Wheat—An Exceptional Crop

Wheat—An Exceptional Crop

Botanical Features, Chemistry, Utilization, Nutritional and Health Aspects

Herbert Wieser

Peter Koehler

Katharina A. Scherf

ELSEVIER

WP
WOODHEAD
PUBLISHING
An imprint of Elsevier

CEREALS
& GRAINS
ASSOCIATION

Woodhead Publishing is an imprint of Elsevier
The Officers' Mess Business Centre, Royston Road, Duxford, CB22 4QH, United Kingdom
50 Hampshire Street, 5th Floor, Cambridge, MA 02139, United States
The Boulevard, Langford Lane, Kidlington, OX5 1GB, United Kingdom

Library of Congress Cataloging-in-Publication Data
A catalog record for this book is available from the Library of Congress

British Library Cataloguing-in-Publication Data
A catalogue record for this book is available from the British Library

ISBN: 978-0-12-821715-3 (print)
ISBN: 978-0-12-821716-0 (online)

For information on all Woodhead publications
visit our website at https://www.elsevier.com/books-and-journals

Publisher: Charlotte Cockle
Acquisitions Editor: Megan R. Ball
Editorial Project Manager: Andrea Dulberger
Production Project Manager: Debasish Ghosh
Cover Designer: Alan Studholme

Typeset by SPi Global, India

Contents

Author biographies ix
Preface xi
Frequent abbreviations xiii

1. Botanical features of wheat

1.1 Phylogeny and species 1
1.2 Grain structure 5
1.3 Breeding 8
References 11

2. Chemical composition of wheat grains

2.1 Carbohydrates 13
2.1.1 Starch 16
2.1.2 Nonstarch polysaccharides 18
2.2 Proteins 21
2.2.1 Storage proteins 25
2.2.2 Gluten 31
2.2.3 Metabolic proteins 32
2.2.4 Protective proteins 35
2.2.5 Miscellaneous proteins 35
2.3 Lipids 36
2.4 Vitamins and minerals 39
2.5 Phytochemicals 41
References 44

3. Cultivation and milling of wheat

3.1 Farming 47
3.2 Storage of grains 54
3.3 Milling process and products 55
References 59

4. Wheat-based food and feed

4.1 History 61
4.2 Baked products 62
4.2.1 Bread 63
4.2.2 Fine bakery goods 80
4.2.3 Dry baked products 84
4.3 Paste products 86
4.3.1 Pasta 86
4.3.2 Other durum products 91
4.3.3 Noodles 92
4.3.4 Miscellaneous products 94
4.4 Breakfast cereals 95
4.5 Beverages 96
4.6 Animal feed 97
4.7 Economics 98
References 101

5. Wheat-based raw materials

5.1 Starch 103
5.1.1 Starch production 104
5.1.2 Native starch 105
5.1.3 Amylose, amylopectin 107
5.1.4 Chemically modified starch 108
5.1.5 Starch-derived sweeteners 110
5.2 Gluten 113
5.2.1 Gluten production 114
5.2.2 Chemical composition 115
5.2.3 Physical properties 117
5.2.4 Modification of gluten 119
5.2.5 Food applications 121
5.2.6 Nonfood applications 124
5.3 Wheat germ oil 126
5.4 Bran 128
5.5 Straw 129
5.6 Ethanol production 129
References 130

6. Nutritional value of wheat

6.1 Nutrition and health 133
6.2 Wheat as a staple food 134
6.3 Starch 136
6.4 Dietary fiber 137
6.4.1 Colonic functions 138
6.4.2 Cancer 139
6.4.3 Cardiovascular diseases 140

6.4.4 Post-prandial glycemic response 141
6.4.5 Obesity and weight loss 141
6.5 Proteins 142
6.6 Lipids, minerals, and vitamins 143
6.7 Phytochemicals 145
6.8 Wheat processing 146
References 147

7. Principles of wheat hypersensitivities

7.1 Human immunity 149
7.2 Wheat proteins 152
7.3 Digestion of proteins 153
7.4 Intestinal permeability 154
7.5 Blood-brain barrier 157
References 157

8. Celiac disease

8.1 History 161
8.2 Epidemiology 163
8.3 Genetics 167
8.4 Causative proteins 168
8.4.1 Chemistry of gluten proteins 169
8.4.2 Testing toxicity 170
8.4.3 Toxicity of cereals 172
8.4.4 Toxicity of cereal proteins and peptides 173
8.5 Environmental factors 177
8.6 Serology and pathology 179
8.7 Clinical features 181
8.7.1 Gastrointestinal complaints 183
8.7.2 Manifestations of malabsorption 183
8.7.3 Skin, hair, and nail symptoms 185
8.7.4 Liver, kidney, and cardiovascular diseases 187
8.7.5 Infections and reproductive disorders 188
8.7.6 Neurological disorders 189
8.7.7 Psychiatric disorders 191
8.7.8 Malignancy and mortality 193
8.8 Associated diseases 195
8.9 Pathomechanism 198
8.9.1 Adaptive response 200
8.9.2 Innate response 202
8.10 Diagnosis 203
8.10.1 Serological tests 205
8.10.2 Intestinal histology 207
8.10.3 Genetic tests 208
8.10.4 Effects of a gluten-free diet 209
8.10.5 Mass screening 209
References 210

9. Wheat allergy

9.1 Food allergies 221
9.2 Wheat food allergy 223
9.3 Wheat-dependent exercise-induced anaphylaxis 226
9.4 Respiratory allergy 228
9.5 Contact urticaria and contact dermatitis 229
References 230

10. Non-celiac gluten sensitivity

10.1 General characteristics 233
10.2 Clinical features 234
10.3 Pathogenesis and triggering factors 235
10.4 Diagnosis 238
10.5 Conclusions 241
References 241

11. Irritable bowel syndrome

12. Treatment of dietary wheat hypersensitivities

12.1 Gluten-free diet 249
12.2 Gluten-free products 251
12.3 Gluten analysis 255
12.3.1 Extraction 255
12.3.2 Reference material 256
12.3.3 Immunochemical methods 256
12.3.4 Nonimmunochemical methods 260
12.4 Life on a gluten-free diet 261
12.5 Alternative therapies 263
References 265

13. Future research on wheat hypersensitivities

13.1 Epidemiology 269
13.2 Genetics 270
13.3 Triggering factors 270
13.4 Environmental factors 271
13.5 Pathogenic mechanisms 272
13.6 Diagnosis 273
13.7 Therapies 274
13.8 Gluten analysis 275

Index 277

Author biographies

Herbert Wieser studied Food Chemistry at the University of Munich and completed his M.S. in 1970. He obtained a Ph.D. degree in food chemistry in 1974 from the Technical University of Munich. In 1973, he joined the German Research Center for Food Chemistry and was member, head and scientific advisor of the working group "structure/function relationships of biopolymers" until his retirement in 2017. Herbert Wieser's major research interest focused on the chemistry of cereal proteins, in particular of gluten proteins, and structure/function relationships of these proteins. He had a special interest in celiac disease with focus on toxic protein structures. He published 330 original scientific papers and book contributions. He was awarded the Max-Rubner Award of the German Nutrition Society in 1984, the Osborne Medal of the AACC International in 2007, and the Neumann Medal of the German Association of Cereal Research in 2009.

Peter Koehler is professor for Food Chemistry at Technische Universität München and chairman of the Working Group on Prolamin Analysis and Toxicity. His research was focused on basic and applied topics in cereal science. Typical examples are structure and functional properties of wheat gluten and other cereal proteins as well as enzymes and emulsifiers in breadmaking. In the field of gluten hypersensitivities, he worked on gluten analysis and gluten detoxification by means of enzymes. He developed a novel reference material for competitive ELISAs for gluten quantitation and coordinated collaborative studies on the validation of ELISA kits. Koehler was awarded with the Scientific Award of the German Bread Industry in 1994, Fellow of the ICC Academy in 2013, "Bread Senator" of the German Association of Industrial Bakers in 2013 and Fellow of the Cereals & Grains Association. In 2017, Koehler joined biotask AG, an analytical laboratory specialized in cereals and cereal products.

Katharina A. Scherf leads the Department of Bioactive and Functional Food Chemistry at the Institute of Applied Biosciences, Karlsruhe Institute of Technology (KIT), Karlsruhe, Germany. Together with her team, she studies the complex interplay between structure, functionality and bioactivity of food biopolymers and uses these fundamental insights to improve food security, quality, and safety. One of her main interests are analytical, immunological and biochemical aspects of celiac disease, non-celiac gluten sensitivity, and wheat allergy.

Having studied Food Chemistry, Katharina Scherf obtained her Ph.D. degree from the Technical University of Munich (TUM) and was a research group leader at the Leibniz-Institute for Food Systems Biology at the TUM. She was awarded with several scientific prizes including the Young Scientist Research Award of the Cereals & Grains Association in 2018, the Gerhard-Billek-Prize of the German Chemical Society in 2015 and the Research Award of the German Coeliac Society in 2014 and 2019.

Preface

Wheat is one of the most important crops cultivated by mankind. Wheat-based foods have been staple foods since thousands of years and are essential for food security worldwide. They are the source of important essential nutrients such as carbohydrates, proteins, dietary fiber, vitamins, minerals, and phytochemicals and supply up to 20% of the energy intake of the global population. Numerous health benefits are associated with the consumption of wheat-based foods, especially made of whole grains. Moreover, they are considered to be among the safest and cheapest foods available and have an impressing global economic value. The unique combination of chemical and physical properties of grain constituents such as starch and gluten contributes to the outstanding importance of wheat for non-food applications. Altogether, the benefits of wheat were commonly unquestioned for a long period of time.

In the last decades, however, wheat has received an increasingly negative image due to the raised awareness of disorders caused by wheat-based foods. A small portion (around 5–10%) of people suffers from wheat hypersensitivities (e.g., celiac disease, wheat allergy, and non-celiac gluten sensitivity) and have to avoid or reduce the intake of wheat-containing foods. Several pseudoscientific books and numerous media reports fueled the overall perception that wheat consumption makes people sick, stupid, fat, and addicted culminating in recommendations that the general population should avoid wheat consumption. The commingling of sound, controversial, and wrong statements have made it virtually impossible for people to disentangle scientific facts from false claims. Thus more and more Western individuals were made unsecure and decided to adopt a wheat-free diet even without any medical advice or sign of health problems.

The general aim of this book is to present the exceptional position of wheat among food crops regarding botanical features, chemistry, utilization, nutritional, and health aspects. A special aim is to demonstrate objectively the benefits and pitfalls of wheat for mankind based on a profound knowledge of wheat science and technology as well as wheat-mediated diseases. The book is organized in 13 chapters. The first part of the book (Chapters 1–3) presents basic knowledge of wheat. Chapter 1 provides the botanical features of wheat including phylogeny, classification into different species, morphological structure of grains, and breeding strategies. Chapter 2 reviews the chemistry of grain constituents and is focused on the structures and quantities of carbohydrates, proteins, lipids, and micronutrients. Chapter 3 deals with farming, optimal crop management, grain storage, and milling process.

The second part (Chapters 4–6) concerns the overall benefits of wheat for the general population. Chapter 4 reviews the wide range of food products made from wheat grains and flours with special attention to baked goods and paste products. Chapter 5 describes the utilization of wheat constituents such as native and modified starch and gluten as raw materials for food and non-food applications. Chapter 6 presents the importance of wheat grain constituents for a healthy and diversified human nutrition with focus on proteins, dietary fiber, vitamins, minerals, and phytochemicals.

The third part (Chapters 7–13) is attended to wheat-related disorders that affect a small portion of the population. Chapter 7 provides an overview on the principles of human immunity in the special field of wheat hypersensitivities. Chapter 8 considers the overall features of celiac disease including clinical pictures, causative proteins, environmental factors, pathomechanisms, and diagnosis. Chapter 9 summarizes current knowledge of wheat-related allergies such as dietary wheat allergy, wheat-dependent exercise-induced anaphylaxis, respiratory allergy, and contact urticaria. Chapter 10 is dedicated to non-celiac gluten sensitivity, a newly identified type of wheat hypersensitivities with unclear pathogenesis and difficult diagnosis. Chapter 11 provides information on irritable bowel syndrome being among the most frequent bowel complaints and in parts caused by the intake of wheat-containing foods. Chapter 12 describes the conventional and alternative treatments of wheat-mediated disorders and presents information on gluten-free products and gluten analysis. Chapter 13 finally indicates future tasks that should be adopted by researchers involved in the investigation of wheat hypersensitivities.

This book will be a useful resource for all qualified persons engaged in cultivation and utilization of wheat and management of wheat-related diseases such as agronomists, breeders, food chemists and technologists, physicians, nutritionists, and dietitians. Experts in one discipline will find this book helpful to keep track of the science in other disciplines. It will also be interesting for public and industrial research centers, universities, and hospitals. Moreover, it will be an excellent introduction for those new to these fields and supply students and interested laypersons with a basic background. Last but not least, this book will provide a useful source of information for patients suffering from wheat-related diseases to learn more about the clinical features, triggering factors, diagnosis, and treatment of their disease.

We wish to express our thanks to Mrs. Sabine Bijewitz (Leibniz-Institute for Food Systems Biology at the Technical University of Munich) for designing the illustrations and to the people, who have allowed reproduction of some of the illustrations included in this book. Finally, we want to thank the editorial and production teams at Elsevier and Cereals & Grains Association for their time, effort, expertise, and advice.

Herbert Wieser
Peter Koehler
Katharina A. Scherf
January 2020

Frequent abbreviations

AGA	antigliadin antibody
Amino acid codes	see Table 2.3
APC	antigen-presenting cell
ASA	acetylsalicylic acid
ASD	autism spectrum disorder
ATI	amylase/trypsin inhibitor
AX	arabinoxylan
CBP	chorleywood bread process
CD	celiac disease
CVD	cardiovascular disease
DF	dietary fiber
DGPA	deamidated gliadin peptide antibody
DH	dermatitis herpetiformis
DNA	deoxyribonucleic acid
ELISA	enzyme-linked immunosorbent assay
EMA	antiendomysium antibody
ESFA	European Food Safety Authority
FAO	Food and Agriculture Organization of the United Nations
FODMAP	fermentable oligo-, di- and monosaccharides and polyols
GA	gluten-associated ataxia
GFD	gluten-free diet
GS	glutenin subunit
HLA	human leukocyte antigen
HMW	high-molecular-weight
HWP	hydrolyzed wheat protein
IBS	irritable bowel syndrome
ICC	International Association for Cereal Science and Technology
IEL	intraepithelial lymphocyte
IFN	interferon
Ig	immunoglobulin
IL	interleukin
LC	liquid chromatography
LMW	low-molecular-weight
MS	mass spectrometry

NCGS	non-celiac gluten sensitivity
NSAID	nonsteroidal antiinflammatory drug
NSP	nonstarch polysaccharide
PCR	polymerase chain reaction
PEP	prolyl endopeptidase
RCD	refractory celiac disease
RNA	ribonucleic acid
SDS	sodium dodecylsulfate
T1D	type 1 diabetes
T2D	type 2 diabetes
TCR	T-cell receptor
TG	transglutaminase
TGA	antitransglutaminase antibody
TLR	Toll-like receptor
VCE	video capsule endoscopy
WA	wheat allergy
WDEIA	wheat-dependent exercise-induced anaphylaxis
WHO	World Health Organization

Chapter 1

Botanical features of wheat

Cereals belong to the most important agricultural crops cultivated by mankind. They are grown on nearly 60% of total farmland in the world and are widely used for human nutrition, animal feed, industrial products, and generation of renewable energy. The main cereals are wheat, corn, rice, barley, rye, sorghum, millet, and oats. Botanically, cereals are grasses and belong to the plant family *Poaceae*. Cultivated wheat comprises five species: einkorn, emmer, durum wheat, spelt, and common wheat. They are annual plants and produce dry, one-seeded fruits (grains) with a fairly uniform anatomy consisting of fruit and seed coat, aleurone layer, starchy endosperm, and germ. Grains represent the most valuable and nutritive parts of cereals. Their morphological structure fulfills all criteria for a successful reproduction of the plant. Common wheat and durum wheat, the economically most important species, have been the target of intense breeding. Systematic wheat breeding started in the late 19th century resulting in currently around 25,000 varieties worldwide. They have been bred to specific purposes such as high grain yield, resistance against diseases, and high quality of end products. The present chapter covers some basic aspects of the phylogeny, grain structure, and breeding strategies of wheat.

1.1 Phylogeny and species

Botanically, wheat (*Triticum*) belongs to the monocot family *Poaceae* ("grass" family) and is closely related to rye (*Secale*) and barley (*Hordeum*) as members of the subfamily *Pooideae* and the tribe *Triticeae* (Fig. 1.1) [1]. The other well-known cereals are either distant relatives of wheat (oats) or show separate evolutionary lines (rice, corn, sorghum, and millet). Currently, cultivated wheat plants belong to five species with different numbers of subgenomes and chromosomes: diploid einkorn (*Triticum monococcum* L.), tetraploid emmer [*Triticum dicoccum* (Schrank) Schübler] and durum wheat (*Triticum durum* Desf.), and hexaploid spelt (*Triticum spelta* L.) and common wheat (*Triticum aestivum* L.) (Table 1.1). Traditionally, wheat has been classified into winter wheat sown in late summer or autumn and spring wheat sown in spring (see Section 3.1). The phylogenetic development of wheat from diploid einkorn to hexaploid spelt and common wheat occurred by natural hybridization and by human selection. With the beginning of the 20th century, intense breeding did not only lead to numerous different varieties within a species but also to a specific crossing of different

Wheat—An Exceptional Crop. https://doi.org/10.1016/B978-0-12-821715-3.00001-0

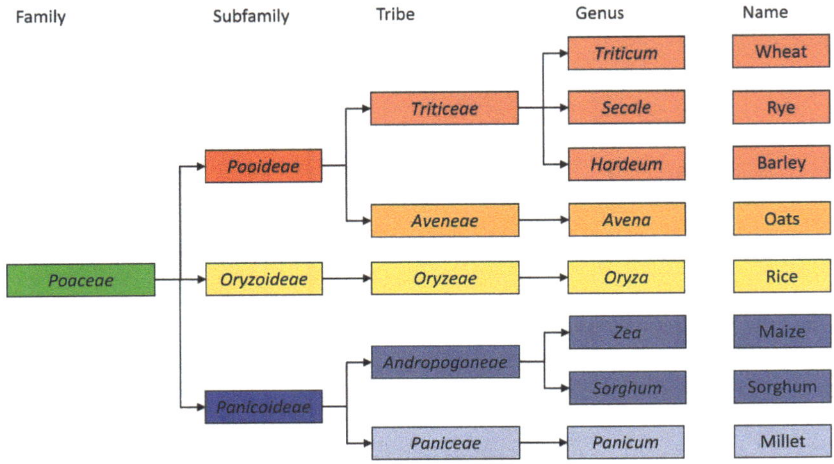

FIG. 1.1 Phylogeny of cereals belonging to the monocot family *Poaceae*.

TABLE 1.1 Classification of cultivated wheat [*Triticum* (T.)] species.

Species	Systematic name	Genome	Chromosomes	Husk
Einkorn	*T. monococcum*	AA	2×7=14	Hulled
Emmer	*T. dicoccum*	AABB	4×7=28	Hulled
Durum wheat	*T. durum*	AABB	4×7=28	Naked
Spelt	*T. spelta*	AABBDD	6×7=42	Hulled
Common wheat	*T. aestivum*	AABBDD	6×7=42	Naked

species and even different cereals (see Section 1.3). Crossing of spelt and common wheat, for example, resulted in hybrids that combined the advantages of both species. Exchange of a wheat half chromosome by a rye half chromosome led to so-called wheat/rye translocation lines and these wheat hybrids have the disease resistance of rye. Furthermore, durum wheat and rye were crossed to obtain triticale (*Triticum* + *Secale*), a man-made hexaploid hybrid.

The botanical origin of the wheat plant is not known with certainty, but a good deal of evidence indicates that einkorn (genome AA) developed from a type of wild grass native to the arid lands of Asia Minor [2]. The first evolutionary event, leading to polyploid wheats, was the hybridization of two diploid wild grasses: *Triticum urartu* (genome AA), closely related to einkorn, and a yet unknown *Aegilops* species (genome BB) related to *Aegilops speltoides* (goat grass). Their crossing resulted in the tetraploid emmer (genome AABB). The cultivation of emmer provided a range of subspecies, some of which developed a naked, free-threshing tetraploid wheat known as durum wheat. When emmer reached the region south of the Caspian See around 5000 BC and hybridized with diploid *Aegilops tauschii* (genome DD), hexaploid common wheat (*Triticum aestivum*, genome AABBDD) developed [3]. The origin of hexaploid spelt is currently not clear. The hybridization of emmer and *Aegilops tauschii* might have taken place in Asia Minor around 6000 BC. The much later appearance of spelt in Europe probably was the result of the hybridization of emmer and common wheat. Thus it is unclear whether spelt has a single origin in Asia or two separate origins in Asia and Europe.

Einkorn (from German "Einkorn" = single grain) is a diploid wheat species carrying the genome AA with 14 chromosomes (Table 1.1). It was the first wheat to be domesticated around 10,000 years ago in the Middle East (in the "Fertile Crescent"). It was widely cultivated in the Neolithic Age but was gradually replaced first by emmer and then by common wheat and durum wheat. Today, einkorn is grown in small farmlands in Western Turkey, Balkan countries, Italy, Spain, and Central Europe. Einkorn belongs to the hulled wheats, i.e., grains do not break free from the husk with threshing and therefore it is difficult to remove the husk from the seed. Einkorn tolerates poor and dry soils and exhibits a pronounced resistance against cereal diseases like rust and mildew. However, low yields and poor dough and bread-making qualities have led to the almost complete disappearance of einkorn.

Emmer is a tetraploid wheat with the genome AABB and 28 chromosomes. Along with einkorn it was one of the first crops cultivated in the Middle East, but today it is a marginal crop preferably grown in mountainous areas. Like einkorn, emmer is a hulled wheat and characterized by modesty regarding soil, climate, and fertilization as well as by highly expressed resistance to diseases. Due to economic reasons and poor product (pasta and bread) quality, emmer was widely replaced by durum wheat and common wheat.

Durum wheat is tetraploid like emmer with the genome AABB and 28 chromosomes. In contrast to emmer, durum wheat is a "naked" cereal, i.e., free-threshing. It was developed from domesticated emmer and is now produced in many parts of the world especially in warm areas with low precipitation. Among wheat species, durum wheat grains have the highest hardness (durum in Latin means "hard"). Due to this hardness, grains are mostly milled into semolina that is preferably used for making pasta ("pasta wheat"). Most durum varieties have been bred for a yellow endosperm that gives pasta its color. Khorasan wheat is a tetraploid hybrid derived from durum wheat and a wild wheat form (*Triticum polonicum*). One variety is marketed under the trademark "Kamut". Khorasan wheat is quite resistant against diseases, has modest requirements regarding fertilization, and is therefore most suitable for extensive agriculture. It needs a warm and dry climate for cultivation and is mainly produced in North America and Southern Europe.

Spelt is a hulled hexaploid wheat with the genome AABBDD and 42 chromosomes. It was an important staple food in the Middle East and Europe from the Bronze Age to the Middle Age but was then successively replaced by common wheat that had higher yields, lower processing costs, and better baking performance. Only in some regions of Southern Germany, Austria, and Switzerland spelt remained the most important cereal until 100 years ago. Since some years, spelt is again increasingly cultivated notably in organic farming and water-protected areas because of its robustness, modesty with regard to climate and soil, and fewer requirements for the use of fertilizers and pesticides. The pleasant taste and flavor of spelt products and a possibly low potential to trigger hypersensitivities increased the popularity of spelt. To compensate for the disadvantages of spelt (e.g., increased height of plants, lower yield, and poorer bread-making quality compared to common wheat) and to preserve its desirable properties (e.g., resistances to disease and unfavorable soil and climate), cross-breeding of spelt and common wheat was started at the beginning of the 20th century. Since that time, spelt/wheat cross-breeds have been cultivated and processed into food products similar to pure spelt. The degree of common wheat crossing can be determined by the analysis of storage protein patterns using high-performance liquid chromatography for instance [4]. Unfortunately, information on the degree of crossing does not have to be labeled on spelt products, which would be of special interest for persons sensitive to common wheat and tolerant to spelt.

Common wheat, known as bread wheat, represents around 90% of cultivated wheat worldwide. Common wheat is hexaploid with the genome AABBDD and 42 chromosomes that carry around 100,000 genes. Using recent advances in sequencing techniques, the International Wheat Genome Sequencing Consortium (IWGSC) published a detailed description of the genome of common wheat, the result of 13 years of collaborative international research [5]. The research article, authored by more than 200 scientists from 73 research institutions in 20 countries, presents an annotated reference genome of the common wheat variety Chinese Spring. The 21 chromosome-like sequence assemblies contain both coding and noncoding elements across the A, B, and D subgenomes and give access to 107,891 high-confidence genes including their genomic context of regulatory sequences. This community resource establishes the foundation for accelerating wheat research and application through improved understanding of wheat biology and genomic-assisted breeding.

Like durum wheat, common wheat belongs to the free-threshing species. Approximately 25,000 different commercial varieties are grown around the world. "Modern" common wheat refers to varieties that were developed after the incorporation of so-called dwarfing genes (Rht-B1 and Rht-D1) in the 1950s (see Section 1.3). "Old", so-called heritage wheat varieties were developed before this time. In North American terminology, common wheat is additionally divided into hard and soft wheat (according to the hardness or softness of the grain) and into white and red wheat (according to the color of the bran).

The classical usage of common wheat flour is the production of bread and other baked goods but additionally comprises an extremely wide range of other food and nonfood applications. Compact wheat (*Triticum compactum*) like club wheat is regarded as a subspecies of *Triticum aestivum* and closely related to common wheat but has more compact ears. Their shorter rachis segments lead to spikelets packed closer together. Compact wheat is not well suited for bread production but is excellent for the production of certain types of cake and pastry.

Common wheat plants are quite demanding with regard to growing conditions specifically concerning soil, fertilization, and climate. The weather should be rainy during the growing stage and dry during the maturing stage. Moreover, common wheat is much more sensitive to plant diseases than rye for instance. In order to enhance wheat resistance to rust and mildew, the short arm of wheat 1B chromosome was replaced by the short arm of rye 1R chromosome that carries the resistance genes (1BL/1RS translocation) [6]. These translocation lines now comprise a part of the variety assortments in many countries. Unfortunately, their usages are often associated with problems concerning processing and product quality such as dough stickiness and poor bread properties.

Triticale (*Triticosecale*) is a cereal species hybridized by man, in which the genomes of tetraploid durum wheat (AABB) and diploid rye (RR) have been combined. Durum wheat is used as the female partner and rye as the male pollen donor. The sterile hybrid has to be treated with colchicine to achieve fertility. Triticale combines favorable properties of wheat (high yield and grain quality) with desirable properties of rye (disease tolerance and modesty on soil and climate). The main producers of triticale are Poland, Germany, France, and Russia. Triticale is used as food (baked products) and livestock feed and its biomass is used for the production of energy and bioethanol.

1.2 Grain structure

Wheat produces dry one-seeded fruits botanically defined as caryopses, in which the fruit coat is fused to the seed coat at maturity. Some different terms are used to describe the caryopsis and its component parts. Grain is another term for the caryopsis and is used throughout this book. The frequently used terms seed and kernel are botanically not strictly correct. Whole wheat grain is in fact not a seed but a fruit, in which the seed is surrounded by an adherent pericarp. The term kernel has different botanical meanings (e.g., it is applied to nuts) and is therefore a less accurate description. The term germ, used to denote the embryo of the grain, and the term bran, comprising fruit coat, seed coat, nucellar epidermis, and aleurone layer (see below), are deduced from the terminology of millers and adopted in this book.

Wheat grains represent the most nutritive and commercially most valuable part of the plant. Grain sizes vary widely depending upon the species, variety, and environmental conditions, but limits of 4–10 mm long and 2.5–4.5 mm wide are unlikely to be exceeded. Correspondingly, single grain weights can be quite different (30–60 mg). The "thousand kernel weight" (TKW), the weight of 1000 grains, is an important quality criterion of harvested wheat. It can vary from variety to variety, from harvest year to year, and even from field to field. TKW is the simplest single index of the ratio of starchy endosperm to outer grain layers, an important factor for millers: the higher the TKW, the higher the portion of starchy endosperm and the yield of white flour.

Morphologically, the grain is formed as caryopsis, i.e., the fruit coat (pericarp) is strongly bound to the seed coat (testa). In einkorn, emmer, and spelt, the husk is fused with the fruit coat ("hulled" wheat) and cannot be removed simply by threshing as with the free-threshing "naked" durum wheat and common wheat. When hulled wheats are threshed, the ears break up into spikelets and not into naked grains and a dehusking process is needed to remove the husks. The anatomy of wheat grains is fairly uniform: Fruit and seed coats enclose the germ and the endosperm that consists of the aleurone layer and the starchy endosperm [7] (Fig. 1.2). The starchy endosperm makes up 80–85%, the aleurone layer 6–9%, the pericarp and testa 5–6%, and the germ around 3% of dry grain weight [8]. The color of the grains is usually termed "white" or "red" related to pigments in the testa. Purple or black grains also exist but are not common.

The pericarp surrounds the entire seed and consists of several layers (Fig. 1.3). The outer epidermis is composed of long narrow cells forming a complete layer. The inner layers consist of intermediate cells, cross cells, and tube cells. Intermediate cells are variable in shape and do not completely cover the grain. Cross cells are located further to the interior of the grain beneath the intermediate cells. Their form is long and cylindrical and their long axis lies at a right angle to the axis of the grain. They are tightly packed with little or no intercellular space. Tube cells are elongated and knobbly in shape, are not packed

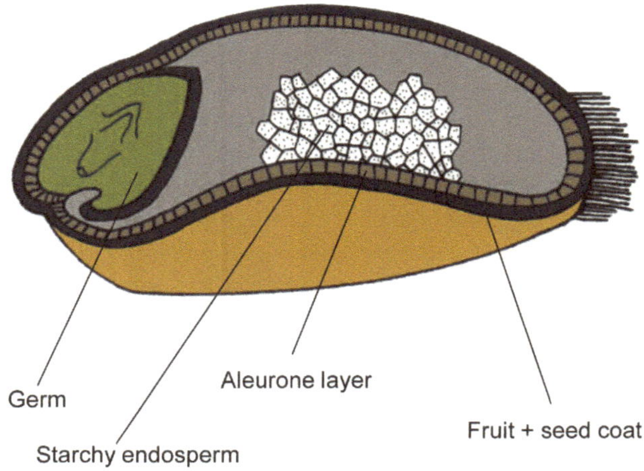

FIG. 1.2 Longitudinal cross-section of a wheat grain showing the main botanical feature.

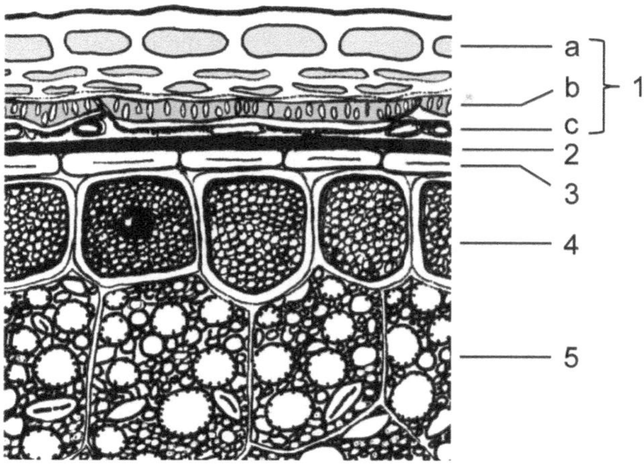

FIG. 1.3 Longitudinal section of the outer layers of a wheat grain. 1 Pericarp: a, epidermis; b, cross cells; and c, tube cells; 2, testa; 3, nucellar epidermis; 4, aleurone layer; and 5, starchy endosperm.

tightly, and have many intercellular spaces. Their long axis runs parallel to the grain's own axis.

The testa is firmly attached to the tube cells on the outer side and to the nucellar epidermis on the inner side. It consists of three layers: a thick outer cuticle, a layer containing pigments (e.g., for red-colored wheat), and a thin inner layer. The nucellar epidermis, represented by a single crushed layer of epidermal cells, is located between the testa and the aleurone layer. Morphologically, pericarp, testa, and nucellar epidermis are summed up as bran. This botanical definition differs from the technological definition of millers: The term bran stands for the mixture of pericarp, testa, nucellar epidermis, and the aleurone layer; the remaining starchy endosperm corresponds to white flour (see Section 3.3). The endosperm consists of the outer aleurone layer and the inner starchy endosperm. The aleurone layer is a one-cell layer and covers both the starchy endosperm and the germ. Those aleurone cells that cover the starchy endosperm are thick-walled and contain large nuclei and a large number of granules. The aleurone cells covering the germ are thin-walled and do not contain granules. The starchy endosperm consists of three types of cells that vary in size, shape, and location within the grain. The peripheral cells form the first row neighboring to the aleurone layer; they are small and equal in diameter in all directions or slightly elongated. Next to the peripheral cells are several rows of elongated prismatic cells. They extend inward to the center of the cheeks. Central cells occur in the center of the cheeks and are similar in shape being rounded or polygonal and vary considerably in size. Starch granules occur in prismatic and central cells but not in peripheral cells.

The germ is located on the lower dorsal side of the caryopsis and consists of two major components: the embryonic axis (rudimentary root and shoot) and the scutellum (storage organ). The embryonic axis can be divided into three regions: the shoot or epicotyl, the mesocotyl, and the radicel. The scutellum, a massive storage organ providing proteins, phytic acid, and lipids, is bordered on the dorsal side by the embryonic axis and on the ventral side by the starchy endosperm thus enabling close contact between the storage site (endosperm) and the embryo.

Grains are the reproductive organs of wheat. The hour of birth occurs after self-pollination of the ovules by pollens of the flowering plant. Within around 30 days of maturation, the different compartments (outer layers, endosperm, and germ) are assembled, formed to a caryopsis and gradually filled with a number of various compounds including starch, proteins, dietary fiber, and lipids. Drying on the field to less than 15% moisture before harvesting yields grains that are stable and fertile for a long time. When water penetrates the grain usually after sowing into the wet soil, the grain starts to germinate. Based on the activation of the germ (embryonic axis and scutellum), a new generation of the wheat plant begins to develop. Building materials, necessary for the development of the plant at the beginning of growth, are stored in different compartments of the grain. Lipids and phytic acid are stored in the scutellum of the germ. After enzymatic hydrolysis, they provide the seedling with fatty acids and phosphorus, respectively. Enzymes, necessary for degradation and synthesis of compounds (e.g., hydrolases and synthases), are enriched in the aleurone layer. Starch and storage proteins, the major components of the grain, are located in the starchy endosperm. Amylose and amylopectin, present in the starch granules, are degraded to glucose by amylases. Glucose serves as source of energy and prime material for other compounds. The degradation of storage (gluten) proteins by peptidases produces free amino acids and other nitrogen-containing compounds (due to the high glutamine content). After around 2 weeks of germination, roots and shoots have grown and take over their tasks, namely uptake of minerals by the roots and photosynthesis by the first green leaves, respectively. The reproductive cycle is completed, when ripe grains are present on the ears.

1.3 Breeding

In general, wheat breeding is essential to human welfare to ensure staple food for an increasing population by providing new varieties with improved properties. Common (bread) wheat and durum (pasta) wheat are the commercially most successful wheat species. Intense breeding worldwide led to the existence of around 25,000 different varieties that usually have been bred with a specific purpose in mind such as high grain yield, resistance against diseases and insects, tolerance regarding extreme temperatures, drought or salinity, and high quality of end products [9]. Breeding has been practiced by farmers and gardeners for thousands of years by simply selecting plants with particular desirable properties and using them as progenitors for subsequent generations. Systematic wheat breeding started in the late 19th century, when the practice of hybridization of different wheat parents followed by selection was initiated and when breeding was closely linked to the development of Mendelian genetics. The standard method was crossing two lines of the same species using hand emasculation, then inbreeding the progeny and selection of the desired hybrid. After World War II, a number of techniques were developed that allowed breeders to hybridize also more distantly related species. Recently, genetic engineering was used to produce a favorable phenotype by adding specific desirable genes or knocking down undesirable genes. In summary, wheat breeding strategies and technologies have become complex and now have a broad scientific basis including agronomy, plant science, genetics, molecular biology, and chemistry.

Modern breeding methods comprise hybridization, genetic mutation or genetic transformation [10]. Most classical breeding programs include hybridization of two promising parents (father F and mother M) with different desirable properties. For example, a high-yield line may be crossed with a line resistant to stem rust. Wheat is a self-pollinating plant, i.e., it carries both male and female parts. Therefore the selected female parent F has to be emasculated by removing unripe anthers followed by pollination with selected male parent M. This hand-operated method can be modified by a procedure known as cytoplasmic male sterility to produce so-called hybrid wheat. It consists of using wheat with shrunken anthers that do not release pollen. Therefore there is no need to remove anthers from the female parent F and the pollination with male parent M can be performed directly. Independent of the breeding method, the new plant is usually crossed again with one of its parents ("backcrossing"). The advantage is that the developed useful traits are introduced into a plant without loss of favorable traits that have been previously constructed.

Doubled haploid techniques (e.g., wheat-by-maize system and anther culture system) are used for doubling chromosomes of haploid hybrids of the first generation in order to generate homozygous lines and to speed up the process to reach homozygosity of the plants. Another method for doubling the number of chromosomes is the application of the alkaloid colchicine, the toxin of autumn crocus. A well-known example is the development of triticale, a cross between durum wheat and rye (see Section 1.1). Normally, distantly related plants do not reproduce each other. When tetraploid durum wheat is crossed with diploid rye, the triploid first generation, containing an uneven number of chromosomes, is usually sterile. Doubling the number of chromosomes by colchicine results in the fertile hexaploid triticale.

The most famous examples for classical breeding are the so-called semidwarf or dwarf wheats. N. E. Borlaug (awarded a Nobel Peace Prize in 1970) from the International Maize and Wheat Improvement Center (CIMMYT) in Mexico, engineered the hybridization of the Japanese wheat Norin-10, possessing dwarfing Rht genes, with the US wheat Brevor-14, an event that has been called "Green Revolution". The new genotype became the main source of dwarf and semidwarf wheat plants with significantly shorter stem length. The introduced Rht-B1 and Rht-D2 genes were shown to be responsible for the inactivation of gibberellic acid, a plant hormone that lengthens stems. Short stems have the advantage that they prevent lodging (collapse of the stems in the fields) of the plant, which is favorable for grain yield, grain quality, and modern harvesting techniques. The low-stature wheats have been introduced into the breeding programs of most wheat-producing countries and have dramatically altered height and yield of modern commercial wheat. Reduced plant height is an important trait in wheat line breeding but is of even greater importance in hybrid wheat breeding [11]. Almost 90% of the world`s wheat cultivation now comprise varieties with the dwarf or semidwarf alleles.

Genetic mutation is another technique for breeding new wheat varieties. Mutagenesis, i.e., the changes in DNA sequences, is the original source of variation of genes constantly occurring in nature. For example, gene mutations are naturally initiated by copying errors during cell division. Using γ-rays, X-rays, ultraviolet light or reactive chemicals, the obtained mutations can artificially generate new traits of wheat. These treatments have been used for breeding since some decades ("mutation breeding"). For example, the transfer of high iron and zinc contents from *Aegilops kotschyi* into common wheat was enabled by γ-ray irradiation [12]. Colchicine and ethyl methyl sulfate are one of the most frequently applied chemical compounds for mutation breeding.

Genetic modification can be also achieved by adding or knocking down specific genes (called genetic engineering) in order to develop a new wheat plant. Unlike traditional breeding that affects the crossing of thousands of genes to a hybrid plant, genetic engineering involves only one or a few genes. Desirable genes can be inserted into the genome by using specific bacteria or by direct methods like the gene gun or microinjection. For example, the ectopically expressed sucrose transporter HvSUT1 gene from barley was used to transform wheat, which resulted in increased grain yield (+28%) and increased grain protein content (+5%) [13]. RNA interference (RNAi) has become an extremely valuable tool for the specific silencing of unfavorable genes. The discovery of RNAi won the Nobel Prize in Physiology and Medicine in 2006 and RNAi is now used for gene engineering in many fields of interest. For instance, RNAi led to the successful silencing of α-gliadins, an important trigger of celiac disease, without any major negative effects on the baking quality [14]. However, to remove all celiac-toxic epitopes from wheat without worsening bread-making quality is a formidable challenge and not surprisingly, such celiac-safe wheats have not yet been developed despite over a decade of research [15].

The so-called CRISPR (Clustered Regularly Interspaced Short Palindromic Repeats)-system in combination with CRISPR-associated (Cas) genes is a relatively new technique to insert, edit or delete gene sequences. The system has been detected as an immune mechanism used by prokaryotes that transfers resistance to foreign genetic elements and therefore provides a specific immunity. Based on these findings, a technique was developed that allowed the removal of existing whole genes or gene sections and the addition of new genes or gene sections. In recent years, numerous research papers on the applications of CRISPR-Cas have been published [16]. Applications on wheat modifications are en route. For example, gene editing using CRISPR/Cas can precisely remove or modify the DNA sequences coding for immunogenic gluten peptides involved in celiac disease and wheat allergy [17]. One genetically modified wheat line named E82 showed outstanding nutritional properties with very low immunogenic gluten and a low stimulation capacity of T cells from celiac disease patients [18]. In next future, transgenic wheat made by CRISP-Cas may become commercially available in the United States, Canada, Argentina and other countries but not in Europe because of strict regulation.

Taken together, genetic engineering offers an efficient alternative to traditional breeding. However, consumer and regulatory acceptance especially in Europe is widely low. Although every commercial transgenic plant is heavily regulated and must undergo extensive testing, uncertainties on the ecological impact of genetically modified plants, the safety of foods, and the concepts used for safety evaluation still remain. Remarkably, corresponding strict controls for plants, obtained by artificial mutation breeding (see above), have never been considered, although numerous changes in DNA structures might occur.

The next step after hybridization, genetic mutation or transformation is selection, i.e., selecting plants with propagated desirable traits and eliminating those with less desirable traits. The time of selection (number of generations) depends on the method for evaluating traits. In cases, when single genes or gene complexes are known to be closely related to specific traits, these genes or their corresponding expressed proteins can be identified by DNA or protein analysis in grains from the first generations (F1–F3). The evaluation of agronomics, disease resistance, and grain quality traits, for instance, needs higher amounts of grains and is usually performed within the next generations (F4 and onwards) during several years of harvesting and sowing. The last stages before releasing a promising new wheat line are extensive field evaluation and production of seedlings. Commercial new wheat varieties are then available after 10–12 years of breeding time, but this period may be shortened using new approaches like genomic selection across multiple breeding cycles [19] or genome-based hybrid wheat breeding [20].

References

[1] Bouchenak-Khelladi Y, Salamin N, Savolainen V, Forest F, van der Bank M, Chase MW, et al. Large multi-gene phylogenetic trees of the grasses (Poaceae): progress towards complete tribal and generic level sampling. Mol Phylogenet Evol 2008;47:488–505.

[2] Feuillet C, Langridee P, Waugh R. Cereal breeding takes a walk on the wild side. Trends Genet 2008;24:24–32.

[3] Avni R, Nave M, Barad O, Baruch K, Twardziok SO, Gundlach H, et al. Wild emmer genome architecture and diversity elucidate wheat evolution and domestication. Science 2017;357:93–7.

[4] Koenig A, Konitzer K, Wieser H, Koehler P. Classification of spelt cultivars based on differences in storage protein compositions from wheat. Food Chem 2015;168:176–82.

[5] International Wheat Genome Sequencing Consortium (IWGSC). Shifting the limits in wheat research and breeding using a fully annotated reference genome. Science 2018;361. eaar7191.
[6] Zeller FJ. 1B/1R wheat-rye chromosome substitutions and translocations. In: Missouri Agricultural Experimental Station, editor. Proceedings of the fourth international wheat genetics symposium. Columbia, MO, USA; 1973. p. 209–21.
[7] Delcour JA, Hoseney RC. Structure of cereals. Principles of cereal science and technology. 3rd ed. St. Paul, USA: AACCI; 2010. p. 1–9.
[8] Van de Kamp JW. Paving the way for innovation in enhancing the intake of whole grain. Trends Food Sci Technol 2012;25:101–7.
[9] Mondal S, Rutkoski JE, Velu G, Singh PK, Crespo-Herrera LA, Guzmán C, et al. Harnessing diversity in wheat to enhance grain yield, climate resilience, disease and insect pest resistance and nutrition through conventional and modern breeding approaches. Front Plant Sci 2016;7:991.
[10] Baenziger PS, DePauw RM. Wheat breeding: procedures and strategies. In: Carver BF, editor. Wheat science and trade. Hoboken, USA: Wiley-Blackwell; 2009. p. 275–308.
[11] Würschum T, Liu G, Boeven PH, Longin CF, Mirdita V, Kazman E, Zhao Y, et al. Exploiting the Rht portfolio for hybrid wheat breeding. Theor Appl Genet 2018;131:1433–42.
[12] Verma SK, Kumar S, Sheikh I, Malik S, Mathpal P, Chug V, et al. Transfer of useful variability of high grain iron and zinc from Aegilops kotschyi into wheat through seed irradiation approach. Int J Radiat Biol 2016;92:132–9.
[13] Saalbach I, Mora-Ramirez I, Weichert N, Andersch F, Guild G, Wieser H, et al. Increased grain yield and micronutrient concentration in transgenic winter wheat by ectopic expression of a barley sucrose transporter. J Cereal Sci 2014;60:75–81.
[14] Becker D, Wieser H, Koehler P, Folck A, Mühling KH, Zörb C. Protein composition and techno-functional properties of transgenic wheat with reduced α-gliadin content obtained by RNA interference. J Appl Bot Food Qual 2012;85:23–33.
[15] Shewry PR, Tatham AS. Improving wheat to remove coeliac epitopes but retaining functionality. J Cereal Sci 2016;67:12–21.
[16] Ledford H. CRISPR, the disruptor. Nature 2015;522:20–4.
[17] Jouanin A, Boyd L, Visser RG, Smulders MJ. Development of wheat with hypoimmunogenic gluten obstructed by the gene editing policy in Europe. Front Plant Sci 2018;9:1523.
[18] Garcia-Molina MD, Gimenez MJ, Sanchez-Leon S, Barro F. Gluten free wheat: are we there? Nutrients 2019;11. pii: 487.
[19] Michel S, Ametz C, Gungor H, Epure D, Grausgruber H, Löschenberger F, et al. Genomic selection across multiple breeding cycles in applied bread wheat breeding. Theor Appl Genet 2016;129:1179–89.
[20] Zhao Y, Li Z, Liu G, Jiang Y, Maurer HP, Würschum T, et al. Genome-based establishment of a high-yielding heterotic pattern for hybrid wheat breeding. PNAS 2015;112:15624–9.

Chapter 2

Chemical composition of wheat grains

Much of our understanding of wheat properties comes from the knowledge of the chemical composition of grain constituents. The 1970s and 1980s were the grand era of cereal analysis and book series such as "Advances in Cereal Science and Technology" (vols. I–X) and "Wheat—Chemistry and Technology" (vols. I and II), both edited by the unforgettable Yeshajahu Pomeranz and the American Association of Cereal Chemists (AACC International), have provided numerous analytical data on the composition of cereals that supply basic knowledge up to this day. A lot of information, included in this chapter, is derived from that time. The basic chemical composition of mature wheat grains varies in a relatively small range and depends on both genetic factors (species and variety) and growing conditions (e.g., soil, climate, and fertilization). Most data, presented in the following sections, are derived from common wheat [1]. An average composition of basic constituents for grains from common wheat is given in Table 2.1 [2]. Carbohydrates, mainly in the forms of starch and dietary fiber (nonstarch polysaccharides), are by far the predominant class of constituents followed by proteins [3]. Lipids, minerals, vitamins, and phytochemicals belong to the minor constituents. This chapter aims at providing an overview on the chemical composition of wheat grains regarding both structures and quantities of important compound classes and single compounds.

2.1 Carbohydrates

Carbohydrates are chemically defined as compounds that consist of polyhydroxy-aldehyde (aldose) and polyhydroxy-ketone (ketose) monosaccharide units. Almost all monosaccharide units, occurring in wheat grains, are compounds with five C-atoms (pentoses) or six C-atoms (hexoses). Depending on the C-chain length, a variable number of chiral C-atoms are present leading to different isomers. Usually, the structures are displayed in the so-called Fischer projection. The positions of the OH-groups at the chiral C-atoms (right or left from the C-chain) determine the type of monosaccharide. The position of the OH-group at the chiral C-atom with the highest numbering (e.g., C5 in glucose) is used to assign the monosaccharide to the so-called D- or L-series (Fig. 2.1A and B). If the OH-group is oriented to the right, the monosaccharide belongs to

Wheat—An Exceptional Crop. https://doi.org/10.1016/B978-0-12-821715-3.00002-2

TABLE 2.1 Average contents (%) of basic constituents of grains of common wheat [2].[a]

Compound	Content
Starch	58.2 (66.7)
Dietary fiber	13.3 (15.2)
Water	12.7
Protein	10.6 (12.1)
Lipids	1.8 (2.1)
Minerals	1.7 (1.9)
Sugars[b]	0.6 (0.7)

[a] *Values in brackets based upon dry weight.*
[b] *Saccharose, fructose.*

the D-series and with the OH-group to the left, an L-monosaccharide is present. D- and L-forms of each type of monosaccharide are enantiomers ("mirror image" forms). Monosaccharide units of wheat carbohydrates mainly occur in the D-form.

By intramolecular reaction of hydroxyl groups with the aldehyde (C1) or keto group (C2), pentoses and hexoses cyclize into five- or six-membered rings (furanoses or pyranoses) including one O-atom and four or five C-atoms, respectively. The schematic structures of hexopyranose and pentopyranose are shown in the Haworth projection in Fig. 2.1C and D. This intramolecular reaction generates an additional chiral C-atom providing two ring forms that are called α- or β-anomers (Fig. 2.1E and F). Structures of the most important monosaccharide types present in wheat are shown in Fig. 2.1E and G–J, namely α-D-glucopyranose, β-D-xylopyranose, β-D-galactopyranose, α-D-fructofuranose, and α-L-arabinofuranose. Both pyranose and furanose rings are not planar. The preferred conformation of pyranose is the so-called chair conformation (see Fig. 2.4), while the envelope and the twist conformations are the most stable conformational isomers of furanose.

Monosaccharide units can be covalently linked by so-called glycosidic bonds yielding oligo- and polymers. For example, glucose units in α-(1,4) or α-(1,6) linkages form amylose and amylopectin molecules, respectively (see below). Based on the number (n) of units, carbohydrates are conveniently classified into monosaccharides ($n = 1$), oligosaccharides ($n = 2–10$), and polysaccharides ($n > 10$).

Carbohydrates are the most abundant components of wheat grains and constitute around 70–75% of the grain. Major carbohydrates are polysaccharides (amylose, amylopectin, and nonstarch polysaccharides), while oligo- and monosaccharides represent minor components [4]. Dry grains contain low

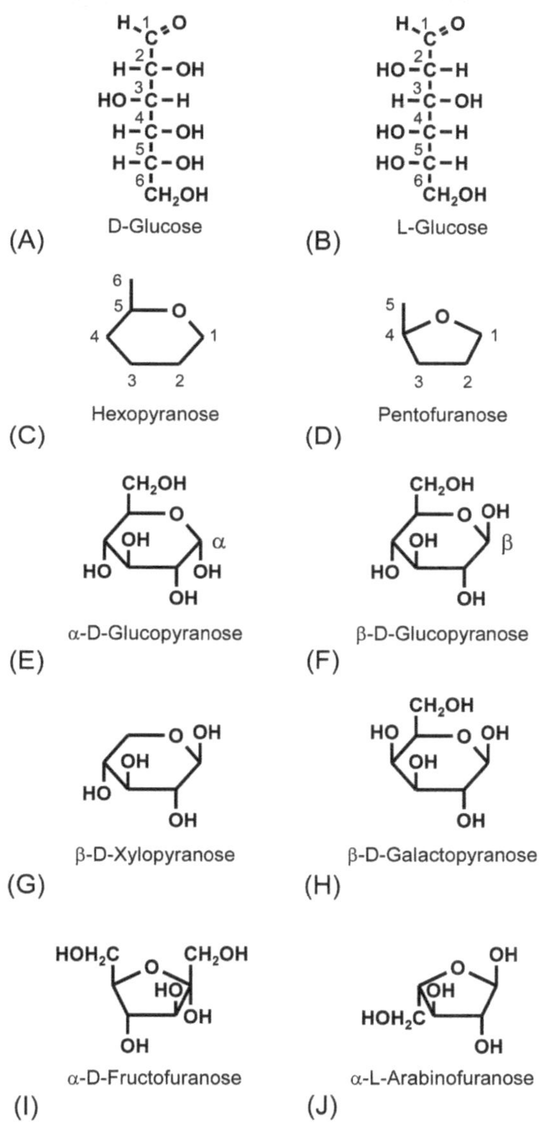

FIG. 2.1 General structures of pentoses and hexoses and examples of different monosaccharide types present in wheat grains.

levels of monosaccharides (glucose ≈0.06%, fructose ≈0.07%), disaccharides (sucrose ≈1.0%, maltose ≈0.09%), trisaccharides (raffinose ≈0.3%), and fructo-oligosaccharides (< 1%). The highest concentrations of mono- and oligosaccharides are present in the germ (≈250 mg/g) followed by pericarp and testa (40–60 mg/g). The starchy endosperm is poor of these carbohydrates (≈10 mg/g). Fermentable oligosaccharides (fructo-oligosaccharides and galactans), disaccharides (lactose), monosaccharides (fructose), and polyols (sorbitol, mannitol, xylitol, and maltitol) present in foods have been grouped into the so-called FODMAPs (fermentable oligo-, di-, monosaccharides, and polyols). They are poorly absorbed in the small intestine and can lead to gastrointestinal complaints (see Chapters 10 and 11). FODMAPs are found in a wide variety of foods with wheat grains having relatively high contents (≈2%).

Polysaccharides, the main components of wheat grains, can be structurally divided into starch and nonstarch polysaccharides. They all are polymers built by monosaccharide units that form linear and branched chains. Major contributors to the polysaccharide structures are glucose, xylose, arabinose, fructose, and galactose (Fig. 2.1E and G–J).

2.1.1 Starch

Starch is the major component of wheat grains with contents of around 58% based on grain weight and around 67% based upon dry weight. It occurs only in endosperm cells, where it is present in a granular form [5]. The water-insoluble starch granules contain the plant's main storage carbohydrates and provide the germ with energy in the form of glucose during germination. Starch granules consist of two types and sizes. The large, lenticular granules (A-starch) are 20–35 μm in the long dimension and the small spherical granules (B-starch) are 2–10 μm in diameter. The large granules constitute only 3–4% of the total number of granules, but they contribute 50–75% of the total weight of granules. Apart from minor constituents, both granule types consist of two polysaccharide compounds: amylose and amylopectin. Their quantitative distribution is around 25% and 75%, respectively. Both amylose and amylopectin are built up by glucose units. Amylose consists of α-(1,4)-linked D-glucopyranosyl units (Fig. 2.2A) and its structure is almost linear. Small parts of the molecules have α-(1,6)-linkages that provide slightly branched structures. The degree of polymerization ranges from 500 to 6000 glucose units corresponding to a molecular mass of 80,000–1 million. Amylose is able to form helical structures that include polar lipids (see Section 2.3) and this occurs in both native and gelatinized starch.

Like amylose, amylopectin consists of α-D-glucose primarily linked by α-1,4-bonds. However, it is branched to a much greater extent than amylose with 4–5% of glucose units additionally linked by α-(1,6)-glycosidic bonds (Fig. 2.2B). Structural models suggest that amylopectin also has double helices organized in parallel. Due to variable glucose linkages, amylopectin has three types of chains. Chain A is composed of α-(1,4)-linked glucose, chain B consists of glucose both α-(1,4)- and α-(1,6)-linked and chain C contains α-(1,4)- and (1,6)-linked glucose plus the terminal reducing group. All chains are interconnected and form clusters with molecular masses as high as 100 million. Thus amylopectin is one of the largest molecules found in nature. Due to the lower molecular mass, the number of amylose molecules exceeds those of amylopectin.

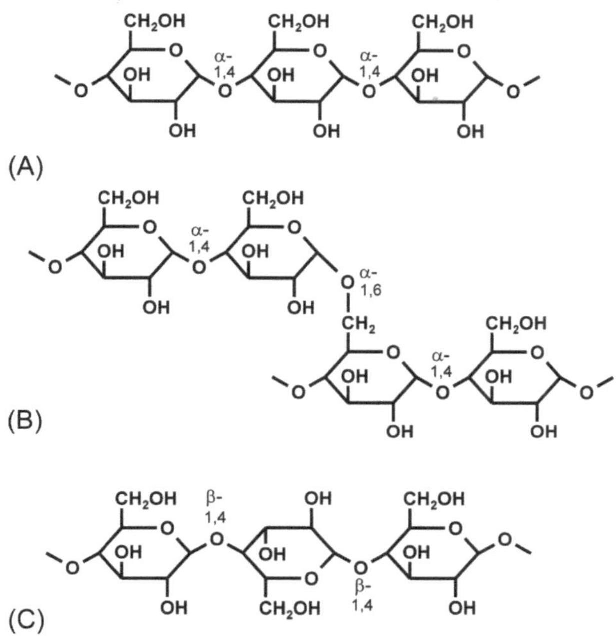

FIG. 2.2 Partial chemical structures of (A) amylose; (B) amylopectin; and (C) cellulose.

At the microscopic level, amylose and amylopectin are radially organized and form growth rings. They reflect alternating amorphous and semicrystalline shells. The amorphous shells are less dense and contain amylose and less-ordered amylopectin. The semicrystalline shells are composed of alternating amorphous and crystalline lamellae. The latter are formed by amylopectin double helices, while the former consist of the amylopectin branching regions.

Starch granules absorb water up to 50% of their dry weight, when heated in aqueous suspension, and swell [6]. Swelling is accompanied by a limited increase of the size and volume of the granules. At the end of swelling, they lose their structure and form gels, i.e., they gelatinize. The molecular order is completely transformed to an amorphous state. Upon subsequent cooling, the amorphous system reassociates to a more ordered crystalline state (the so-called retrogradation state). Retrogradation processes can be divided into two subprocesses. The first has been termed amylose retrogradation and occurs in a time range of minutes to hours. The second is amylopectin recristallizaton and takes place within hours or days. Therefore amylose retrogradation is responsible for the initial hardness of the gel, whereas amylopectin recrystallization determines the long-term structure and crystallinity of the gels. Gelatinization and retrogradation properties of starch are important factors for the digestibility of starch and the quality of baked goods and starch-derived products (see Sections 4.2 and 5.1).

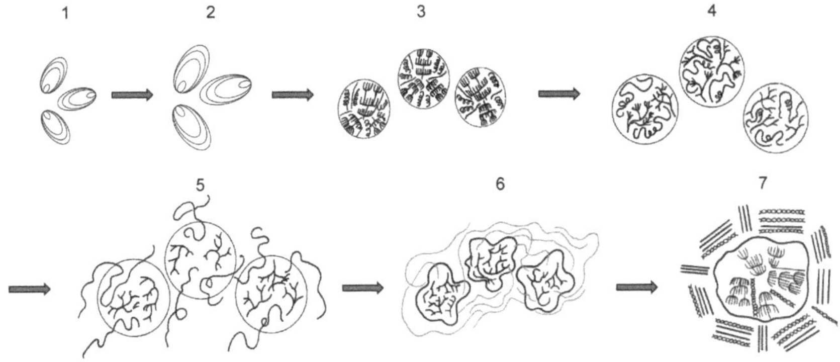

FIG. 2.3 Schematic changes of starch granules during heating and cooling of a starch-water mixture. (1) Native starch granules; (2) swelling; (3)–(5) gelatinization; (6) amylose retrogradation; and (7) amylopectin recrystallization.

Fig. 2.3 shows the processes that take place upon heating of a starch suspension and subsequent cooling and storage. In cold water, native starch granules are not soluble and form a suspension (1). Upon heating, the granules take up a limited amount of water. This process is referred to as swelling (2). Further heating leads to gelatinization associated with loss of crystalline structures (3), amylose leaching, and partial resolution of granules (4). At the end of heating, a starch paste with partially dissolved amylose and dispersed amylopectin molecules is present (5). Upon cooling, retrogradation starts: Firstly, an amylose network forms around the granule remnants (amylose retrogradation) (6). Secondly, ordered crystalline amylopectin structures form during storage (amylopectin recrystallization) (7).

Lipids and proteins are minor constituents of starch granules. The level of lipids in the granules is around 8–12 mg/g. They mainly consist of lysophospholipids (lysolecithin) and in smaller amounts of glycolipids and free fatty acids. As so-called starch lipids they are associated with amylose as well as with the outer branches of amylopectin. These lipid complexes lead to a delay of the onset of gelatinization and affect the properties of the starch especially in baking applications. The protein content of starch granules is about 5 mg/g. A considerable part stems from granule-bound starch synthase, a starch-synthesizing enzyme associated with amylose, and also from friabilin, a protein complex containing puroindolines that are responsible for the endosperm softness of common wheat.

2.1.2 Nonstarch polysaccharides

Wheat grains contain around 13% of nonstarch polysaccharides (NSPs). Major NSPs are the cell wall polysaccharides arabinoxylans, cellulose, and β-glucans; further NSPs are fructans and arabinogalactan peptides. NSPs differ from the starch components amylose and amylopectin in the nongranular state, the structure of their constituent saccharide units, and the nature of their unit linkages. The noncellulosic NSPs are sometimes named "hemicelluloses" and the pentose-containing NSPs are named "pentosans". NSPs belong to the group of dietary fiber [7], because human gastrointestinal enzymes are not able to cleave the predominant β-glycosidic linkages and therefore they are considered to be beneficial by regulating stool frequency, adding fecal bulk, and promoting the growth of gut bacteria (see Section 6.4).

The contents of the different NSP components, analyzed in grains of 129 winter wheat varieties, are included in Table 2.2 [8]. Arabinoxylans (AX) that accounts for 6–7% are the predominant NSPs of the grain. Based on solubility, AX can be divided into a water-extractable fraction (WEAX) and a water-unextractable fraction (WUAX) occurring in a ratio of about 1:3. They have a common general structure, in which WEAX act as a building block or precursor of WUAX. Both consist of linear β-(1,4)-D-xylopyranosyl chains that can be substituted at the C(O)2- and/or C(O)3-positions with α-L-arabinofuranose (Fig. 2.4). A minor component of AX is ferulic acid (Fig. 2.4) that is linked to arabinose by an ester bond at the C(O)5-position. Ferulic acid residues can form cross-links between AX molecules and are responsible for the insolubility

TABLE 2.2 Contents (%) of nonstarch polysaccharides (NSP) in winter wheat varieties [8].

Compound	Range	Mean
Total NSP	11.5–15.5	13.4
Arabinoxylan	5.5–7.4	6.5
Cellulose	1.7–3.1	2.1
Lignin	0.7–2.0	1.3
Fructan	0.8–1.9	1.3
β-Glucan	0.5–1.0	0.7

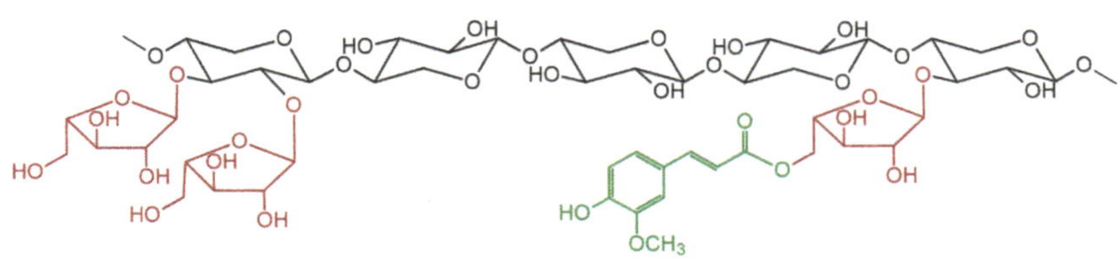

FIG. 2.4 Structural elements of arabinoxylan: β-D-xylpyranosyl residues substituted at C(O)-3 or C(O)-2 and C(O)-3 with an α-L-arabinofuranosyl residue (violet) and additionally with ferulic acid (green).

of WUAX. The amount of ferulic acid is much greater in arabinoxylans of the outer layers than in those of the starchy endosperm. WEAX are able to absorb 15–20 times more water than their own weight and form highly viscous solutions thus improving gas holding capacity of wheat doughs via the stabilization of gas bubbles during fermentation. WUAX also have a high water-holding capacity and assist in water-binding during dough mixing. However, they are considered to have a negative impact on bread-making, as they interfere with the formation of a continuous gluten network and thus destabilize the gas bubbles formed during fermentation.

Cellulose is the major structural polysaccharide of many plants. The structure is quite simple: The carbohydrate polymers are composed of β-(1,4)-linked D-glucose units and differ from amylose only in the β-linkage instead of the α-linkage (Fig. 2.2C). The degree of polymerization can range from 1000 to 14,000 units. Cellulose crystallizes as monoclinic, rod-like crystals. Because of its high molecular mass and crystalline structure, cellulose is insoluble in water and its ability to absorb water is poor. Wheat grains contain about 2–3% cellulose including 1–2% lignin. It is concentrated in the pericarp and testa and strongly lignified, whereas the content of cellulose in the less lignified starchy endosperm is rather low. Lignin fills the space between the cellulose polymers and is composed of complex cross-linked phenolic polymers with molecular masses in excess of 10,000.

β-Glucans, also called lichenins, are made up of linear D-glucopyranosyl chains joined by mixed β-(1,3)- and β-(1,4)-glycosidic linkages. The content of β-glucans in wheat grains is 0.5–1.0%; the major portion is present in cell walls of the aleurone layer and the starchy endosperm. β-Glucans of wheat are less soluble in water than those of barley and oats. Nevertheless, they are mucous substances that provide increased viscosity in an aqueous medium.

Fructans, called inulin, levan or phlein, are fructose polymers of different length with or without glucose. The monosaccharide units (up to 60 units) are linked by glycosidic β-(2,1)-bonds (inulin) or by β-(2,6)-bonds (levan or phlein) and occur both linear and branched. The content of fructans in the grain is 0.8–1.9%. They are enriched in the outer layers and to a lesser degree in the germ. Fructans with only few fructosyl units are termed fructo-oligosaccharides.

Arabinogalactan peptides occur in wheat grains in concentrations of approximately 0.3% and are concentrated in the starchy endosperm. They have molecular masses of approximately 23,500 and consist of a carbohydrate and a peptide moiety. The backbone of the carbohydrate part contain β-(1,6)-linked D-galactopyranose units substituted in the C(O)3-position with an α-L-arabinofuranosyl or a β-D-galactopyranosyl residue. The latter can also be substituted in its C(O)3-position with an α-L-arabinofuranosyl residue. The peptide has 15 amino acid residues including three hydroxyproline residues each linked to a carbohydrate chain.

2.2 Proteins

Proteins are linear polymers formed by carboxamide linkages (termed peptide bonds) of amino acids. The basic structure of amino acids includes an α-C-atom linked to an amino group, a carboxyl group, an H-atom, and a variable residual group (side chain) R (Fig. 2.5A). Usually, 20 amino acids (Table 2.3), different in the side chain R, participate in the formation of linear linkages of proteins. R can be a second H-atom (glycine) or aliphatic, aromatic, and heterocyclic residues. All amino acids present in proteins have configurations corresponding to the L-form (Fig. 2.5A). The structure of side chains determines the physical and chemical properties of amino acids and proteins. For example, they can be polar, charged and polar or nonpolar. Linkages of amino acids are formed by a peptide bond between the amino group and the carboxyl group (Fig. 2.5B). Protein structures can be modified by other molecules attached to the amino acids such as sugars (glycoproteins) or lipids (lipoproteins).

The type of amino acids (amino acid composition), their number (molecular mass), and their order (amino acid sequence), together termed primary structure, determine the individual properties of proteins. The primary structure decides on the steric dimension of a specific protein segment (secondary structure) and of the whole protein (tertiary structure) as well as on the ability to form aggregates with other protein molecules (quaternary structure). Disulfide linkages, formed by cysteine residues either within a protein chain (intrachain) or between proteins (interchain), complete both tertiary and quaternary structures. Aside from proteins defined as polymers with hundred and more amino acid residues ($n \geq 100$), polypeptides ($n = 10–99$) and oligopeptides ($n = 2–9$) exist.

Proteins make up the second important compound class in wheat grains. Their content may vary from less than 6% to more than 20%, although most commercial grain samples are between 10% and 15% depending on both genotype (species and variety) and growing conditions (soil, fertilization, and climate).

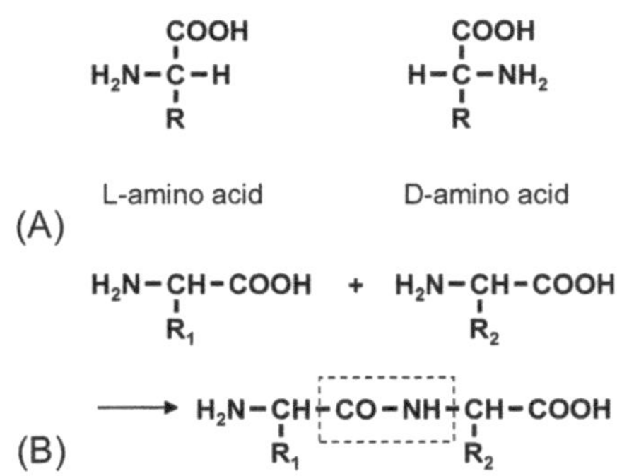

FIG. 2.5 (A) Configurations of amino acid enantiomers and (B) linkage of two amino acids by a peptide bond.

TABLE 2.3 Common names, abbreviations, properties of side chains, and molecular masses of standard amino acids.

Name	Three-letter code	One-letter code	Side chains	Molecular mass
Alanine	Ala	A	Nonpolar	89.1
Arginine	Arg	R	Charged, polar	174.2
Asparagine	Asn	N	Polar	132.1
Aspartic acid	Asp	D	Charged, polar	133.1
Cysteine	Cys	C	Nonpolar	121.2
Glutamic acid	Glu	E	Charged, polar	147.1
Glutamine	Gln	Q	Polar	146.1
Glycine	Gly	G	Nonpolar	75.1
Histidine	His	H	Polar	155.2
Isoleucine	Ile	I	Nonpolar	131.2
Leucine	Leu	L	Nonpolar	131.2
Lysine	Lys	K	Charged, polar	146.2
Methionine	Met	M	Nonpolar	149.2
Phenylalanine	Phe	F	Nonpolar	165.2
Proline	Pro	P	Nonpolar	115.1
Serine	Ser	S	Polar	105.1
Threonine	Thr	T	Polar	119.1
Tryptophan	Trp	W	Nonpolar	204.2
Tyrosine	Tyr	Y	Polar	181.2
Valine	Val	V	Nonpolar	117.2

Durum wheat and spelt plants appear to produce the highest and common wheat plants the lowest amount of grain proteins within the *Triticum* species [9]. Regarding fertilization, the supply of the growing plant with nitrogen specifically determines the final protein content of the grain [10]. Remarkably, the continuing increase in atmospheric carbon dioxide concentration will lead to a severe decrease of the grain protein content [11].

Proteins are distributed over the whole grain, their concentrations within each compartment however are considerably different. According to a study of Jensen and Martens [12], who investigated separated compartments, the germ contains 341 mg/g (7.5% of total proteins on a dry weight base), the aleurone layer 229 mg/g (14.2%), pericarp and testa 108 mg/g (3.8%), and the starchy endosperm 102 mg/g (74.5%). Considering the proportions of the compartments, most proteins (74.5%) are located in the starchy endosperm.

Mutation of few ancestral genes during thousands of years resulted in numerous modifications of protein structures so that wheat grains actually contain hundreds of different protein components that are in parts closely related. They have been traditionally classified into four fractions (albumins, globulins, gliadins, and glutenins, named "Osborne fractions") according to different solubilities and residual proteins in addition [13,14]. Albumins are soluble in water and dilute salt solutions. Globulins are insoluble in pure water but soluble in dilute salt solutions. Due to structural and functional similarities, albumins and globulins are frequently dissolved together with a dilute salt solution and combined in one fraction (albumins/globulins). Gliadins are insoluble in water and salt solutions but soluble in aqueous alcohols (e.g., 60% ethanol or 50% propanol). Native glutenins are insoluble in water, salt solutions, and aqueous alcohols. Like gliadins, they are soluble in aqueous alcohols after cleavage of disulfide bonds by reducing agents resulting in monomeric glutenin subunits. The Osborne fractions can be obtained from flour by subsequent extraction with the four solvents mentioned. After exhaustive extraction, a small portion of residual proteins such as membrane proteins and lipid proteins remains insoluble. According to distinct differences in amino acid sequences, gliadins can be subdivided into four types (ω5-, ω1,2-, α-, and γ-gliadins) and glutenins into two types [high-molecular-weight glutenin subunits (HMW-GS) and low-molecular-weight glutenin subunits (LMW-GS)].

The contents and proportions of Osborne fractions and single protein types vary considerably and depend on species, variety, and growing conditions. Table 2.4 gives an overview on the range of the proportions (%) of extractable protein fractions (albumins/globulins, gliadins, and glutenins) and single gliadin and glutenin types of whole grain flours from different wheat species (mean values for eight varieties each grown under the same conditions) [9].

The botanical relationship of the different wheat species is well reflected by the protein compositions: They all contain albumins and globulins, all types of gliadins, and all types of glutenins. Gliadins are generally predominant proteins in wheat grains (50–71% of total extractable proteins) followed by albumins/globulins (15–32%) and glutenins (6–26%). Among single protein types, α- and γ-gliadins belong to the major types, LMW-GS are in a medium range and ω5- and ω1,2-gliadins as well as HMW-GS are minor components. The data of Table 2.4 clearly demonstrate that varieties within each species show broad ranges of protein proportions, even when they are cultivated under identical conditions. Under different growing conditions, the range of variations may be even broader. In conclusion, highly significant general differences between wheat species, regarding the qualitative and quantitative protein compositions, are missing and only some distinct trends, indicated in the following paragraphs, can be recognized.

TABLE 2.4 Range of proportions (%, based upon total extractable proteins) of protein fractions and types in grains from different wheat species [9].[a]

		GLIA				
Species	**ALGL**	**Total**	**ω5**	**ω1,2**	**α**	**γ**
Einkorn	21.5–25.9	62.5–71.0	2.4–11.9	1.0–2.2	36.1–40.4	13.8–22.2
Emmer	22.4–25.3	60.4–65.7	1.4–4.5	3.7–6.2	25.1–31.2	23.2–30.6
Durum wheat	16.7–24.8	56.1–66.1	2.8–7.2	1.5–5.0	25.1–38.5	19.7–29.5
Spelt	14.5–26.7	57.1–65.9	1.8–4.0	3.6–5.8	25.5–31.2	21.4–26.8
Common wheat	20.3–32.0	50.1–60.5	1.9–3.7	2.5–4.6	21.5–27.7	20.1–25.3

	GLUT			
Species	**Total**	**HMW**	**LMW**	**GLIA/GLUT**
Einkorn	5.9–14.4	0.9–2.0	4.4–12.2	4.2–12.0
Emmer	9.7–17.0	2.0–4.2	7.2–13.1	3.6–6.7
Durum wheat	12.6–26.0	2.2–4.0	9.6–21.2	2.2–5.3
Spelt	14.7–21.7	2.5–5.0	11.5–16.1	2.8–4.0
Common wheat	16.3–25.7	3.6–6.7	13.2–18.9	2.0– 3.2

[a] *ALGL, albumins + globulins; GLIA, gliadins; GLUT, glutenins; HMW, high-molecular-weight glutenin subunits; LMW, low-molecular-weight glutenin subunits; ω5, ω1,2, α, γ, gliadin types.*

Einkorn and common wheat show the highest and lowest proportions of gliadins, respectively and corresponding low and high proportions of total glutenins. Among gliadin types, einkorn is significantly different from the other species due to low proportions of ω1,2- and γ-gliadins and high proportions of ω5- and α-gliadins. Among glutenin subunits, there is a distinct trend in increasing proportions of HMW-GS from einkorn to common wheat. The proportions of LMW-GS are clearly decreased in einkorn and emmer and tend to be increased in durum wheat, spelt, and common wheat. The ratio of gliadins to glutenins is the most characteristic feature of the species: high in einkorn (4.2–12.0%) and low in common wheat (2.0–3.2%). Altogether, one of the most important aims of wheat breeding and selection—high bread-making quality—is well reflected by the lowest ratio of gliadins to glutenins and the highest proportions of HMW-GS of common wheat varieties, two parameters that are decisive for dough properties and bread quality [15].

The proportions of Osborne fractions as well as single gluten protein types are strongly influenced by nitrogen and sulfur fertilization. High levels of nitrogen fertilization increase the proportions of total gliadins, ω5- and ω1,2-gliadins, and HMW-GS, while those of total glutenins, γ-gliadins, and LMW-GS decrease [10]. In case of sulfur deficiency, the gliadin/glutenin ratio increases, ω5- and ω1,2-gliadins become major components and the proportions of γ-gliadins and LMW-GS decrease distinctly [16].

The amino compositions of the Osborne fractions are summarized in Table 2.5 [17]. Albumins and globulins are closely related and characterized by relatively balanced "ordinary" compositions. They differ significantly from gliadins and glutenins that show unbalanced "extraordinary" compositions. Gliadins present the highest values for glutamine (37%) and proline (17%), whereas especially the essential amino acids lysine (0.8%), methionine (1.1%), and tryptophan (0.4%) occur only in small amounts. Compared with gliadins, glutenins (plus residual proteins) show a more balanced composition with less glutamine (30%) and proline (12%) and increased values for most of the other amino acids, in particular of glycine (8%).

Aside from Osborne classification, wheat proteins can be alternatively grouped into four classes according to their different functions: (1) Storage proteins occurring in the starchy endosperm (gliadins and glutenins); (2) metabolic proteins (e.g., enzymes); (3) protective proteins (e.g., enzyme inhibitors) both mainly present in the aleurone layer and the germ and belonging to the albumin and globulin fractions; and (4) miscellaneous proteins with further specific functions. These groups are described in more details in the following Sections 2.2.1–2.2.5.

2.2.1 Storage proteins

The storage protein fraction amounts to about 8% of the whole grain and consists of gliadins and glutenins (together termed gluten proteins). They are essential for the baking performance of wheat flour (see Section 4.2) but are also

TABLE 2.5 Amino acid composition (mol%) of Osborne fractions [17].[a]

AA	ALBU	GLOB	GLIA	GLUT[b]
Ala	6.9	7.5	2.8	4.4
Arg	4.0	6.4	1.7	2.7
Asx	9.7	7.7	2.7	3.7
Cys	3.2	3.6	2.2	1.4
Glx	20.9	15.2	37.1	30.1
Gly	6.9	8.3	2.9	7.9
His	1.8	2.4	1.7	1.8
Ile	3.3	3.8	4.1	3.5
Leu	6.4	7.3	6.9	6.9
Lys	3.0	4.0	0.8	2.1
Met	1.6	2.0	1.1	1.3
Phe	3.1	3.1	4.6	3.6
Pro	9.3	6.9	16.6	11.9
Ser	6.2	6.6	5.9	7.3
Thr	3.8	4.6	2.3	3.6
Trp	1.1	0.9	0.4	0.6
Tyr	2.8	2.9	2.0	2.4
Val	6.0	6.8	4.2	4.8

[a] *AA, amino acids (code see Table 2.3; Asx, asparagine or aspartic acid; Glx, glutamine or glutamic acid); ALBU, albumins; GLOB, globulins; GLIA, gliadins; GLUT, glutenins.*
[b] *Residual proteins included.*

precipitating factors of several wheat hypersensitivities (see Chapters 8–10). Gliadins and glutenins are encoded on the chromosomes 1A (einkorn), 1A and 1B (emmer and durum wheat), and 1A, 1B, and 1D (spelt and common wheat). Closely related genes lead to distinct homologies between proteins of all wheat species. Storage proteins are major proteins of the grain (corresponding to around 70–80% of total proteins) and are located exclusively in the starchy endosperm. Their only function is to provide the germ with amino acids and nitrogen during the first stage of germination. Their synthesis starts around ten days after flowering of the plants and ends at around 30 days after flowering.

The storage protein fraction contains a complex mixture of different components. According to significant differences in amino acid compositions and sequences, they can be classified into four gliadin types (ω5-, ω1,2-, α-, and γ-gliadins) and two glutenin types (HMW- and LMW-GS) (see above). Typical of all these proteins are the extremely unbalanced amino acid compositions showing high contents of glutamine (up to 53 mol%) (Table 2.6) [14]. In contrast to most amino acids, glutamine (the amidated form of glutamic acid) contains two nitrogen (N) atoms (Fig. 2.6A) that are important for the nitrogen supply of the germ. Proline is the second predominant amino acid with contents up to 29 mol%. Unique to this amino acid is the secondary amino group (Fig. 2.6B) that causes kinks within the protein chains and therefore allows a dense packing of the protein strands in the starchy endosperm. Moreover, peptide bonds that contain a proline residue are resistant to most cleaving enzymes and prevent the storage proteins from degradation by external enzymatic attacks. The relatively high proportions of hydrophobic amino acids such as leucine, phenylalanine, and tyrosine are of further importance for the function of storage proteins.

TABLE 2.6 Amino acid composition (mol%) of gliadin and glutenin types [14].[a]

AA	ω5	ω1,2	α	γ	HMWx	HMWy	LMW
Ala	0.2	0.6	2.2	2.9	2.9	3.8	3.9
Arg	1.0	0.5	2.2	1.3	1.2	2.7	2.5
Asp	0.2	0.0	0.4	0.7	0.5	0.6	0.3
Cys	0.0	0.0	2.2	2.6	0.5	1.1	2.8
Gln	53.1	41.6	35.9	36.1	35.8	32.2	32.3
Glu	1.9	1.4	1.8	0.7	2.1	2.7	1.8
Gly	0.7	0.8	2.6	2.9	20.0	17.9	3.2
His	1.4	0.5	2.2	1.3	0.5	2.4	1.1
Ile	4.3	1.6	4.7	4.9	0.5	1.1	4.3
Leu	3.1	4.0	8.0	7.2	4.4	3.8	8.1
Lys	0.7	0.5	0.4	0.7	0.7	1.1	0.3
Met	0.0	0.0	0.7	1.6	0.3	0.6	1.8
Phe	9.3	8.3	3.7	4.9	0.3	0.3	4.3
Pro	19.8	29.0	14.3	17.5	13.0	11.0	12.8
Ser	2.9	5.9	5.1	5.5	6.0	6.6	8.9
Thr	0.5	2.4	1.8	2.0	2.9	3.6	1.4
Trp	0.0	0.3	0.4	1.0	1.0	0.9	0.7
Tyr	0.7	1.6	3.7	0.3	5.7	5.2	3.5
Val	0.2	0.5	5.1	4.6	1.7	2.4	5.3

[a] *AA, amino acids (code see Table 2.3); ω5, ω1,2, α, γ, gliadin types; HMW, LMW, glutenin types.*

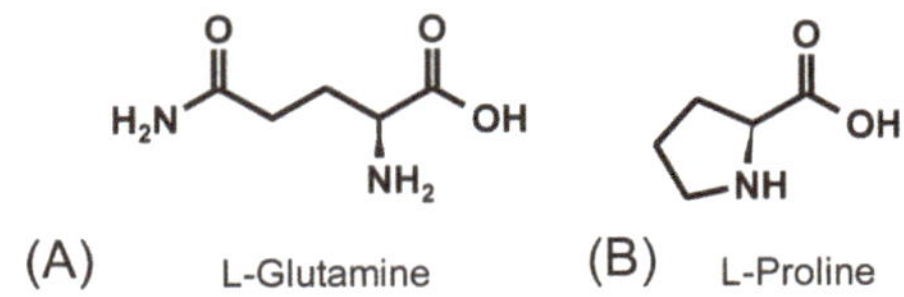

FIG. 2.6 Structures of the amino acids (A) L-glutamine and (B) L-proline.

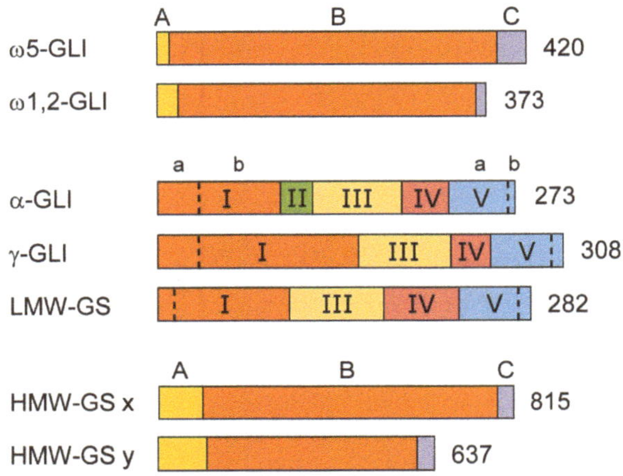

FIG. 2.7 Schematic architecture of wheat storage proteins (GLI, gliadins; GS, glutenin subunits; HMW, high-molecular-weight; LMW, low-molecular-weight).

They guarantee the insolubility of storage proteins in water and thus prevent the leakage of proteins out of the grains during germination. Additional common features of all storage proteins are low contents of the essential amino acids lysine (0.3–1.1 mol%), methionine (0.0–1.8 mol%), and tryptophan (0.0–1.0 mol%).

Gliadins, the alcohol-soluble storage proteins, are mainly present as monomers. Initially, they were classified into α-, β-, γ-, and ω-gliadins (in order of decreasing mobility) on the basis of acid gel electrophoresis. Later studies on amino acid sequences however demonstrated that α- and β-gliadins fall into one type (summed up as α/β- or α-type) and that ω-gliadins can be differentiated into the ω5- and ω1,2-types. Each type comprises several protein components that structurally differ only in the modification of a few amino acid sequences (see database Uni Prot KB).

ω5-Gliadins have molecular masses of around 50,000 and are characterized by the highest contents of glutamine, proline, and phenylalanine among gliadin types that together amount for around 80% of the total compositions (Table 2.6). Due to the lack of cysteine residues, ω5-gliadins are not able to form disulfide cross-links. The amino acid sequences can be subdivided into a short N-terminal segment A, a long central segment B, and a short C-terminal segment C (Fig. 2.7). Segments A and C contain nonrepetitive sequences with balanced amino acid compositions, whereas segment B contains repetitive sequences consisting mainly of glutamine, proline, and phenylalanine. Typical repetitive units of segment B are pentapeptides such as QQQFP (one-letter-code for amino acids, see Table 2.3) repeated up to fifty times. The structures of ω1,2-gliadins are related to those of ω5-gliadins, but they contain less glutamine and more proline and have lower molecular masses (≈40,000) and a shorter segment B (Fig. 2.7). The repetitive units consist of heptapeptides such as PQQPFPQQ repeated up to thirty times.

The structures of α- and γ-gliadins are related and differ significantly from those of ω-gliadins. They have molecular masses in a range from 28,000 to 35,000 and the proportions of glutamine and proline are much lower (Table 2.6). Both α- and γ-gliadins consist of N- and C-terminal domains. The N-terminal domain of α-gliadins can be divided into two segments (Fig. 2.7). Segment I contains short nonrepetitive sequences (a) and repetitive sequences (b) with dodecapeptide units such as QPQPFPQQPYP, repeated up to five times. Unique to α-gliadins is segment II containing poly-glutamine sequences with up to 18 glutamine residues. The N-terminal domain of γ-gliadins consists of segment Ia with nonrepetitive sequences and segment Ib with repetitive sequences containing heptapeptide units such as QPQQPFP repeated up to fifteen times. Segment II is missing. Within the C-terminal domains, α- and γ-gliadins are homologous (segments III–V). They present sequences that are nonrepetitive and possess more usual amino acid compositions. With a few exceptions, α- and γ-gliadins contain six and eight cysteine residues, respectively, located in the C-terminal domain; they form three and four homologous intrachain disulfide bonds (Fig. 2.8).

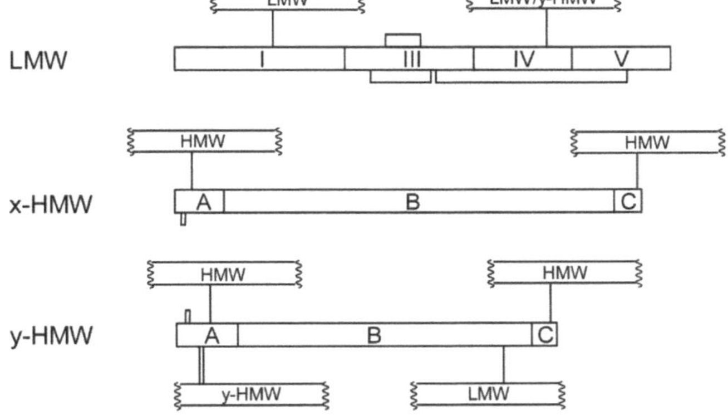

FIG. 2.8 Schematic design of disulfide structures of gluten proteins.

A portion of around 15–20% of the gliadin fraction does not consist of monomeric but oligomeric proteins with molecular masses of around 100,000–600,000 that have been named as HMW-gliadins but also as aggregated gliadins or ethanol-soluble glutenins. They are formed by modified gliadins with an odd number of cysteine residues that are linked together as well as linked to LMW-GS and HMW-GS by interchain disulfide bonds yielding alcohol-soluble oligomers. Thus HMW gliadins contain all types of gluten proteins. Their potential polymerization to glutenins via disulfide bonds is interrupted by the presence of so-called terminators (e.g., compounds with a free thiol group like glutathion) during biosynthesis resulting in oligomers that are still soluble in aqueous ethanol [18].

The glutenin fraction comprises polymeric proteins linked by interchain disulfide bonds. The polymers have varying molecular masses ranging from around 600,000 to more than 10 million. Thus a part of glutenin polymers belongs to the largest proteins in nature. After cleavage of disulfide bonds, the resulting monomers (glutenin subunits) are alcohol-soluble like gliadins and can be differentiated into two types: LMW-GS and HMW-GS. The structural architecture of LMW-GS is similar to those of α- and γ-gliadins (Fig. 2.7). The N-terminal domain (segment I) consists of glutamine- and proline-rich repetitive units such as QQQPPFS repeated up to thirteen times. The C-terminal domain (segments III–V) is homologous to those of α- and γ-gliadins within segments III and V, whereas segment IV is unique to LMW-GS. Eight cysteine residues are present in LMW-GS; six residues are in positions homologous to α- and γ-gliadins and form three intrachain disulfide bonds (Fig. 2.8). Two additional cysteine residues are located in segments I and IV and form disulfide bonds with different protein types resulting in oligomeric HMW gliadins and polymeric glutenins.

HMW-GS consist of three structural domains (Fig. 2.7): a nonrepetitive N-terminal domain A, a repetitive central domain B, and a nonrepetitive C-terminal domain C. Domains A and C are characterized by a balanced amino acid composition and the presence of almost all cysteine residues. Two cysteine residues located in the A and C domains have been proposed to form interchain disulfide bonds resulting in large polymers. Domain B contains repetitive hexapeptides such as QQPGQG forming a backbone with inserted hexapeptides (e.g., YYPTSP) and tripeptides (e.g., QQP or QPG). According to different amino acid sequences HMW-GS are divided into two subtypes: the x- and the y-type. Compared to the x-type, y-type HMW-GS show a shorter segment B (about 50 repeats compared to about 70 repeats) and an additional cysteine residue in segment B that is linked to LMW-GS. HMW-GS form both intra- and interchain disulfide bonds, the latter being involved in head-to-tail polymerization (Fig. 2.8).

Disulfide bonds play an important role in determining the structures and properties of gluten proteins. They are formed between sulfhydryl groups of cysteine residues, either within a single protein (intrachain) or between proteins (interchain).

Intrachain bonds stabilize the steric structures of the proteins and interchain bonds provoke the generation of oligomers and polymers. The formation of disulfide bonds starts rapidly after synthesis of proteins within the lumen of endoplasmatic reticulum as integral part of protein folding. Intrachain links are generated more rapidly than interchain links. To participate in growing polymers like glutenins, proteins need at least two cysteines forming interchain disulfide bonds. HMW- and LMW-GS usually fulfill this requirement and generate linear polymers via cysteines of the N-terminal and C-terminal domains (HMW-GS) and segments I and IV (LMW-GS) (Fig. 2.8). Polymerization is differently stopped by so-called terminators such as glutathione or modified gliadins with an odd number of cysteines resulting in a wide range of molecular masses. The overriding importance of disulfide bonds for dough properties can be demonstrated by the addition of reducing agents that weaken the dough by the cleavage of disulfide bonds and by the addition of oxidizing agents that form disulfide bonds and strengthen the dough.

In summary, the composition of storage proteins is characterized by a complex mixture of monomers, oligomers, and polymers comprising molecular masses from around 30,000 to more than 10 million. They are characterized by long unique repetitive sequences that are rich in glutamine, proline, and hydrophobic amino acids and by specific intra- and interchain disulfide bonds that stabilize the steric structure. On the one hand, they optimally fulfill their function as storage proteins: insoluble in water, tightly packed, and providing the germ with nitrogen and amino acids during germination. On the other hand, their ability to form a cohesive, viscoelastic network (called gluten), when mixed with water, determines the unique baking quality of wheat.

2.2.2 Gluten

The term "gluten" has experienced an increased awareness during the last decades mainly due to the enhanced interest in gluten hypersensitivities and gluten-free diet [19]. However, when people are asked what gluten actually is, the vast majority of persons only has some vague and often wrong ideas. Even among many scientific specialists, the meaning of gluten is not exactly known. Different definitions of the term gluten may have contributed to confusions. Originally, gluten (from Latin for "glue") has been defined as the rubber-like proteinaceous mass that remains, when wheat dough is washed with water or salt solutions to remove soluble constituents and starch granules (Fig. 2.9). The starch industry, using corn as raw material, expanded the meaning of gluten with the term "corn gluten", the proteinaceous byproduct of corn starch production. Corn gluten mainly consists of the storage proteins of corn (zeins). In the field of hypersensitivities, all storage proteins of wheat (gliadins and glutenins), rye (secalins), barley (hordeins), and oats (avenins) were summarized within the terms gluten or gluten proteins, first in the field of celiac disease and afterwards in the fields of other gluten hypersensitivities (see Section 8.4 and Chapters 9–11). According

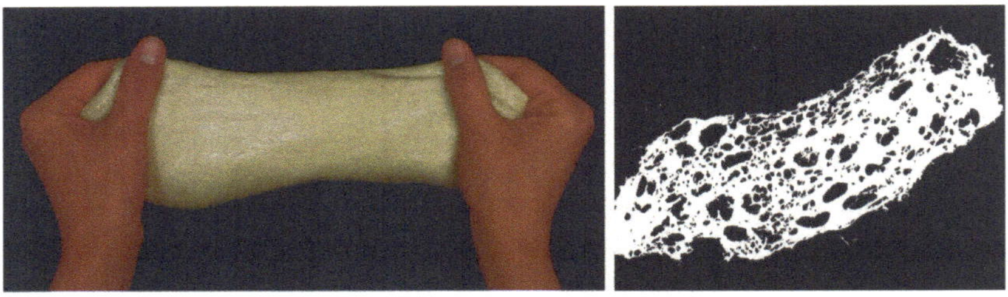

FIG. 2.9 Gluten, a rubberlike proteinaceous mass isolated from wheat flour (left); gluten network in a hydrated flour particle (starch granules have been removed by amylase digestion) (right).

to the Codex Alimentarius (ALINORM 08/31/26, Appendix III, WHO, Rome), gluten is defined as "a protein fraction from wheat, rye, barley, oats or their cross-bred varieties and derivatives thereof, to which some persons are intolerant and that is insoluble in water and 0.5 mol/L NaCl".

The preparation of gluten from wheat flour was first described by the Italian biochemist Jacopo Beccari in 1745. Gluten was recognized as first protein of plant origin; before that time, it was believed that proteins were present only in materials of animal origin. Today, Beccari's procedure is widely used for the industrial preparation of so-called vital gluten, a byproduct of wheat starch production (see Section 5.2). Chemically, wheat gluten mainly consists of gliadins and glutenins (see above) that form a three-dimensional network linked by covalent and noncovalent bonds, when flour is mixed with water (Fig. 2.9). Gluten proteins decisively account for the unique physical properties of wheat dough such as cohesivity, viscosity, extensibility, and elasticity. Moreover, the gluten network mediates the gas-holding properties of dough during the fermentation and baking process (see Sections 4.2.1.2 and 4.2.1.3).

2.2.3 Metabolic proteins

Wheat grains contain many proteins with metabolic functions. They primarily occur in the albumin and globulin fractions. The class of metabolic proteins is extremely heterogeneous and most of them are located in the germ and aleurone layer, others are distributed throughout the endosperm. Among metabolic proteins, enzymes are predominant and play specific degrading, synthetic, and regulatory roles [20,21]. Most of them are produced during grain maturation, are nonactive in the mature and dry grain and are fully active, when the grain is penetrated with water during the germinating phase. Enzymes are proteins that catalyze reactions with high specificity. According to their different specificities, wheat enzymes can be assigned to degrading (hydrolyzing), synthesizing, and other metabolic functions and ordered according to their substrate specificities. The most important enzymes present in wheat grains are described in the following paragraphs.

Enzymes that cleave polysaccharides (glycoside hydrolases) are most abundant in wheat grains. Amylases hydrolyze starch and other α-(1,4)-glucans and release either oligosaccharides of 6–7 glucose units (α-amylase) or the disaccharide maltose (β-amylase). Glucoamylases (γ-amylases) supplementarily cleave α-(1,6) glycosidic bonds. β-Amylases are present in the endosperm in relatively large amounts. They occur in several isoforms that differ in the electrophoretic behavior and pH optimum. β-Amylases accumulate during starch and gluten protein deposition; they are free during early grain development and bound to other proteins by disulfide bonds at maturity. Activation occurs upon disulfide bond reduction that releases active monomers with four thiol groups and one disulfide group and molecular masses of around 64,000. Starting from the nonreducing end, β-amylases catalyze the hydrolysis of each second 1,4-glycosidic bond and release maltose. In comparison to amylose, amylopectin is hydrolyzed to a less degree, because reaction is stopped, before branch points are reached.

α-Amylases are present in mature grains only in trace amounts but increase greatly upon germination. Numerous isoforms have been isolated and characterized. They are similar in that they are acidic, contain no free cysteine, have two disulfide groups and molecular masses between 42,000 and 45,000. α-Amylases attack amylose inside the helix, hydrolyze glycosidic bonds that are one turn distant and mainly release oligosaccharides. Amylopectin is hydrolyzed at random and the branch points are overleapt yielding oligosaccharides, maltose, and glucose. A mixture of α- and β-amylases does not completely degrade starch, as neither enzyme can break α-(1,6)-glycosidic bonds present in amylopectin. Glucoamylase, the third main type of amylases, completes the total conversion of starch to glucose. It hydrolyzes α-(1,4)-linkages at the nonreducing ends of the partially hydrolyzed polysaccharides and, although slower, α-(1,6)-linkages.

Enzymes that hydrolyze proteins are called "peptidases" (formerly "proteinases"). They cleave the peptide bond between two adjacent amino acid residues in a protein. Proteolytic enzymes can be differentiated into endo- and exopeptidases. Endopeptidases hydrolyze peptide bonds somewhere along the protein chain. The cleavage site is dependent on the amino acid side chains next to the peptide bond. Exopeptidases remove single amino acids from the N-terminal end (aminopeptidases) or the C-terminal end (carboxypeptidases) of proteins. Wheat grains contain numerous different peptidases with unique properties and specificities. They may vary considerably in time of development, distribution within the grain, concentration, and specificity. In sound, dry grains, peptidase activity is low but increases significantly with the time of germination. The most important peptidases are those endo- and exopeptidases that hydrolyze the storage proteins to provide the germ with amino acids and nitrogen during the first stage of development. Endopeptidase, cleaving those peptide bonds that include a proline residue (so-called prolyl endopeptidases), are of special importance. They hydrolyze peptide bonds on the C-terminal side of prolyl residues and thus are complementary to many other peptidases that are not able to degrade the proline-rich storage proteins.

Various enzymes called lipases affect grain lipids. Lipases are a subclass of esterases and hydrolyze the ester bonds of triglycerides releasing mono- and diglycerides and long-chain free fatty acids. Lipases are present in all tissues of the grain including pericarp, testa, germ, and aleurone layer with high activities and the endosperm with low activity. Phospholipases are special types of lipases that liberate fatty acids from phospholipids. Phospholipase A removes one fatty acid from diacylphospholipids, while phospholipase B removes both fatty acids.

Some other enzymes, important for the physiology and technological properties of wheat, are shortly described in the following. Lipoxygenase catalyzes the peroxidation of certain polyunsaturated fatty acids and mainly occurs in the germ. The enzyme is important for bleaching pigments as well as for the mixing tolerance of dough. Phytase hydrolyzes phytic acid to inositol and free orthophosphate. The enzyme contributes to improved uptake of phosphate from the diet and prevents minerals from binding to intact phytic acid that would lead to mineral deficiencies. Polyphenol oxidases oxidize phenols to quinones and thus can cause browning of dough from whole grain flour. These enzymes are formed very early during grain development and decrease upon maturation. At maturity, the largest amount of activity is located in the outer layers of the grain. The enzyme peroxidase is a hemoprotein that catalyzes the oxidation of a number of aromatic amines and phenols by hydrogen peroxide. It is present in the grain at early stages of maturation and is believed to affect the browning of doughs by destruction of carotenoid pigments. Ascorbic acid oxidase catalyzes the oxidation of L-ascorbic acid to L-dehydroascorbic acid and glutathione dehydrogenase catalyzes the oxidation of glutathione in the presence of dehydroascorbic acid. Both enzymes essentially contribute to the well-known technological effect of ascorbic acid as dough improver.

Developing grains have numerous enzymes that are responsible for the synthesis of constituents. Especially important are enzymes that synthesize starch and proteins. The first step of starch synthesis is the conversion of glucose-1-phosphate to ADP-glucose by the enzyme glucose-1-phosphate adenyltransferase. Afterwards, the enzyme starch synthase catalyzes the reaction between ADP-glucose and a growing chain of glucose residues linked by 1,4-α-glycosidic bonds, whereby amylose is created. Starch branching enzymes hydrolyze internal 1,4-glycosidic bonds of α-glucans and introduce 1,6-glycosidic bonds between these chains generating amylopectin. Protein synthesis is associated with the enzyme amino acyl-tRNA synthetase. It plays an important role in RNA translation, i.e., the expression of genes to create proteins in developing wheat. The enzyme catalyzes the esterification of an amino acid to its compatible transfer-RNA to form an amino acyl-tRNA. The specificity of the reaction is ensured by multiple forms of the enzyme, one for each amino acid of the genetic code. Once the tRNA is charged, ribosomes can transfer the amino acid onto a growing peptide chain.

2.2.4 Protective proteins

A number of proteinaceous inhibitors found in wheat protect the grain constituents from the enzymatic attack of external enemies such as fungi, worms, and insects. Those inhibitors that prevent degradation of starch and storage proteins by the hydrolyzing enzymes amylases and peptidases are most important [21]. These inhibitors occur mainly in the starchy endosperm and the germ. The group of amylase inhibitors comprises monomeric, dimeric, and tetrameric forms that contain subunits with molecular masses from about 12,000 to 16,000. They inhibit mammalian and insect amylases but not the endogenous wheat α- and β-amylases, which reflects their protective role very well. Peptidase inhibitors such as serpins are active against external peptidases like trypsin and chymotrypsin but do not inhibit endogenous exo- and endopeptidases that are activated during germination [22]. Some wheat inhibitors are bifunctional, i.e., they are capable of simultaneously inhibiting amylase and peptidase activity. A prominent group of bifunctional inhibitors comprises amylase/trypsin inhibitors representing up to 17 homologous proteins with molecular masses of around 15,000. Many inhibitors present in wheat are involved in the initiation of wheat hypersensitivities, for example, amylase inhibitors in baker's asthma or amylase/trypsin inhibitors in nonceliac gluten sensitivity (see Sections 9.4 and 10.3).

2.2.5 Miscellaneous proteins

The group of purothionins occurs as lipid complexes in the endosperm and consists of low-molecular-weight, high-cysteine basic proteins. They are extractable from flour with petroleum ether. Single components (e.g., α- and β-purothionins) are closely related; they contain around 45 amino acid residues corresponding to a molecular mass of around 5000. In their native state, purothionins are dimeric and the binding to the lipid components (phospholipids or glycolipids) occurs by electrostatic forces. Their physiological function is unknown, but it seems to be of protective nature, because purothionins are toxic to bacteria, yeast, vertebrates, and insect larvae.

Wheat endosperm contains another type of lipoproteins, the so-called CM-proteins. They are extractable by a mixture of chloroform and methanol (CM) and have molecular masses ranging from 13,000 to 17,000. They can be made water-soluble after dialysis in a solution of urea that dissociates lipids. Therefore they are sometimes classified as albumins or globulins, but they do not exactly fit into Osborne's solubility classes. Some CM-proteins are homologous to inhibitors of cereal α-amylases and to trypsin inhibitors (see above) as well as to legume storage proteins. Thus they are members of a superfamily of seed proteins with still unknown functions.

Puroindolines a and b are proteins with molecular masses of around 13,000 that are bound to the surface of starch granules. They are the molecular basis of grain hardness and texture: When both puroindolines are present, grain texture is soft ("soft wheat"). When one of the puroindolines is absent, the result is a hard texture ("hard wheat"). Soft wheat grains are more friable, require less energy for milling and produce flours with small particle size. Hard wheat flours are coarser and have more broken and damaged starch granules. Durum wheat lacks both puroindolines and the texture is very hard. Due to this hardness, durum wheat is usually milled into semolina and not into flour.

Agglutinins are lectin-like proteins that occur in the germ. One of the wheat germ agglutinins was characterized as a glycoprotein having glucose as its major carbohydrate constituents and a molecular mass of 23,500. Another agglutinin was identified as protein without glucose and an apparent molecular mass of 17,800. Agglutinins are known to cause mammalian cells in culture to "agglutinate" by binding to carbohydrate receptors on the cell membrane surface. Therefore they are often called "phytohaemagglutinins".

So-called triplet proteins or triticins are tetrameric proteins with molecular masses of 150,000, 155,000, and 160,000 and occur in the endosperm. Upon reduction of disulfide bonds, they are broken down into di- and monomers. Amino acid analysis indicated similarity to legume globulins. Farinins are surface-associated proteins of the starch granule. They show structures like avenins (the storage proteins of oats) with molecular masses of around 30,000 and are composed of two protein chains linked by a disulfide bond.

2.3 Lipids

Lipids may be defined as chemical compounds with a pronounced hydrophobicity, soluble in nonpolar organic solvents and insoluble in water. Wheat grains contain around 1.5–2.5% lipids. The majority of wheat lipids are derivatives of fatty acids linked to glycerol by ester bonds (glycerides) (Fig. 2.10A); fatty acids with 16 or 18 C-atoms (Fig. 2.10C) are the major constituents of wheat lipids. Their average proportions on a molar base are the following: 21% palmitic acid, 1% stearic acid, 11% oleic acid, 62% linoleic acid, and 5% linolenic acid. According to different chemical structures, the main wheat lipids can be classified into nonpolar lipids (simple glycerides) and polar lipids (glycolipids and phospholipids). Additionally, the lipid fraction of the grain contains small amounts of free fatty acids, sterols, sphingolipids, diol lipids, waxes, and hydrocarbons. Carotenoids and tocopherols, also belonging to wheat lipids, are included into Section 2.5.

The group of glycerides consists of mono-, di-, and triglycerides (glycerol linked to one, two or three fatty acids, respectively). The major glycerides in wheat are triglycerides; their content in the whole grain is 0.9–1.2%. Diglycerides and monoglycerides are only present in traces. Glycolipids consist of glycosyl glycerides as major components and steryl glycosides as well as glycosyl ceramides as minor components. The basic structure of glycosyl glycerides is glycerol linked to one or two sugar moieties and one or two fatty acids, respectively. The sugar residue is usually galactose (Fig. 2.10B) and less commonly glucose. Glycosyl glycerides mainly are derivatives of 1,2-diacyl-3-β-D-galactopyranosyl-L-glycerol

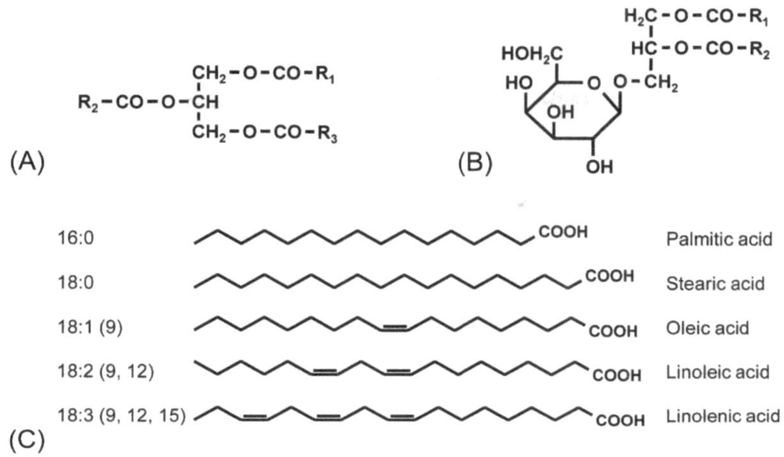

FIG. 2.10 Structures of (A) triglycerides, (B) monogalactosyl diacylglycerol, and (C) fatty acids.

with up to three additional galactosyl residues attached through α-1,6-glycosidic bonds. Digalactosyl diglycerides are most frequent in wheat followed by monogalactosyl diglycerides. They are predominantly found in the nonstarch lipid fraction [23] (Table 2.7). The principal steryl glycosides contain sitosterol and campesterol. The sugar bound to steryl glycosides is usually glucose and the fatty acid is linked to the C6-atom of glucose. Glycosyl ceramides consist of long-chain (mostly C18) bases that are amide-linked to fatty acids. Different sugars are joined to the primary hydroxyl group of ceramides by β-glycosidic bonds.

The major phospholipids are glycerol derivatives esterified with two fatty acids and one phosphatidic acid linked to various compounds. Most frequent are phosphatidyl choline (lecithin), phosphatidyl ethanolamine, and phosphatidyl inositol. They are enriched in the starch lipid fraction (Table 2.7). Minor phospholipids are lysophopholipids with only one fatty acid; they occur as starch lipids.

Most quantitative data on lipids are available for common wheat. The quantitative distribution of lipids among grain compartments is quite different [23]. Germ (≈200 mg/g) and aleurone layer (≈100 mg/g) have the highest and pericarp and testa (≈10 mg/g) the lowest total lipid concentrations. The contributions of aleurone layer, germ, and nonstarch endosperm to total lipids are similar (≈25% each), that of starch is somewhat lower (≈18%), whereas pericarp and testa contribute only around 4% (Table 2.7). The average proportions of the major lipid classes, found in common wheat and presented in Table 2.7, amount to around 50% nonpolar lipids, 36% phospholipids, and 12% glycolipids.

Lipids of the germ and aleurone layer almost exclusively consist of nonpolar lipids. The lipids of the starchy endosperm have to be subdivided into starch-associated lipids and nonstarch lipids, because they have a strongly different composition: Starch-associated lipids are highly enriched in lysophospholipids, whereas nonstarch endosperm lipids are characterized by well-balanced

TABLE 2.7 Range of amounts (μg) of lipids in a single grain and their average distribution (% of total lipids) [23].[a]

Compartment	TL	NL	GL	PL
Whole grain	916–1244	454–686	96–175	289–516
	(=100.0%)	(51.4%)	(12.3%)	(36.3%)
Starchy endosperm				
Nonstarch	215–321	71–115	68–93	50–113
	(23.1%)	(8.4%)	(7.3%)	(7.4%)
Starch	139–256	6–10	2–17	131–229
	(17.8%)	(0.7%)	(0.9%)	(16.2%)
Aleurone layer	220–387	159–321	8–22	39–57
	(27.3%)	(21.6%)	(1.4%)	(4.3%)
Germ	270–319	214–267	0–10	40–47
	(26.2%)	(21.7%)	(0.5%)	(4.0%)
Pericarp/testa	43	37	3	3
	(3.9%)	(3.3%)	(0.3%)	(0.3%)

[a] *GL, glycolipids; NL, nonpolar lipids; PL, phospholipids; TL, total lipids.*

proportions of nonpolar lipids, glycolipids, and phospholipids. Nonpolar lipids represent the major lipid class of pericarp and testa.

2.4 Vitamins and minerals

Wheat grains are important sources of a number of vitamins, notably B-vitamins and vitamin E. B-vitamins are water-soluble and have quite different chemical structures (Fig. 2.11), whereas vitamin E is fat-soluble and belongs to the group of tocopherols (see Section 2.5). The average amounts of single vitamins in 100 g of whole grains and white flour from common wheat are compared in Table 2.8 [2]. In whole grains, nicotinamide (5100 μg) is by far the predominant vitamin followed by α-tocopherol (1400 μg) and pantothenic acid (1200 μg). The contents of thiamine (455 μg) and pyridoxine (269 μg) are in a medium range and those of riboflavin (94 μg) and folic acid (87 μg) are in a low range. Biotin (6 μg) occurs in traces. Vitamins are not distributed evenly throughout the wheat grain. For example, the starchy endosperm contains less than 5% of total thiamine and more than 40% of total pantothenic acid [24]. The aleurone layer contains more than 80% of total nicotinamide, more than 60% of pyridoxine and around 40% of both riboflavin and pantothenic acid. Thiamine and α-tocopherol are strongly enriched in the germ. Altogether, most of the vitamins are concentrated in the aleurone layer and the germ. Pericarp and testa are relatively poor in vitamins.

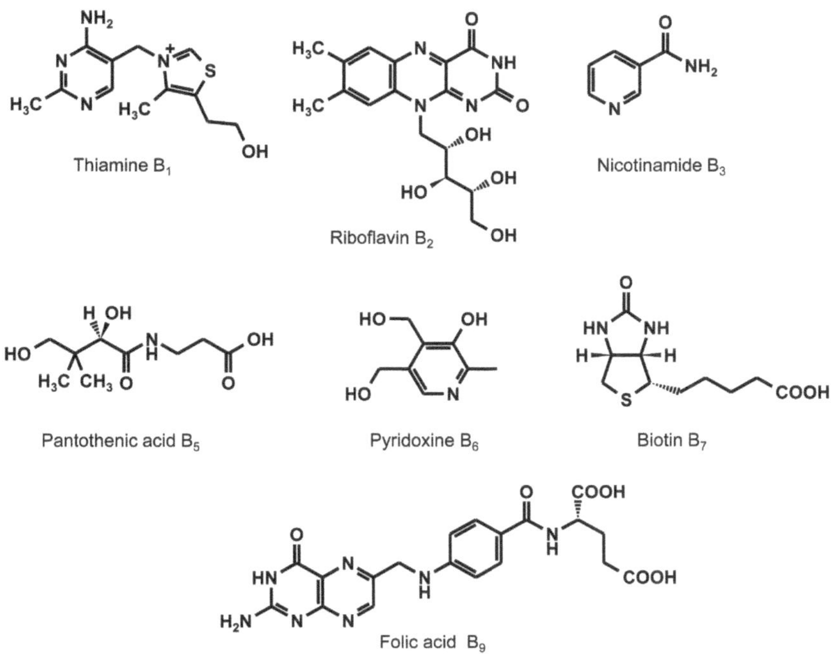

FIG. 2.11 Structures of B-vitamins.

TABLE 2.8 Average contents of vitamins and minerals in whole grains and white flour [2].

Compounds	Whole grains	White flour
Vitamins (μg/100 g)		
Nicotinamide (B_3)	5100	700
α-Tocopherol (E)	1400	n.v.[a]
Pantothenic acid (B_5)	1200	210
Thiamine (B_1)	455	60
Pyridoxine (B_6)	269	180
Riboflavin (B_2)	94	30
Folic acid (B_9)	87	10
Biotin (B_7)	6	2
Minerals (mg/100 g)		
Potassium	380	108
Phosphorus	342	74
Magnesium	97	n.v.[a]
Calcium	33	15
Sodium	7.7	2.0
Iron	3.2	1.4
Manganese	3.1	0.6
Zinc	2.6	0.7
Copper	0.4	0.1

[a] *n.v., no value available.*

The contents of vitamins in white flour are considerably lowered in a range from −33% (pyridoxine) to −89% (folic acid) as compared to whole grains.

The total mineral (ash) content of whole grains (100 g) amounts to around 2000 mg. Major minerals are potassium (380 mg) and phosphorus (342 mg) followed by magnesium (97 mg) and calcium (33 mg) (Table 2.8). Minor minerals are zinc (2.6 mg), manganese (3.1 mg), and iron (3.2 mg). Copper and selenium are trace minerals (< 1 mg). The aleurone layer contains 61%, the starchy endosperm 20%, the germ 12%, and pericarp and testa 7% of total minerals [24]. The contents in white flour are lowered in a range from −55% (calcium) to −81% (manganese) as compared to whole grains.

2.5 Phytochemicals

Phytochemicals are nonnutritive chemicals of plants that are responsible for organoleptic properties such as flavor and color and/or have biological significance like antioxidant activity. Detailed analyses of phytochemicals were carried out under the EU FP6 HEALTHGRAIN program on 150 common wheat lines and 50 lines of other wheat species and other cereals grown on the same site [25]. Most quantitative data presented in the following are taken from the HEALTHGRAIN program and are based on dry weight of whole grain flour from common wheat. Wheat grains contain several main classes of phytochemicals: carotenoids, flavonoids, terpenoids, and phenolics. The contents of phytochemicals are dependent on wheat species and varieties as well as on growing conditions.

The major colored compounds are carotenoids (carotenes and xanthophylls) and flavones. Carotenoids are defined as polyene hydrocarbons that consist of eight isoprene units. β-Carotene (Fig. 2.12A), known as provitamin A, is the major carotene of the wheat grain (≈0.2 μg/g). Xanthophylls are hydroxylated carotenes and includes lutein (≈1.1 μg/g) (Fig. 2.12B) and zeaxanthin (≈0.3 μg/g). They do not possess vitamin A activity. β-Carotene and lutein

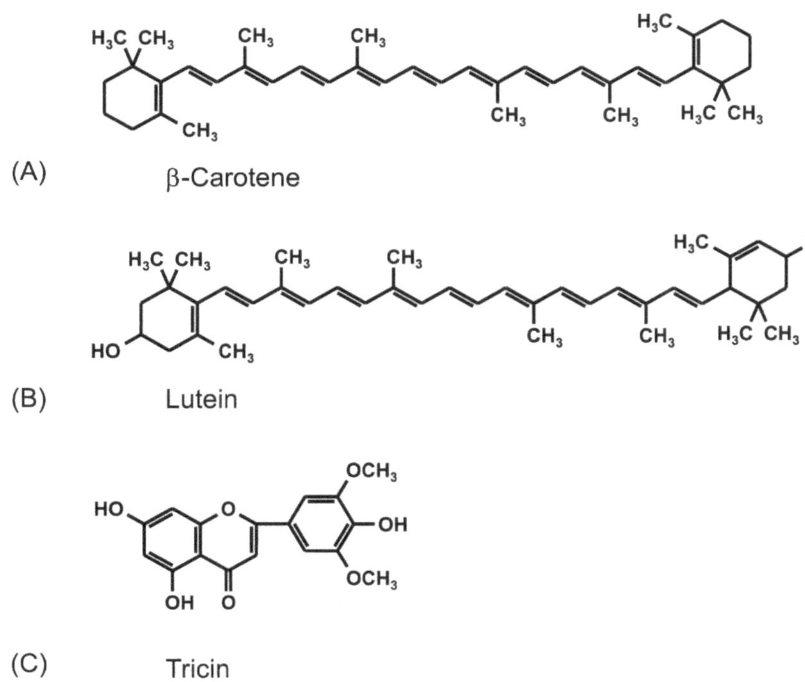

FIG. 2.12 Structures of the most important pigments in wheat grain. (A) β-carotene; (B) lutein; and (C) tricin.

are the major determinants of the yellow color of durum wheat endosperm. Carotenoids are particularly enriched in the germ. They are insoluble in water and soluble in organic solvents and oils. They are sensitive to light and are easily oxidized to colorless compounds.

The structures of flavones are based on the backbone of 2-phenylchromen-4-one. The predominant flavone of wheat is tricin (4,5,7-trihydroxy-3,5-dimethoxy flavone) (Fig. 2.12C). In addition, different glucosides of tricin (4-O-, 5-O-, and 7-O-glycosides) occur. Tricin is insoluble in water, but its glycoside variants are soluble. Tricin and related compounds are present in largest quantities in the germ (≈50 μg/g) and pericarp and testa (5–10 μg/g) and least in the starchy endosperm (< 1 μg/g).

The most important terpenoids in wheat grains are phytosterols and tocopherols. Phytosterols are plant steroids made up of four condensed C6-rings with a hydroxyl group and an aliphatic side chain and thus they are structurally related to cholesterol. Two types of phytosterols can be found: sterols with a double bond in the sterol ring structure and stanols without a double bond. The most common sterols occurring in wheat grains (≈800 μg/g) are β-sitosterol (Fig. 2.13A), campesterol, and stigmasterol. Stanols (≈200 μg/g) include sitostanol and campestanol (Fig. 2.13B) as major compounds. The structure of tocopherols is based on a chromane ring substituted with an aliphatic chain with 14 C-atoms. Major tocopherols in wheat grains (≈50 μg/g) are compounds with vitamin E activity such as α-tocopherol (Fig. 2.13C), β-tocopherol, and tocotrienols. α-Tocopherol (≈14 μg/g) has the highest biological activity, i.e., in vivo antioxidative effect.

Phenolic compounds are the most abundant components within phytochemicals. Their chemical structure contains at least one aromatic ring with at least one hydroxyl group. Phenolic acids are the main compounds of phenolics in wheat grain (≈660 μg/g). They consist of a phenolic ring and a carboxylic group and are derived from either cinnamic acid or benzoic acid. Minor amounts occur in a free form or are linked to carbohydrates, while major amounts are bound to nonstarch polysaccharides. Among phenolic acids, ferulic acid (Fig. 2.4B) and to a lesser extent *p*-coumaric acid are predominant components. They are bound to arabinoxylans (see Section 2.1.2) and are released by fermentation in the colon.

Lignans are a group of polyphenolic compounds derived from phenylalanine via dimerization of substituted cinnamic alcohols to a dibenzylbutan skeleton. They usually occur free or are glycosidically linked to sugars; diglucosides are most common. Lignans amount to about 10 μg/g grain and mainly occur in the outer layers of the grain. They are metabolized by the intestinal microbiota to enterolignans ("mammalians lignans") such as enterodiol (Fig. 2.13D) and enterolactone. Both compounds have been associated with antioxidative activities and estrogen-like properties.

Alkylresorcinols like bilobol (Fig. 2.13E) are phenolic lipids consisting of 1,3-dihydroxybenzene substituted with an alkyl chain (C15–C25) in position 5.

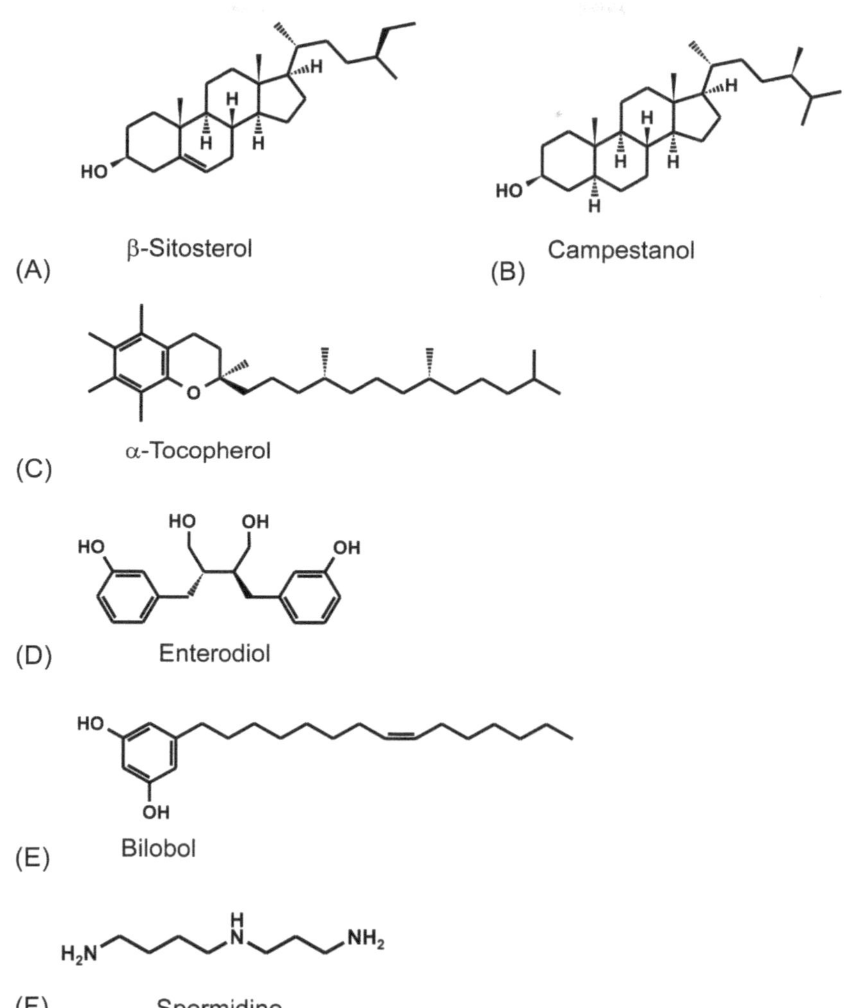

FIG. 2.13 Structures of selected phytochemicals. (A) β-sitosterol; (B) campestanol; (C) α-tocopherol; (D) enterodiol; (E) bilobol; and (F) spermidine.

Their contents in whole grains are in a range of around 200–700 μg/g. Because alkylresorcinols almost exclusively occur in the outer layers of the grains, their content can be used as marker of both presence and amount of bran in flour or other wheat-based products. Moreover, alkylresorcinols and their plasma and urine metabolites are used as selective nutritional biomarkers for the intake of whole grain wheat (and rye).

Spermidine (*N*-(3-aminopropyl)butan-1,4-diamine) (Fig. 2.13F) belongs to the polycationic aliphatic amines and occurs in most living organisms. It is known to be an essential regulator of various cellular processes regarding DNA stability as well as growth, differentiation, and apoptosis of cells. Its physiological functionality plays an important role in counteracting aging and promoting longevity. Wheat germs are the best known dietary source of spermidine (243 mg/kg) and corresponding extracts have been used for long-term intervention studies in humans to investigate the impact of spermidine treatment on cognition and brain integrity [26].

References

[1] Pomeranz Y. Chemical composition of kernel structures. In: Pomeranz Y, editor. Wheat—chemistry and technology. 3rd ed., vol. 1. AACC, St. Paul; 1988. p. 97–158.

[2] Souci SW, Fachmann W, Kraut H. Lebensmitteltabelle für die Praxis. 5th ed. Stuttgart: Wissenschaftliche Verlagsgesellschaft; 2011. p. 241–3.

[3] Koehler P, Wieser H. Chemistry of cereal grains. In: Gobbetti M, Gänzle M, editors. Handbook on sourdough biotechnology. Berlin, Heidelberg: Springer; 2012. p. 11–46.

[4] Lineback DR, Rasper VF. Wheat carbohydrates. In: Pomeranz Y, editor. Wheat—chemistry and technology. 3rd ed., vol. 1. AACC, St. Paul; 1988. p. 277–372.

[5] Delcour JA, Hoseney RC. Principles of cereal science and technology. 3rd ed. St. Paul: AACC; 2010. p. 23–51.

[6] Dengate HN. Swelling, pasting, and gelling of wheat starch. In: Pomeranz Y, editor. Advances in cereal science and technology. vol. 6. St. Paul: AACC; 1984. p. 49–82.

[7] Wisker E, Feldheim W, Pomeranz Y, Meuser F. Dietary fiber in cereals. In: Pomeranz Y, editor. Advances in cereal science and technology. vol. 7. St. Paul: AACC; 1985. p. 169–238.

[8] Andersson AA, Andersson R, Piironen V, Lampi AM, Nyström L, Boros D, et al. Contents of dietary fibre components and their relation to associated bioactive components in whole grain wheat samples from the HEALTHGRAIN diversity screen. Food Chem 2013;15:1243–8.

[9] Geisslitz S, Wieser H, Scherf KA, Koehler P. Gluten protein composition and aggregation properties as predictors for bread volume of common wheat, spelt, durum wheat, emmer and einkorn. J Cereal Sci 2018;83:204–12.

[10] Wieser H, Seilmeier W. The influence of nitrogen fertilisation on quantities and proportions of different protein types in wheat flour. J Sci Food Agric 1998;76:49–55.

[11] Wieser H, Manderscheid R, Erbs M, Weigel H-J. Effects of elevated atmospheric CO_2 concentrations on the quantitative protein composition of wheat grain. J Agric Food Chem 2008;56:6531–5.

[12] Jensen SA, Martens H. The botanical constituents of wheat and wheat milling fractions. II. Quantification by amino acids. Cereal Chem 1983;60:170–7.

[13] Wrigley CW, Bietz JA. Proteins and amino acids. In: Pomeranz Y, editor. Wheat—chemistry and technology. 3rd ed., vol. 1. AACC, St. Paul; 1988. p. 159–275.

[14] Wieser H, Koehler P, Konitzer K. Celiac disease and gluten—multidisciplinary challenges and opportunities. London, Waltham, San Diego: Academic Press Elsevier; 2014. p. 99–119.

[15] Wieser H, Kieffer R. Correlations of the amount of gluten protein types to the technological properties of wheat flours determined on a micro-scale. J Cereal Sci 2001;34:19–27.

[16] Wieser H, Gutser R, von Tucher S. Influence of sulphur fertilisation on quantities and proportions of gluten protein types in wheat flour. J Cereal Sci 2004;40:239–44.

[17] Belitz HD, Grosch W, Schieberle P. Food chemistry. 4th ed. Berlin: Springer; 2009. p. 678–9.

[18] Schmid M, Wieser H, Koehler P. Isolation and characterization of high-molecular-weight (HMW) gliadins from wheat flour. Cereal Chem 2016;93:536–42.

[19] Shewry P. What is gluten—Why is it special? Front Nutr 2019;6:101.

[20] Kruger JE, Reed G. Enzymes and color. In: Pomeranz Y, editor. Wheat—chemistry and technology. 3rd ed., vol. 1. AACC, St. Paul; 1988. p. 441–500.

[21] Brijs K, Courtin CM, Goesaert H, Gebruers K, Delcour JA, Shewry PR, et al. Enzymes and enzyme inhibitors endogenous to wheat. In: Khan K, Shewry PR, editors. Wheat—chemistry and technology. St. Paul: AACC; 2009. p. 401–35.

[22] Altenbach SB, Vensel WH, Dupont FM. The spectrum of low molecular weight α-amylase/protease inhibitor genes expressed in the US bread wheat cultivar Butte 86. BMC Res Notes 2011;4:242–54.

[23] Morrison WR. Lipids. In: Pomeranz Y, editor. Wheat—chemistry and technology. vol. 1. St. Paul: AACC; 1988. p. 373–439.

[24] Betschart AA. Nutritional quality of wheat and wheat foods. In: Pomeranz Y, editor. Wheat—chemistry and technology. vol. 2. St. Paul: AACC; 1988. p. 91–130.

[25] Shewry PR, Hawkesford MJ, Piironen V, Lampi AM, Gebruers K, Boros D, et al. Natural variation in grain composition of wheat and related cereals. Agric Food Chem 2013;61:8295–303.

[26] Schwarz C, Stekovic S, Wirth M, Benson G, Royer P, Sigrist S, et al. Safety and tolerability of spermidine supplementation in mice and older adults with subjective cognitive decline. Aging 2018;8:19–33.

Chapter 3

Cultivation and milling of wheat

The starting point of wheat cultivation around 8000 BC was the so-called Fertile Crescent, an area of Minor Asia, where grains of wheat grasses were collected for consumption. As time went on, wheat cultivation spread over the whole globe and together with rice, wheat is now the predominant food crop worldwide [1]. Wheat products are staple foods around the world and decisively contribute to the diet of a large portion of mankind. Therefore farming of wheat belongs to one of the most important agricultural tasks. Farmers have to make fundamental decisions for successful farming including the selection of suitable varieties, planting areas, and time tables as well as effective crop management. Mature grains, the most valuable part of wheat plants, are harvested once a year in a short period of time but are processed and consumed throughout the year. Therefore they must be stored soundly and safely, so that they retain their end-use properties. Moreover, proper storage gives humans an appropriate buffer against crop failure and starvation. To transform grains into those products that can be easily processed and have more desirable properties, milling has traditionally been the most applied technology. Modern milling enables the production of huge amounts of different types of flour, farina, and grist as well as separated grain fractions like bran and germ. The detailed description of farming, storage of grains, and milling is given in the ensuing sections.

3.1 Farming

The first signs of cereal cultivation were found in the Fertile Crescent of the Middle East (including Anatolia, Lebanon, Syria, Palestine, and Iraq), where wild wheat and barley plants were successfully cultivated due to the favorable environmental conditions there. Some human tribes changed from nomadic to stable settlement, because cereal cultivation permitted food storage and sowing stored seeds for the next season. Because dry cereal grains could be stored for relatively long periods of time, they provided humans with security against food failure and starvation. Thus humans successively supplemented hunting and gathering by farming, which marked the beginning of agriculture, one of the greatest revolutions mankind ever experienced. Einkorn and emmer were the first wheat species that were widely cultivated and became main crops in the Fertile Crescent during the Neolithic. From there, ancient wheat spread eastward through India to China and westward to Egypt, Greece, and Italy

Wheat—An Exceptional Crop. https://doi.org/10.1016/B978-0-12-821715-3.00003-4

and reached Northern Europe including England, Ireland, and Scandinavian countries. Einkorn and emmer were supplemented and stepwise replaced by spelt, durum wheat, and common wheat. Wheat did not evolve naturally on the American continent but was introduced by Europeans at the beginning of the Modern Age. Christopher Columbus' crew planted a few grains in Puerto Rico in 1493 and somewhat later, Spanish explorers introduced wheat to Mexico and the southeast of North America. Bartholomew Gosnold brought wheat to New England in 1602 and shortly afterwards, pilgrims on the Mayflower transported wheat to Massachusetts. Since that time, wheat has been gradually cultivated and utilized throughout the world.

Farming of the different wheat species is similar. They all are annual plants and consequently, one planting yields one harvest per year. They grow best in a moderate climate. Rainfall during the growing season is strongly related to yield: The drier regions—parts of Africa, Russia, and South America—have the lowest yield, whereas regions with more rainfalls like much of Europe achieve yields up to five times those of the driest regions. Winter wheat requires vernalization by low temperature (lower than 5 °C). It is sown in late summer or fall to germinate and to develop into young plants that afterward remain in the vegetative phase during winter and resume growth in early spring. Winter wheat is predominant throughout the world, where the soil does not freeze sufficiently to winter-kill the young plants. Spring wheat does not tolerate very low temperature and is usually sown in spring and matures in late summer. Because yields are generally lower than for winter wheat, the production of spring wheat has gradually fallen throughout the world.

Wheat plants are cultivated almost all over the world. According to the FAO Production Statistics of the Food and Agriculture Organization (FAO) of the United Nations [2], more than 100 countries produce economically significant amounts of wheat sown on an averaged 221 million hectares of arable land and yielding around 740 million tons of grains per year. India, Russia, and China are, by far, leading in the areas of land used for wheat production followed by USA and Australia (Table 3.1). The annual global economic value of harvested wheat is around 100 billion US dollars. About 95% of the wheat grown globally is common wheat and most of the remaining 5% is durum wheat. The other species (einkorn, emmer, and spelt) are economically unattractive, because they are hulled wheats with a though glume that has to be removed before milling the grains. Moreover, their mean grain yields are only 35% (einkorn), 45% (emmer), and 63% (spelt) compared to common wheat (= 100%) [3].

The total grain yield may vary considerably from year to year and from country to country and is highly influenced by the weather. Because the wheat plant is dependent on moisture at sowing time and throughout the growing season, droughts can have marked negative effects on the production and economy. On the other hand, widespread and prolonged rainfall during maturing and harvest time damages grains, which renders them suitable only for cheap animal feed. The widespread drought in many regions of Europe in 2018 and the subsequent

TABLE 3.1 Annual production of wheat grains (average of 2014–2016) [2].

Country	Tons	Hectares	Tons/Hectare
China	129,360,812	24,185,532	5.349
India	91,960,000	30,723,333	2.993
Russia	64,930,583	25,696,963	2,527
USA	57,948,570	18,530,620	3.127
France	37,068,228	5,446,657	6.806
Canada	29,166,800	9,439,700	3.090
Germany	26,266,000	3,234,700	8.120
Pakistan	25,690,262	9,182,096	2.798
Ukraine	25,581,633	6,351,967	4.027
Australia	23,773,370	12,093,034	1.966
World	739,993,141	221,176,028	3.346

crop loss of wheat drastically demonstrated the influence of weather and climatic changes on yield. In spite of these global problems, the average wheat production worldwide has been rather stable during the last decade, which is highly important for national and international economics and human nutrition. China and India are leading in the production of wheat grains (17% and 12% of world production, respectively) followed by Russia and USA (Table 3.1). The yield of grains per hectare of cultivated land is highly different. European countries like Germany and France achieve 7–8 tons per hectare, while Australia and Russia harvest only around 2.0–2.5 tons per hectare. Improving the conditions of farming may help to increase grain yield, which would be important for both the economic success and the alimentation of world population in future.

The wheat plant is quite adaptable to environmental conditions, for instance growing in Palestine far below the sea level and in Himalayan regions up to 3000 m above sea level. It grows in areas far north at 65° north latitude (Norway) and far south at 45° south latitude (Argentina). Wheat is harvested somewhere in the world in any given month. Examples are in January: New Zealand and Chile; February: India; March: Egypt; April: Asia Minor and Mexico; May: Central Asia and North Africa; June: Southern United States and Southern Europe; July: Northern United States and Central Europe; August: England and Central Russia; September: Scandinavia and Canada; October: Northern Russia and Finland; November: Southern Africa; December: Argentina and Australia.

The selection of qualified sites for wheat farming is important, because type and nutrient and moisture levels of the soil as well as weather and climate are important factors for the thriving of wheat plants. Similarly, suitable wheat varieties, regarding growing conditions and end-product quality, should be selected. Optimal crop management over the whole growing period is essential for the development of the plant and for grain yield. Each stage of development may require specific applications of fertilizers, fungicides, insecticides, herbicides, and growth regulators. However, making unnecessary applications is bad for both economic and environmental reasons. Thus it is in the farmer's best interest to make applications, when they are needed but only when needed ("good agricultural practice"). Moreover, it should be noted that some targets of applications (e.g., fungi, viruses, insects, and even weed) become resistant, when the same pesticide is applied over a longer period. Field control and treatment by means of aerial drones may be a future option to optimize the applications of fertilizers and pesticides. For ecological reasons, organic farming should be taken in consideration whenever possible.

Most of the following measures are referred to intensive cultivation that is predominant worldwide. Proper fertilization is a critical factor in producing optimum grain yields, which is not only a matter of applying the correct amount of the right fertilizer but also at the right time. Levels of fertilizers are best based on representative soil tests, because soil testing is the most reliable tool for making good economic and environmental decisions about applying fertilizers. Nitrogen (N) is usually the most limiting nutrient for wheat production [4]. The rate of N application determines the concentrations and proportions of grain proteins that are, for example, crucial for the bread-making quality of flour. The time of N application can be as important as the amount used. There a several potential times for N application: before, during or after planting, either as single or split applications. It is likely best to apply N on winter wheat in the spring and on spring wheat before planting. Split (two or three) applications tend to improve yield. Most N fertilizer sources are urea, ammonium nitrate, and ammonium sulfate. Another way to correct N deficiency in soil is the use of pasture rotations using legumes, lucernes or clovers that precede farming of wheat for a season.

Second to N, sulfur (S) is an important limiting nutrient for wheat production [5]. S deficiencies in soils are now fairly common. Formerly, wheat did not show any evidence of S deficiency, because air pollution from industry and traffic provided sufficient amounts of S in the soil. However, the massive decrease in the input of S from the atmosphere over the last decades reduced the S availability in soils dramatically. This has consequently led to a severe S deficiency in many crops that drastically affected protein concentrations and proportions in the grains and worsened dough properties and baking quality of wheat flours. The S requirement of the wheat plant is about one-tenth of the N requirement. Ammonium sulfate is a preferred S (and N) source for application. Phosphorus (P) and potassium (K) applications are recommended, when soil tests indicate deficiencies. Fertilizers are most frequently applied in form of P_2O_5 and K_2O, respectively.

Pest management is a key component of wheat production. Major target pests in wheat farming include microorganisms, insects, and weeds and a number of different pesticides (fungicides, insecticides, and herbicides) are used to protect the plant. There are numerous different wheat diseases mainly caused by fungi and viruses. They can be divided into seed- and soil-borne diseases (e.g., *Stagonospora* blotch), leaf and head-blight diseases (e.g., powdery mildew and leaf rust), crown and root diseases (e.g., head scab), and viral diseases (e.g., yellow mosaic). Most diseases are best managed through the use of combined tactics, for example, use of resistant varieties, crop rotation, and application of fungicides. The latter are important tools in killing parasitic fungi and their spores. There is a long list of fungicides on the market; some of them have a broad spectrum of disease control activity and others have very specific uses. The main chemical classes are strobilurins and triazoles. Mixtures and alternations of fungicides are often most effective to reduce the likelihood that fungicide resistance will develop in target pathogens. Seed treatment may help, when conditions favor common soil- and seed-borne fungi (e.g., *Fusarium*, *Pythium*, and *Septoria*). Foliar fungicides are used to manage common foliar and head diseases (e.g., leaf, stripe, and stem rust, powdery mildew, speckled leaf blotch, and *Stagonospora* leaf and glume blotch). Effects of foliar fungicides are greatest, when they are applied between full extension of the flag leaves and anthesis (begin of flowering).

The application of fungicides against fungi that produce mycotoxins is of special importance. Fungi develop under certain conditions of humidity and temperature and are responsible for preharvest contamination of wheat plants with mycotoxins [6]. Improper storage of grains may even increase toxin production. Primary fungi producing mycotoxins are members of the genera *Fusarium*, *Aspergillus*, *Penicillium*, and *Claviceps*. Examples of the best known mycotoxins are trichothecenes, ochratoxin A, and aflatoxins. Deoxynivalenol (DON) is one of the most widely distributed *Fusarium* mycotoxins occurring in cereals, particularly found in wheat and corn. Like many other mycotoxins, DON enters the food chain and appears in products made from contaminated wheat; the baking process does not destroy toxicity.

Among vermin, aphids, mites, caterpillars, and the larvae of some butterflies and moths species feed upon wheat crops. The management of infestation depends on vermin population and includes cultural practices (e.g., tillage, crop rotation, and date of planting), wheat variety, seed treatment, and application of insecticides. Insecticides are substances that are used to kill or inactivate vermin by two different principles: Systemic insecticides, incorporated and distributed throughout the whole plant, destroy the insect, when parts of the plants are consumed. Contact insecticides are harmful to vermin upon direct contact. Major chemical classes of insecticides are organochlorines, organophosphates, carbamates, pyrethroids, ryanoids, and neonicotinoids.

Herbicides are used to prevent the growing of weeds; both preemergence and postemergence treatments are possible. Winter annual weed species such as dandelion, purple deadnettle, and wild garlic emerge in fall and become competitive and troublesome during spring. Summer annual weeds such as ragweed will be of less concern in early spring and will be outcompeted by the growing wheat crop. The determination of the severity of weed infestations in wheat fields is key in regarding the necessity of herbicide application. In particular, fields that contain uniform infestations should be taken into consideration for herbicide application to avoid yield loss and harvest interference problems. The right time of application is important, because the majority of herbicides is labeled for application at certain wheat growth stages. Frequently, they have very short time windows, in which they can be applied.

Glyphosate (*N*-(phosphonomethyl)glycine) is the most widely used herbicide worldwide (Fig. 3.1). It is absorbed through foliage and inhibits a weed enzyme that catalyzes the synthesis of aromatic amino acids (phenylalanine, tyrosine, and tryptophan) necessary for growth. Studies on the toxicity of glyphosate to humans gave contradictory results; particularly, concerns about its carcinogenic effect on human health persist. In March 2015, the International Agency for Research on Cancer of the World Health Organization (WHO) classified glyphosate as "probably carcinogenic in humans" (category 2A) based on epidemiological studies, animal studies, and *in vitro* studies [7]. In contrast, the European Food Safety Authority (EFSA) concluded in November 2015 that "the substance is unlikely to be genotoxic (i.e., damaging to DNA) or to pose a carcinogenic threat to humans" and later clarified that, while carcinogenic glyphosate-containing formulations may exist, "studies, which looked solely at the active substance glyphosate, did not show this effect" [8]. A report of the WHO and FAO of the United Nations Joint Committee on Pesticide Residues issued in 2016 stated that the use of glyphosate formulations does not necessarily constitute a health risk and gave admissible daily maximum intake limits (1 mg/kg of body weight per day) for chronic toxicity [9]. The European Chemicals Agency (ECHA) classified glyphosate as causing serious eye damage and toxic to aquatic life but did not find evidence implicating it as a carcinogen, a mutagen, toxic to reproduction nor toxic to specific organs [10]. In March 2019, the European Commission announced a plan to set up a group of Member States to act as corapporteurs for the next assessment of glyphosate.

The application of plant growth regulators (PGRs) causes various responses in wheat plants such as shorter and thicker stems and thicker stem cell walls. The plant can be shortened by more than 50%, so that the plant has a better standability and a significantly reduced risk of lodging. The main target of PGRs is gibberellins, substances of the hormonal system of the plant. Gibberellins are

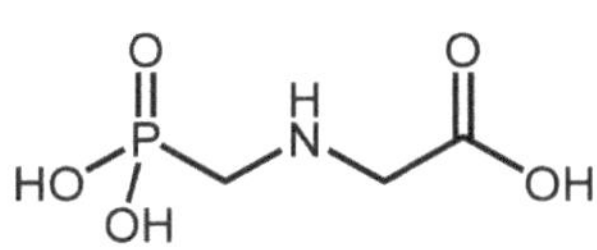

FIG. 3.1 Structure of glyphosate (*N*-(phosphonomethyl)glycine).

tetracyclic diterpene acids with 19 or 20 C-atoms. Acting as plant hormones, they are involved in various developmental processes such as breaking dormancy, germination, and stem elongation. As soon as wheat tiller formation begins, growing tillers initiate the production of gibberellins that trigger stem growth and the process of stem elongation starts. Most PGRs (e.g., chlormequat chloride, mepiquat chloride, and prohexadione calcium) inhibit the biosynthesis of gibberellins, a few PGRs (e.g., 2-chloroethylphosphonic acid) inhibit the production of auxins, another endogenous PGR. Auxins are plant hormones that have an important function in plant growth and development of plant organs as well. Chemically, they are compounds corresponding to indole-3-acetic acid or related derivatives. The best time for PGR application is when stem elongation starts and the first node appears. In particular, application can be a useful tool in situations, where the wheat crop has high yield potential and high risk of lodging. Factors that additionally increase the risk of lodging include wheat varieties with poor standability, high rates of N fertilization, and warm, wet conditions during stem elongation.

Organic farming is an alternative agricultural system that is based on methods imitating the structures of a natural system with the aim to sustain the health of soils, plants, ecosystems, and people [11]. Organic agriculture ranges from standard systems that simply follow organic certification guidelines to strict closed-cycle systems beyond guidelines by limiting external inputs as much as possible. National and international standards regulate the production methods of organic farming and different associations certify organic producers. Although requirements vary slightly between certifying agencies, they promote soil quality, animal and plant diversity, biological processes, and animal welfare. The application of naturally occurring substances is allowed, while with a few exceptions synthetic substances are strictly prohibited. For instance, fertilizers are of organic origin such as compost, green manure, and bone meal; the application of sewage sludge is forbidden. Soil management is accompanied by crop rotation and companion planting. Synthetic pesticides are prohibited, while naturally derived pesticides like pyrethrin and rotenone are permitted. The application of plant growth regulators is generally forbidden. Weed management is based more on suppression and less on elimination. As an example, weed growth can be reduced by turning the soil or by thermal weeding before seedling. A key law of organic farming is the rejection of transgenic plants. An important disadvantage of organic wheat farming is the lower grain yield reduced by up to 30% compared to conventional farming and the price premiums for organic wheat products.

In the next 30 years, 2 billion people will be added to the planet along with the need to feed them. To produce more and better foods around the world will be the main objective and the most urgent and potentially explosive global problem in the near future. To solve this problem, increased food production is necessary despite challenges arising from climate change [12] and additional wheat cultivation might contribute to possible solutions. However, arable land has been lost to urban sprawl, highway constructions, and recreation, for instance, and becomes more and more rare and expensive. Additionally, the cultivation of plants for energy production is a serious competitor for food production. To convert new prairie sods and virgin woods into farmland is questionable not only for environmental reasons. Nevertheless, a number of measures offer the possibility to increase wheat availability.

Achievements of wheat improvement by breeding and genetic engineering may provide hopes for the future. Genetic engineering, if accepted by the population, can be used to tailor wheat properties more selectively and faster than classical breeding (see Section 1.3). Improved wheat plants, resistant to frost, heat, drought, and salty soils, may increase suitable arable land and grain yield significantly to ensure food security [13]. By means of modified wheat plants, agricultural land can become more productive, notably in wheat-producing countries with relatively low average grain yield. Another feasible alternative could be to reduce the cultivation of food plants that need large areas of land and high amounts of valuable water for growth but produce foods with low density of energy and nutrients (water contents far above 90%) such as green salads and a number of vegetables. Much more important, however, might be the reduction of meat production and consumption, particularly in developed countries, which would be additionally desirable for human health, animal welfare, and ecological matters anyway. To feed farm animals like cattle and pigs, high amounts of crop material, in particular corn and soy, are necessary, which requires huge areas of arable land for cultivation. By the reduction of livestock feed production, more land for the cultivation of wheat and other plants, usable for human nutrition, would become available.

3.2 Storage of grains

Once the wheat plants have grown their grains, they die and become brown and dry and harvest can begin. In developed countries, crops are universally harvested by combine harvesters that cut, thresh, and winnow the grains during a single pass across the field. In developing countries, simpler methods may be in use, from combines to hand tools. Harvested wheat grains are relatively easy to store [14]. Storage can vary from simple pouring onto the ground to bulk storage in bins. Today, large grain storage bins are widely common. Hulled wheats are often stored as spikelets, because the husk gives good protection against pests. To ensure safe storage, the moisture content of the grains should be 14% or less. The low moisture content corresponds to low water activity and inhibits microbial spoilage by mycotoxin producing fungi, for instance, and prevents grain metabolism (e.g., pregermination).

Grains usually dry rapidly in the field in a dry season and thus have water contents safe for storage. However, grains have a high moisture content at harvest in rainy years. Such grains spoil rapidly and must be dried effectively and economically. Both low and high temperature drying, frequently done with a full-bin system, are currently in use. Low-temperature drying uses ambient air that is forced through the grain mass. Advantages are that the energy input is low and the functionality of grain components is sustained. A major disadvantage is the relatively long time required to reduce moisture and the need to use air of low relative humidity. High-temperature drying uses air heated to temperatures of around 60–80 °C. The major advantage is saving time, whereas major disadvantages are the cost for heating the air, heat damage of the grains (e.g., shrinkage, increased brittleness, and susceptibility to breakage), and loss of constituent functionality. Stored grains must be protected from insects, birds, and rodents that are able to destroy distinct parts of the harvest. Killing the animals using baits, traps or other instruments is effective only over a short period. Better controls are to keep them away by rodent-proofing of buildings and using good sanitation.

3.3 Milling process and products

Milling of wheat grains is one of the oldest trades and represents now one of the major food industries worldwide. The prime aim of milling is to transfer grains into small particles that can be used as food ingredient. In most cases, white flours are produced that mainly consist of the starchy endosperm separated from the other parts of the grains (pericarp, testa, aleurone layer, and germ). Grains from free-threshing common and durum wheats are delivered to the mills as dehulled grains, whereas grains from hulled spelt, emmer, and einkorn have to be dehulled by special milling or pounding to obtain pure grains. Generally, grains used for milling should be sound, clean, fully matured, and free of foreign material.

In principle, flour production consists of grinding (milling) the grains and sifting (separation) the different flour fractions [15]. However, several operations are necessary, before grains can be subjected to milling. First, the grains must be cleaned by removing impurities. Common equipments are magnetic separators to remove metal, stoners to remove stones, milling separators to remove sticks and other foreign large material, disc separators to remove seeds with longer or shorter grains, and scourer aspirators to remove light material such as dust and straw pieces. In the last years, optical sorting devices have gained widespread occurrence in the milling industry. The equipment scans every single grain for optical properties and effectively removes foreign grains and grains contaminated with ergot fungi. Washing of grains applied for purification has lost importance, because it promotes the growth of microorganisms. The next step of conditioning is grain wetting or steeping in water for some hours, since a raised moisture content (ca. 15–17%) makes the separation of the endosperm from the outer layers and the germ easier. The rate of water uptake can be increased with higher temperatures, however, exceeding 50 °C should be avoided.

The historical development of milling wheat grains started by means of mortar and pestle followed by hand- or animal-driven stone mills that were further developed into water and wind mills. In the last centuries, metal rollers and finally impact mills were introduced. Today, the grinding of grains is most frequently done with roller mills, where pairs of rolls rotate in opposite directions (Fig. 3.2). Rollers can be adjusted to the end-product needed, for example, by modifying surface flutes and gap between rolls and by rotating at different speeds. After each grinding step, sieves in the form of flat sifters allow the separation of the ground stock according to the particle sizes. Roller mill grinding consists of the break system, the purification system, and the reduction system. The brake system includes several breaks (i.e., sets of rolls) combined with sifters in sequence. The breakage function critically depends on the ratio of roll gap to input particle size [16]. The first break rolls open the grain by shearing forces and subsequent sieving separates bran, middlings (large pieces of endosperm), and some flour. Later breaks include a scraping of the bran to remove adhering endosperm (shorts) and the last break flattens the germ, which allows the isolation of germs after sieving. The purification system separates middlings from small particles of bran and attached bran. The purpose of the reduction system is to reduce the particle size of the purified middlings to a particular level of flour fineness by means of a series of rollers and sifters. The products obtained can be classified according to the particle size as follows: > 500 μm for grist, 200–500 μm for farina (common wheat) or semolina (durum wheat), and < 200 μm for flour. Durum wheat grains are so hard that it is difficult to reduce the particle size to flour fineness. Therefore, durum wheat grains are mostly milled into semolina, the best raw material for pasta production.

Pin mills and hammer mills are mainly applied for special purposes. Pin mills are used for grinding wheat grains, when starch damage is undesirable and flour production rates are relatively low. They consist of a revolving rotor with a

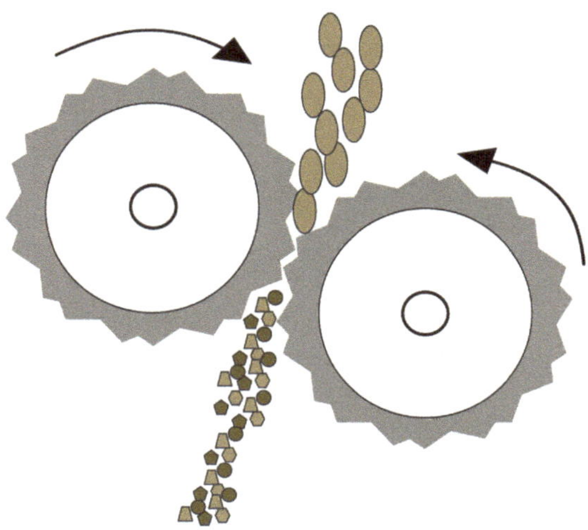

FIG. 3.2 Break rolls of a roller mill.

series of pins attached onto the inside. Grains enter the rotor at the center and are thrown towards the periphery by centrifugal forces. They repeatedly impact the pins and are comminuted yielding flour with the desired particle size. Hammer mills consist of a steel drum equipped with a vertical or horizontal rotating shaft, on which hammers are fixed. The rotor is spun at high speed and fed with material that is shredded by the hammer bars and expelled through screens of a selected size. For example, hammer mills are used for milling grains into coarse flour to be fed to livestock.

Flour dust explosions are the most-feared hazards in mills. Explosions can occur, when flour dust is present in a sufficiently high concentration in the atmosphere of the mill and when there is an ignition source (e.g., electrostatic discharge or electric arcing from machines or other equipment). Even in recent times, gigantic explosions of flour dust completely destroyed mills and killed workers. Therefore, dust control systems are required to keep the atmosphere in mill buildings relatively dust-free.

Flour extraction rates (percentage of produced flour weight related to grain weight applied) are important as an index of milling technique efficiency and millability of different wheats and reflect the quality and properties of flours obtained. Flour yield can vary from around 50% (white flour) up to 100% (whole grain flour). The mineral (ash) contents of flours (range around 0.4–2.2 g/100 g) are highly correlated with flour extraction rates (Table 3.2) [17]. Therefore, flour types are standardized on the basis of their ash contents (ICC 104/1 [18]) in some countries. Flours that do not contain portions of the germ should be stored at moderate temperature (15–20 °C) and dry atmosphere in bins permeable to air or in silos with aeration. Whole grain flours including the germ should be stored under low temperatures such as refrigeration. Most of

TABLE 3.2 Relationship between extraction rate and ash content of wheat flour [17].

Extraction rate (%)	Ash content (g/100 g flour)
40	0.380–0.440
65	0.490–0.580
70	0.600–0.700
75	0.750–0.870
82	1.000–1.150
86	1.160–1.350
91	1.550–1.750
100	1.850–2.200

the flours and byproducts provided by millers are used for manufacturing countless food and non-food products and a small portion as animal feed ingredients (see Sections 4.2–4.6).

Terminology and regulations for "whole grain flour", also called whole wheat flour, wholemeal flour or entire flour, are different from country to country. In the USA and many other countries, "whole wheat flour" and "wholemeal flour", respectively, are defined as flours that contain all parts of the grain (endosperm, pericarp, testa, and germ) in proportions similar to those of the original grain. Canada permits up to 5% of the grain to be removed for making "whole wheat flour". The portions, allowed to be removed, contain much of the germ and some of the bran. In Canada, flour that contains 100% of the original grain, is called "whole grain whole wheat flour". In general, whole grain flour can be made by any grinding method capable of reducing the whole grain to the particle size required by the standard for whole grain flour. For example, stone mills allow the direct production of whole grain flour. In roller mills, flour streams are combined with shorts, germ, and bran that may have been reground by pin or hammer mills. "Graham flour" is similar to whole grain flour, however, it is not sifted during milling and is ground more coarsely. A disadvantage of whole grain flour is that it has a shorter shelf life than white flour, since the higher lipid content, introduced by the germ, leads to rancidity. Therefore, whole grain flour has to be stored under refrigeration for a limited time only.

The influence of the extraction rate on the quantitative composition of flours is shown in Table 3.3 [19]. The composition of whole grain flour (100% extraction rate) corresponds to that of whole grain (Table 2.1). Lowering the extraction rates provokes a considerable change in the quantitative composition of

TABLE 3.3 Influence of the extraction rate on the chemical compositions (g/100 g) of common wheat flours [19].

	Extraction rate			
Compound	**100%**	**82%**	**65%**	**40%**
Starch	66.7	76.2	80.7	81.6
Protein	12.4	13.0	11.2	11.3
Dietary fiber	15.2	6.0	4.9	4.6
Lipids	2.1	2.1	1.3	1.1
Minerals	1.9	1.0	0.6	0.5
B vitamins (mg/100 g)	8.3	3.2	1.3	1.4
Vitamin E (mg/100 g)	1.6	n.v.[a]	0.4	n.v.[a]

[a] *Value not available.*

flour compounds. Flour with an extraction rate of 82% shows a highly significant increase in starch (+ 14%) and drastic decreases in dietary fiber (– 61%), minerals (– 47%), and B-vitamins (– 61%) compared to whole grain flour. These tendencies continue, when the extraction rate is lowered to 65%. Starch increases by + 21% and dietary fiber, minerals, B-vitamins, and vitamin E decrease in a range from – 68% to – 84%. Further refining of flour (40% extraction rate) does not significantly change the chemical composition. Altogether, milling of grains to white flours (40% and 65% extraction rate) generates drastic losses of dietary fiber, minerals, and vitamins, compounds with high nutritive values (see Chapter 6), whereas the starch content distinctly increases.

The discovery of millers in the 1950s that finer flour particles contain more proteins than larger particles has led to a novel milling process that enables separation into subsieve particle classes by so-called air classification. This technique involves a form of impact grinding of white flour followed by centrifugal elutriation by means of air currents. Millers first grind the parent flour by impact in an attrition mill or a pin mill to reduce endosperm particles sufficiently in size. The separation of the flour obtained is then achieved by an air classifier, a machine like an air fan except that the air is drawn from the center of the rotor. Finer particles tend to be carried along with the air current, whereas larger particles are collected along the circumference of the housing. Usually, two cuts are made that yield three fractions: the fine fraction (< 17 μm particle size) with protein contents up to 25%, the medium fraction (17–40 μm) with proteins as low as 5%, and the coarse fraction (> 40 μm) with a protein content similar to that of the parent flour. Fine and medium fractions have special qualities and are therefore suitable for different end uses. The fine fraction is primarily used as a blending flour to increase protein content for bread-making and seems to be advantageous in the manufacture of meat-containing products. The medium fraction flour, for instance, is used for the production of cookies and cakes and as thickener in foods such as soups, gravies, and custards. The coarse fraction is often reprocessed to recover more of the medium fraction.

The most notable recent development in flour milling has been the debranning (or pearling) of grains. In this process, grains pass through abrasion and friction devices before milling. First, the outer thicker layers (pericarp) are removed and then sequentially the thinner inner layers (testa and nucellar tissue). Pearling grains prior to milling has been suggested to yield flours with superior bread-making performance in terms of larger loaf volume, more regular crumb texture, and extended shelf life. Due to the inclusion of the aleurone layer, the nutritional value of the flour is increased. Italian millers extensively apply debranning for durum wheat used in pasta production.

References

[1] Balfourier F, Bouchet S, Robert S, de Oliveira R, Rimbert H, Kitt J, et al. Worldwide phylogeography and history of wheat genetic diversity. Sci Adv 2019;5:eaav0536.

[2] FAOSTAT. Food and Agriculture Organization of the United Nations. 2018.

[3] Longin CF, Ziegler J, Schweiggert R, Koehler P, Carle R, Würschum T. Comparative study of hulled (einkorn, emmer, and spelt) and naked wheats (durum and bread wheat): agronomic performance and quality traits. Crop Sci 2015;56:302–11.
[4] Abrol YP, Kumar A, Nair TV. Nitrate uptake and assimilation and grain accumulation. In: Pomeranz Y, editor. Advances in cereal science and technology. vol. 6. St. Paul: AACC; 1984. p. 1–48.
[5] Randall PJ, Wrigley CW. Effects of sulfur supply on the yield, composition, and quality of grain. In: Pomeranz Y, editor. Advances in cereal science and technology. vol. 8. St. Paul: AACC; 1986. p. 171–206.
[6] De Koe WJ, Juodeikiene G. Mycotoxin contamination of wheat, flour, and bread. In: Cauvin SP, editor. Breadmaking—improving quality. 2nd ed. Oxford, Cambridge, Philadelphia, New Delhi: Woodhead Publishing Ltd; 2012. p. 615–58.
[7] World Health Organization International, Agency for Research on Cancer. Evaluation of five organophosphate insecticides and herbicides. vol. 112. IARC Monographs; 2015.
[8] European Food Safety Authority (ESFA). Glyphosate: EFSA updates toxicological profile. Retrieved May 23. 2016.
[9] WHO/FAO. Report of the Joint Committee on pesticide residues. 2016.
[10] European Chemicals Agency (ECHA). Glyphosate not classified as a carcinogen by ECHA. 2017.
[11] Reganold JP, Wachter JM. Organic agriculture in the twenty-first century. Nat Plants 2016;2:15221.
[12] Asseng S, Ewert F, Martre P, Rötter RP, Lobell DB, Cammarano D, et al. Rising temperatures reduce global wheat production. Nat Clim Change 2015;5:143–7.
[13] Trnka M, Hlavinka P, Semenov MA. Adaptation options for wheat in Europe will be limited by increased adverse weather events under climate change. J R Soc Interface 2015;12:20150721.
[14] Delcour JA, Hoseney RC. Principles of cereal science and technology. 3rd ed. St. Paul: AACCI; 2010107–20.
[15] Bass EJ. Wheat flour milling. In: Pomeranz Y, editor. Wheat—chemistry and technology. vol. 2. St. Paul: AACC; 1988. p. 1–68.
[16] Campbell GM, Bunn PJ, Webb C, Hook SC. On predicting roller milling performance. Part II. The breakage function. Powder Tech 2001;115:243–55.
[17] Schild E. Der junge Bäcker, Band 1. Gießen: Fachbuchverlag Dr. Pfanneberg & Co.; 1985136.
[18] International Association for Cereal Science and Technology: ICC Standard Methods.
[19] Souci SW, Fachmann W, Kraut H. Lebensmitteltabelle für die Praxis. 5th ed. Stuttgart: Wissenschaftliche Verlagsgesellschaft; 2011241–3.

Chapter 4

Wheat-based food and feed

The triumphal rise of wheat as a food crop started around 10,000 years ago in Minor Asia. Settlers collected grains from high-growing wheat grasses, crushed them, and produced porridge for their diet. Heating of porridge on hot stones or wooden embers provided the first flat breads. Later on, the Egyptians developed bread leavened by microorganisms and since that time, bread has successively become a staple food for many people worldwide. Apart from bread, countless assortments of other wheat-based products such as fine bakery goods, dry baked products, pasta, noodles, and breakfast cereals are now globally offered to consumers. Wheat-containing foods are omnipresent in the food markets and restaurants and provide people with a widely varied diet having distinctive organoleptic and nutritive properties. Further appreciated features of bread and other wheat-based products are that they belong to the safest and cheapest foods. The global economic value of wheat is enormous and important for national economies, in particular producers, employees, and consumers. The present chapter covers basic aspects of the manufacture and characteristics of baked goods, paste products, other wheat-based foods, and animal feed. Finally, some remarks on the economic significance of wheat products are given.

4.1 History

For a long time of about 300,000 years, humans mainly lived on meat, fish, fruits, and vegetables acquired by hunting and gathering. Frequently, they were nomads, who followed the herds of moving animals. It was only about 10,000 years ago that cereal crops were introduced as further component of the human diet. The first signs of cereal cultivation were found in the Fertile Crescent of the Middle East (including Anatolia, Lebanon, Syria, Palestine, and Iraq), where wild wheat and barley were successfully grown due to the favorable environmental conditions there. Some human tribes changed from nomadic to stable settlement, because cereal cultivation permitted storage of grains for food production and sowing stored grains for the next season. Because dry cereal grains could be stored for relatively long periods of time, they provided humans with security against food failure and starvation. Thus humans routinely supplemented hunting and gathering by farming, which marked the beginning of agriculture, one of the greatest revolutions mankind ever experienced.

Wheat—An Exceptional Crop. https://doi.org/10.1016/B978-0-12-821715-3.00004-6

Relics, excavated in central Syria, indicated skilled use of tools like sickles and mortars to harvest and grind grains. The crushed material was moistened with water and either eaten as porridge or made into flat cakes that were dried in the sun. At some later period, the practice of placing the flat cakes on hot stones or covering them with hot ashes was developed—the birth of flat bread. Einkorn and emmer were the first wheat species that were widely cultivated and became main crops in the Fertile Crescent during the Neolithic. From there, wheat spread eastward through India to China and westward to Egypt, Greece, and Italy and reached Northern Europe including England, Ireland, and Scandinavian countries. Einkorn and emmer remained popular for thousands of years. This was exemplarily demonstrated by the famous Tyrolean Iceman ("Ötzi"), a mummified Late Neolithic hunter found in the glaciers of the Tyrolean Alps: He had eaten unleavened einkorn flat bread before he died.

Around 5000 BC, Egyptians were the first to learn how to make leavened bread from ancient wheat by using wild yeasts and lactobacilli. Monuments from antiquity show that bread was produced in many different shapes, from small, round loaves to elongated loaves often decorated with seeds. Baking became a most important occupation during the classic period of Greek and Roman domination of Western civilization. There were many large public bakeries during this period, but bread was also baked at home. Wheat played such a dominant role in the Roman Empire that this area has been frequently described as "Wheat Empire", where *panem et circenses* (bread and games) was the maxim of the communities. Throughout the Middle Ages, cultivation of barley and rye was as important as that of wheat. However, barley and rye grains were mainly used as feed, whereas almost all wheat grains were used for food production. The main foodstuff "bread" was considered so fundamental that even prisoners were entitled to getting "bread and water".

Interestingly, wheat and bread have become central to religions. In the Bible, wheat and bread are metaphors for freedom and even salvation. Moses described the Promised Land as "a land of wheat and barley and vineyards". In Genesis (47:19), there is prayer for "And give us seed that we may live and not die, that the land not be desolate". Christians pray "give us this day our daily bread" and consume wheat wavers representing the body of Jesus Christ. Muslims regard unleavened wheat bread ("naan") as sacred insisting it be stored upright and never thrown away in public. Jews celebrate Passover with unleavened wheat bread ("matzo") to commemorate the exodus of the Israelites from Egypt.

4.2 Baked products

As mentioned before, wheat grains were introduced into the human diet around 10,000 years ago. Grains were crushed, moistened with water, and eaten as porridge. Portions of the porridge were dried in the sun and consumed as flat cakes that could be stored longer as compared to porridge. The decisive initial step for the evolution of baked wheat products was heating the cakes on hot stones or in hot ash—the birth of palatable, easily digestible, and durable baked goods. The success of baked products was spurred on by the finding that fermentation with microorganisms like wild yeasts and lactobacilli could leaven flat doughs to provide a loosened and voluminous bread. Since this discovery, bread became a staple food worldwide and many other baked wheat products were developed and refined over time followed by increased popularity. Variations in recipes, additives, dough preparation, and baking processes have resulted in the global leading position of baked wheat products among foods, not only with regard to the consumed amounts but also in their broad variability. In the following, preparation and properties of the most important baked products are described. They include leavened and flat breads, fine bakery products, and dry baked goods.

4.2.1 Bread

Bread is the most important baked wheat product and a staple food worldwide. The slogan "Fiat Panis" ("let there be bread") in the emblem of Food and Agriculture Organization (FAO) of the United Nations may characterize the universal importance of bread among foodstuffs. After its beginning thousands of years ago, the manufacture of bread has spread over the whole globe and remained popular up to this day. That is why the major portion of harvested wheat is utilized for the production of bread. Bread-making consists of several complex processing steps and requires a deep understanding of these processes from the experts engaged in making bread [1]. For many centuries, bread-making has been the domain of small-sized craft bakeries. However, the importance of industrialized production of bread has grown enormously during the last 50 years.

The bread-making quality of wheat is unique among cereals. All other cereal flours yield breads of inferior quality with low volume and non-elastic small-pored crumb (Fig. 4.1). The basis for the special properties of wheat dough and bread are gluten proteins (gliadins and glutenins) (see Section 2.2.1). During mixing of flour with water, the gluten proteins form a three-dimensional network that provides cohesivity, elasticity, and gas-holding

FIG. 4.1 Bread from different cereal flours baked under standardized conditions (1 = wheat; 2 = rye; 3 = barley; 4 = oats; 5 = rice; 6 = corn; 7 = sorghum.

capability to the dough and high volume and regularly pored crumb to the resulting bread. The formation of a gluten network in wheat dough is unique among cereals.

Though countless types of wheat bread exist worldwide that differ in constituents, size, color, texture, and flavor, the bread-making process can be reduced to three basic steps: selection of recipe and raw materials, dough making and handling, and baking. In traditional bakeries, parts of the production steps are done manually or by using only small devices and the number of manufactured bread loaves is limited. In industrial production, processing is automatized as far as possible and thousands and more loaves of bread can be produced daily.

4.2.1.1 Breadmaking recipes

The most important ingredients for making flat bread are flour, water, and salt and additionally yeast or other leavening agents for leavened bread. Flour is the major component of the recipe and is responsible for the overall structure and properties of bread. Tap water (≈60% of flour weight) is added in order to hydrate flour particles, to dissolve or disperse ingredients, and to form a homogeneous dough with optimal consistency during mixing. The basic function of salt (sodium chloride), added at levels of 1–2% (based on flour weight), is to confer a slight salty taste to the bread. In addition, salt promotes gluten network formation during mixing, when the mixer speed is low, and contributes to controlling yeast fermentation. In modern industrial bakeries, salt is added by means of a brine system to ensure rapid distribution. Yeast is the most commonly used leavening agent for bread dough. Sourdough fermentation is only of marginal importance and mainly used for making rye bread. Various nonessential additives supplement the recipe and determine the diversity of bread. The contributions of flour, leavening agents, and nonessential additives to the characteristics of dough and bread are described in the following sections.

Flour

The major portion of flours produced in mills is used for bread-making. Other usages, for instance as thickener for soups, sauces, and dressings and in the manufacture of crumb for coating fish and meat products, are of less importance. Flours (usually from hard wheat in the American terminology) with extraction rates of 70% or lower (see Section 3.3) are preferred for the production of most bread types. White flour largely replaced high extraction flour in the late 19th and early 20th centuries after the introduction of modern mills like roller mills. Today however, breads made from flour with high extraction rates or from whole grain flour are gaining popularity mainly for health reasons (see Chapter 6). Especially in Central, Northern, and Eastern Europe, mixtures of wheat and rye flours in different ratios are common raw materials for breadmaking. Freshly milled flour, sometimes referred to as "green flour", is not suitable for making bread of high quality. It should be stored under ambient conditions for some weeks to stabilize the redox system of the constituents ("flour maturation"). The main reaction induced during maturation is the oxidation of reactive thiol groups by oxygen to obtain stable disulfide groups. Natural maturation may be accelerated using chemical agents like ascorbic acid, chlorine dioxide or azodicarbonamide provided that this complies with legislation.

The baking quality of flour is strongly dependent on several factors including wheat species and variety, environmental and soil conditions under which the wheat was grown, the milling process, and the chemical composition of the flour [2]. Common wheat ("hard wheat" in North America) is usually considered to have the highest quality and therefore breads from other species are only regional niche products. Common wheat varieties however can distinctively differ in their bread-making properties and hence, varieties are graded according to their baking quality, high, medium or low for instance, in many countries. The quality of each variety can be influenced by the conditions of farming, storage, and milling. It is therefore important for both millers as sellers and bakers as buyers to know the quality of flour delivered by the mill. In the following, different chemical and physical methods are described that allow the evaluation of flour quality.

Among flour constituents, proteins are the most important determinants of dough and bread properties [3]. The quantity of flour protein is highly related to bread volume and therefore on the first line of measurements. Analytical methods, specified by the International Association for Cereal Science and Technology (ICC) standard methods [4], are the Dumas combustion method (ICC 167), the Kjeldahl method (ICC 105/2), and near-infrared reflectance (NIR) spectroscopy (ICC 159). Apart from the amount of total flour proteins, both quality (composition) and quantity of gluten influence bread-making properties. Gluten quality can be roughly estimated by relatively simple tests mainly based on the sedimentation or swelling properties of gluten. The most commonly used test is the Zeleny sedimentation test (ICC 116/1). The procedure involves the suspension of flour in dilute lactic acid and volumetric measurement of the amount of sedimented material (mainly hydrated gluten) after a defined time interval. A high sedimentation volume is correlated with high gluten quality. A modification of the Zeleny method is the so-called sodium dodecyl sulfate (SDS) sedimentation test, where lactic acid is replaced by SDS.

Both amount and quality of gluten in flour can be determined by the Glutomatic instrument (ICC 137/1). A piece of dough is prepared under standard conditions and washed with tap water to remove soluble material and starch followed by weighing the remaining wet gluten mass. Amounts lower than 20% (based on flour weight) indicate poor baking quality. To get more information on the quality (strength) of the obtained wet gluten, it is placed onto a special sieve and centrifuged, after which the stronger gluten portion remains on top of the sieve, while the weaker gluten portion passes through the sieve. A high "Gluten Index" (percentage of gluten remaining on the sieve) indicates strong gluten.

Gluten consists of gliadins and glutenins (see Section 2.2.1) and their amounts and proportions are highly specific indicators for dough and bread properties. Examples are the ratio of gliadins to glutenins that influences dough strength and the presence and amounts of specific high-molecular-weight glutenin subunits (HMW-GS) that influence bread volume [5,6]. However, quite sophisticated methods (e.g., column chromatography and electrophoresis) are needed to determine protein quality in this way.

Starch, the main constituent of grains, is the second important determinant of dough and bread properties. Milling of grains results in some damage of starch granules. The degree of damage, usually ranging between 5% and 12%, is primarily influenced by milling process parameters but also by wheat variety, grain hardness, and starch properties. The degree of damage has an important impact on product quality. A small degree has a positive influence on the water absorption of flour, bread volume, and crust browning, for instance. A high degree can result in negative effects such as poor dough properties, large pores of the bread crumb, and smaller bread volume. Therefore it is important to determine the degree of starch damage in flours delivered by the mill. The most frequently used measurements are based on the fact that in an aqueous flour suspension, starch is more subjected to hydrolysis by endogenous α-amylase, when damaged starch granules are present.

The capability of the flour to absorb water and the level of water to be added to the flour for dough preparation are further important indicators for quality. The water absorption capacity is mainly affected by moisture, protein and bran contents, and the content of damaged starch of the flour. The level of added water, necessary for optimal dough consistency, is crucial: Too much water soften doughs and too little water strengthens doughs; both states of consistency are unfavorable for bread-making. Common methods for determining the water absorption capacity of flour are based on instruments like the Farinograph or the Mixograph (see below). They measure the volume of water that is necessary to produce a dough with a defined consistency.

Optimal dough development is a key requirement for the production of bread and other baked goods. Therefore each flour consignment has to be pretested for the properties of dough prepared from the flour. One of the most commonly used instruments is the Farinograph (ICC 115/1). It measures the resistance of the developing dough against the rotating mixer blades from the beginning of dough formation and records it as torque ("Farinogram"). The resistance increases, as the dough develops, reaches a maximum and decreases, when the dough breaks down (Fig. 4.2). The quality parameters recorded are dough development time, dough stability, and mixing tolerance. Similar results can be obtained by different types of Mixographs.

The rheological properties of doughs can be analyzed by an instrument called Extensograph (ICC 114/1). During uniaxial stretching of the dough, the resistance to extension as well as the extensibility are monitored and resulting

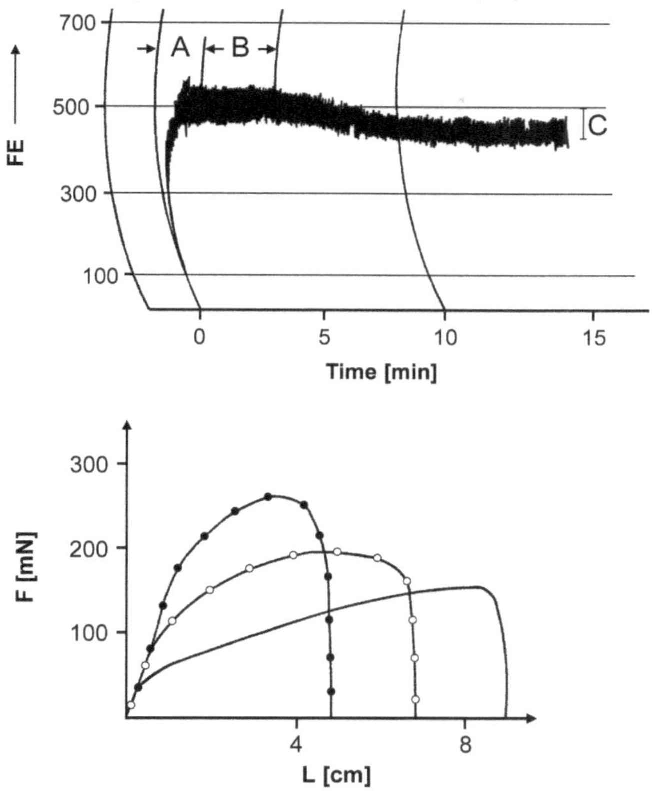

FIG. 4.2 Farinogram; A, dough development time; B, dough stability; C, decrease in dough consistency (top). Extensigrams of strong, medium, and weak doughs (bottom).

shape and area of the curve reflect soft or strong doughs (Fig. 4.2). Moreover, the Extensograph is an important instrument for determining the influence of flour additives like ascorbic acid and enzymes (see below) on dough properties. Another instrument, used for characterizing rheological dough properties, is the Chopin Alveograph (ICC 121). It also measures the resistance to stretching and the extensibility but is based on a different principle. Dough is prepared under standard conditions. Air is blown through a thin sheet of dough to create a growing gas bubble that finally bursts. The pressure, generated during the procedure, is measured manometrically. The characteristics of the monitored curve reflect the stability and extensibility of the dough and are related to gluten strength. Additionally, fundamental rheological tests are used to analyze well-defined physical properties of dough. They include dynamic oscillation, creep and stress relaxation, extensional measurements, and flow viscosimetry. However, the instrumentation is complex and expensive and requires high levels of technical skill.

To obtain final information on the baking quality of a given flour, a standardized baking test is necessary. The test may be carried out using the following recipe (ICC 131): 1000 g flour (14% moisture basis), 18 g dry yeast, 15 g salt, 19 g sucrose, and 0.05 g ascorbic acid. The dough is made by mixing flour and ingredients with water under standardized conditions. Dough pieces are rounded, rested, molded, placed in pans, proofed, and baked. The loaves are evaluated for volume, shape, crust color, and crumb texture.

Leavening agents

Dough, only made from flour, salt, and water, gives a dense, flat product after baking. Flat breads are the most ancient of all breads and still rather common in Africa and Asia. At present, breads from dough leavened by yeast are globally predominant. Wheat breads leavened by acidic fermentation ("sourdough") or by chemical agents are rare. Sourdough breads are leavened by long-lasting fermentation of dough with a piece of prefermented dough ("starter") enriched by strains of lactobacilli and yeasts. They are mainly produced from rye flour or from mixtures of wheat and rye flours. Chemical leavening is achieved by the addition of agents that release gas during dough preparation and baking. It is commonly used, when high osmotic pressure in dough (high concentrations of sugar or salt or low moisture content) inhibits yeast fermentation, in gingerbread for instance. Most used chemical agents are carbon dioxide-generating mixtures of carbonates and organic acids (e.g., tartaric, citric, or adipic acid) or acidic phosphates. The vast majority of chemically leavened baked goods comprises fine bakery goods and dry baked products (see Sections 4.2.2 and 4.2.3) but not bread. So-called aerated bread is leavened by gas (mostly air or carbon dioxide) being forced into the dough under pressure by the usage of pressure-vacuum mixers.

As mentioned before, yeast is by far the most common leavening agent for the production of wheat bread. The manufacture of bread with high volume and porous crumb is usually achieved by the addition of 2–4% of yeast to flour. For this purpose, a special yeast ("Baker's yeast"), mainly consisting of the strain *Saccharomyces cerevisiae*, is applied that metabolizes both maltose (produced from starch by endogenous or added amylases) and sucrose (added to the flour) to carbon dioxide and ethanol. Commercial yeasts are available in a dried form as compressed yeast (around 28–30% dry matter), as cream or as pumpable liquid yeast. The fermentation of sugars by yeast takes place over the whole process of dough preparation and the beginning of the baking process. Single fermentation times depend on the type of dough processing. Typical times for the straight-dough process are 10–30 min during dough bulk fermentation, 5–10 min during dividing and rounding (intermediary proof), 20–50 min during sheeting and molding (final proof), and 5–10 min at the beginning of baking (oven fermentation). Gas bubbles that result from fermentation and mainly consist of carbon dioxide raise the dough and thus increase bread volume. They make up more than 70% of the final bread volume and have a major influence on the textural property of bread. The size and number of bubbles determine the overall appearance of the pored crumb and contribute to the characteristics of different bread types. In addition to gas production, yeast positively affects the rheological properties of dough and produces a variety of flavor compounds typical of bread (see Section 4.2.1.5).

Nonessential ingredients

A number of nonessential ingredients are used to improve both bread-making process and bread quality. They can be added to flour at the mill or at the bakery. Sugar (sucrose) is applied to support yeast fermentation and crust color formation and to provide a sweet taste to some bread types. Rolls and other small baked products may contain up to 15% sugar. Bakery fat (frequently a mixture of oil and solid fat) mainly contains triglycerides (so-called shortening) and is used to improve gas retention within the dough as well as crumb texture and to increase the shelf life of bread. The amount applied varies and depends on the extraction rate of flour; whole-grain flour requires higher levels than white flour. Vital gluten, frequently applied in a range from 3% to 5% of flour weight, is used to fortify weak flours and whole grain flours in order to improve dough strength, crumb structure, and bread volume (see Section 5.2.5).

Many valuable effects can be achieved by using different enzymes such as α-amylase, xylanases, peptidases, and lipases. They lead to improvements of fermentation and dough stability, bread volume and flavor, crumb structure, and shelf life. For example, α-amylase is used to increase maltose production important for fermentation and to extend the shelf life of bread, whereas endoxylanases are used to enhance bread volume. Lipoxygenase-active soy flour improves dough-mixing properties and bread volume and bleaches carotenoids and other colored constituents by oxidation resulting in a white crumb.

Compounds that act as oxidants in bread production include ascorbic acid (via dehydrascorbic acid), potassium bromate, azodicarbonamide, and calcium peroxide. Small levels (≈ 50 mg/kg flour) are sufficient to oxidize low-molecular-weight thiol compounds (glutathione and cysteine) that otherwise would degrade the gluten network in dough. Thereby they improve dough strength and bread size and texture. Except for ascorbic acid (vitamin C) that is claimed to be nondetectable in the final product anymore, the usage of chemical oxidants is frequently restricted by national and international legislations. Reducing agents like cysteine may be used to weaken strong doughs, for example, to reduce resistance to deformation that would otherwise make molding and shaping more difficult and reduce bread volume.

Different emulsifiers, applied at about 0.5% of flour weight, are also used for the improvement of dough properties, bread volume, freshness, and shelf life. They mainly comprise lecithin, mono- and diacylglycerides and their derivatives (esterified with acetic, lactic, tartaric, monoacetyl, or diacetyl tartaric acids). Increased diversity of breads is provided by additional ingredients like milk, eggs, nuts, dietary fiber, and spices as well as by adding other flours, for example from rye, barley, oats, potato, or soy. The addition of spices and oilseeds such as seeds of caraway, coriander, sesame, and sunflower, can be used to confer taste and flavor to the product.

4.2.1.2 Dough preparation and treatment

A number of different processes can be used for dough preparation from wheat flour; examples are the straight-dough system, the sponge-and-dough system, and the short-time system including the so-called Chorleywood bread process (CBP) [7]. For the straight-dough system, all ingredients are mixed into a dough that is then allowed to ferment. The leavened dough is divided into loaf-sized pieces, rounded, and molded into the loaf shape. An additional fermentation period (proof) is applied to increase dough size. In the sponge-and-dough system, about two-thirds of the flour, proportionate water, and yeast are mixed to form a loose dough, called sponge that is allowed to ferment for 4–8 h. The rest of the ingredients is then mixed with the sponge followed by an additional proofing step. The subsequent dough treatment corresponds to the straight-dough system. In comparison to the sponge-and-dough system, the straight-dough system requires less processing time, labor, power, and equipment. However, straight-dough bread is chewier and has a coarser cell structure and less flavor. The CBP is characterized by high-speed dough mixing with high energy input under reduced pressure and short fermentation time. The process requires the addition of higher amounts of water, yeast, oxidants, and fat or surfactants compared to conventional dough preparation methods.

In general, the first step of dough preparation for the purpose of breadmaking includes mixing of flour with water and other ingredients. In this procedure, key factors affecting dough properties are water absorption and mixing conditions. Typical levels of added water are in a range from 55% to 62% of flour weight. The full hydration of flour particles is most important in order to create a cohesive dough mass and the presence of free water allows reactions of flour compounds and ingredients. The optimal amount of water to be added can be determined in pretests by instruments such as the Mixograph and the Farinograph (see above). Proper dough mixing should produce a dough with an optimal degree of development that is strongly influenced by the type and speed of the mixer and the mixing time. Continued mixing after reaching the optimal dough development should be avoided, because it produces wet and sticky doughs, the doughs are "overmixed". The horizontal conventional mixer is the most common commercial mixer. It contains a U-shaped mixer bowl and an agitator drive shaft with two to four mixing arms that rotate at 60–85 rounds per minutes. Heat, generated during the process, is dissipated by refrigerated jackets on the mixer, so that the dough keeps a temperature of 22–30 °C. The mixing time is about 12–16 min. Other mixers in use are high-speed mixers such as the Tweedy mixer (6–8 min mixing time) and the Stephan mixer (2–3 min mixing time).

The CBP procedure, in which mixing and dough development are carried out in a single operation, uses mixing times between 2 and 5 min at an energy input about 5- to 8-times higher than applied by conventional mixing. The process requires a higher protein content (> 13%) of the flour than other processes and it may be necessary to fortify with vital gluten. The recipe for dough preparation requires yeast quantities 50–100% above the normal level and addition of oxidants (e.g., dehydroascorbic acid via ascorbic acid or potassium bromate) and high-melting point fat or emulsifiers. After mixing, only short intermediate and final proofing are required. As a result of the CBP, dough preparation and fermentation time can be drastically shortened ("no-time dough") and bread can be produced very quickly and at low costs. The CBP is widely used for the production of factory bread in the United Kingdom and Australia.

The optimal formation of dough during mixing is evident from its appearance. The mass appears silky and glassy, is coherent and elastic, and does not tend to stick on the wall of the mixer. The completion of mixing can be judged by the baker: When an optimal dough piece is pulled between the fingers, it can be stretched into a thin membrane, in which small bubbles and gluten strands can be seen. Because dough development is related to the power input of the mixer, wattmeters attached to the mixer can be used to shut-off the mixer automatically, when the pretermed energy requirement has been reached.

The next step of dough treatment after mixing is dividing the bulk dough into pieces of appropriate size; the weight can range from less than 100 g to more than 1 kg. Commercial dough dividers perform their function volumetrically. After dividing, dough pieces are rounded. In automatic rounders, pieces are forced to travel against a moving surface continually rotating. Bowl, umbrella, and cone rounders are the most used types. The main objective of rounding is to form a skin around the dough piece to minimize gas loss. Because dividing and rounding operations induce considerable loss of gas, the dough is given a rest period of about 6–8 min (intermediate proof). Pieces, leaving the intermediate proofer, have expanded and become more pliable and extensible. The subsequent molding of dough pieces is the last dough-manipulating stage, during which the gas cells are finally subdivided and uniformly distributed through the dough mass. Molding proceeds in three steps: sheeting, curling, and rolling. Frequently used molder types are drum molders and for special applications curl and bread-twisting molders. Modern molders are often equipped with an automatic panning device. After being molded, self-supported or panned doughs are transported to a proof box for final proofing. Because optimum conditions of proofing depend on many factors (e.g., ingredients, yeast activity, and the type of bread desired), conditions of proofing may vary in a wide range including proofing time of 20–50 min, temperature of 30–45 °C and relative humidity of 80–85%, for instance.

4.2.1.3 Baking process

The baking stage finishes the bread-making process by transforming the foam structure of dough into the sponge structure of bread [8]. Baking mediates the characteristic appearance and properties of the product "bread" by heating the proofed doughs. Strong doughs can be baked without the support of a pan (free-standing loaves), whereas soft doughs have to be baked in a pan. The conditions to be selected for the baking process (e.g., temperature and duration of heating) are highly variable and depend on the size of dough pieces and the desired bread type. In a few remaining traditional baking ovens, wood or coal is used for heating. Modern ovens are heated either by gas, oil, or electricity. In industrial baking plants, two types of ovens are mostly used: the tunnel oven and the traveling-tray oven. The tunnel oven provides a single pass of dough pieces through the baking chamber on a traveling hearth. Doughs enter at one end of the tunnel and the baked loaves are discharged at the opposite end. Tunnel ovens usually consist of several heating zones. Passing time and temperature of single zones can be adapted to the type of bread. For example, a typical one-pound loaf of bread is baked in 15–18 min, when the temperature is set at 174, 218, 227, and 193 °C in a four-zone oven. Traveling tray ovens consist of trays moving over two parallel endless chains that connect two reels. Each tray, holding several dough pieces, is permanently fixed to the conveying chain. The chain pulls the trays from the front of the oven through the hot-air chambers to the back and then to a lower track, where the dough continues to bake on its return to the front of the oven, where it is unloaded. Traveling tray ovens are designed as either single-lap or double-lap ovens. Traveling speed and temperature can be adapted to the type of bread.

The baking process is accompanied by numerous physical and chemical changes that are particularly influenced by the level and time of heating. These changes are different in the inner part (crumb) and the outer part (crust) of the loaf, since heat transfer occurs slowly and produces a gradient of temperature from the crust (>200 °C) to the inner part of the crumb (<100 °C). At the beginning of heating (up to 50 °C), yeast continues to produce gas and increases loaf volume ("oven-spring"). At about 65 °C, starch starts to gelatinize and with increasing temperature, gluten proteins denature and gliadins and glutenins are cross-linked. The surface of the dough, exposed to temperatures of around 200 °C, rapidly forms a crust, where water vaporizes very fast and the crust undergoes browning. This is the result of Maillard reactions between reducing sugars like glucose and free amino groups of proteins, peptides, and amino acids. Moreover, starch degrades to dextrins, disaccharides, and monosaccharides and subsequent caramelization contributes to sweetness, aroma, and color of the crust. The thickness and appearance of the crust are dependent on temperature and baking time and can be influenced by the level of steam added to the oven during baking.

4.2.1.4 Bake-off technology

So-called bake-off technologies have developed quickly over the last decades. The principle usually involves the industrial manufacture of semifinished products that are then finished in baking stores. Freezing of the semifinished material is often a part of the bake-off process. Several different ways of preparing bake-off products are in use. Factories produce material either fully baked and frozen, partially baked and unfrozen, partially baked and frozen, unfermented frozen dough pieces or fermented frozen dough pieces. Goods are finished in baking stations that are equipped with freezing, thawing, proofing, and oven devices. Advantages of the bake-off technology are based on keeping the sophisticated time-consuming operations at the industrial level and the final simple and short operations in vending shops equipped with a minimum of devices and an unskilled staff. The advantages for consumers are the almost ubiquitary occurrence of baking stations and the steady availability of freshly baked products.

4.2.1.5 Bread flavor

The flavor of bread is one of the most important quality characteristics for consumers. Typical flavor compounds are created during three different stages of bread making: dough mixing, dough proofing, and baking. The initial step is mixing. Different flour enzymes (e.g., amylases and peptidases) are activated and produce important precursors of odorants such as sugars and amino acids. The next step is proofing the dough. Apart from the production of ethanol and CO_2, yeast and lactobacilli convert sugars into different alcohols, aldehydes, ketones, esters, and acids and thus provide flavor compounds. Most of these compounds develop the typical crumb flavor. Proofing duration and temperature have a strong effect on crumb flavor. Moreover, prefermentation or a sourdough step may improve flavor quality.

The heat-dependent production of bread flavor during baking is based on two distinctively different thermo-chemical reactions: caramelization and Maillard reactions. Caramelization mainly involves the transformation of sugars (glucose, fructose, maltose, and sucrose), present in the dough after fermentation, into brown complex polymers. Caramelization occurs on the outside of the bread at around 135 °C and proceeds via a sequence of complicated reactions at more than 200 °C, which result in the formation of various oligosaccharides with tastes, ranging from sweet to intensely bitter, and in the coloration of the crust. Odorants such as aldehydes, ketones, and furfural are also formed but seem to play a minor role in the formation of crust flavor.

The second type of heat-dependent flavor-producing reactions is the Maillard reaction, also called nonenzymatic browning. It involves interactions of reducing sugars (glucose, fructose, and maltose) and the amino groups of amino acids, peptides, and proteins. The highly complex reactions start at around 100 °C and result in the formation of colored aldehyde-amine polymers and numerous flavor compounds such as aldehydes, ketones, pyrazines, and pyrroles. They are initially formed in the crust but diffuse from the crust to the crumb during cooling and storage.

White bread has most frequently been studied regarding flavor compounds among bread types. More than hundred odorants, belonging to different chemical classes (e.g., acids, alcohols, esters, furans, aldehydes, ketones, pyrazines, and pyrroles), were identified [9]. However, the mixture of only a few compounds decisively contributes to the overall flavor impressions of crumb and crust, the flavor profiles of which are strongly different: Whereas the crumb is characterized by a "yeasty" flavor note, "roasty", and "caramel" notes predominate in the crust. Some of the most important flavor compounds of crumb and crust of white bread are summarized in Table 4.1. The list indicates that not a single compound by itself but only a complex mixture of different compounds in defined concentrations determines crumb and crust flavor.

The composition of flavor compounds strongly varies between the different bread types and is dependent on many factors such as the kind of flour and ingredients and the conditions of mixing, fermentation, and baking. A number of ingredients, used for dough preparation, affect the overall flavor of bread. For example, the choice of flour has an important influence: bread from white flour differs significantly from bread made from whole grain flour, when aroma is compared. The strain of yeast and its amount can also determine bread flavor and even the amount of salt (e.g., 1.5% or 2.5%) plays a special role. Of course, other additives like dairy products or spices can change the flavor profile dramatically. To enhance the formation of specific flavor compounds, different methods are in use, for example, the addition of specific enzymes that increase sugar and amino acid production during dough mixing and proofing or the addition of flavor precursors directly. Spraying of odorant solutions onto bread or into the package is another method. It is well-known that bread aroma is significantly modified (impaired) after some days of storage. The reasons are the evaporation of volatiles and their diffusion from the crust to the crumb, oxidative reactions, and reaction of flavor components with proteins and polysaccharides.

4.2.1.6 *Bread storage*

During storage, bread readily changes desirable quality characteristics into undesirable ones. The process is called staling and affects the state of both crust and crumb and the flavor profile. Bread staling represents a major problem for

TABLE 4.1 Compounds that determine the flavor of crumb and crust of white wheat bread and their flavor qualities [9].

Crumb	Crust
2.4-Decadienal (fatty, waxy)	2-Acetyl-1-pyrroline (roasty, sweet)
(E)-2-Nonenal (green, tallow)	4-Hydroxy-2,5-dimethyl-3-furanone (caramel)
4,5-Epoxy-2-decanal (metallic)	(Z)-2-Nonenal (green, cucumber)
3-Methyl-1-butanol (malty)	3-Methylbutanal (malty)
Phenylethanol (floral, honey)	3-Methylbutyric acid (sweaty)

consumers and producers and is the main reason for wasting. Under optimal conditions, the shelf life of most white breads is up to 3 days. Due to the absorption of water from the crumb and the surrounding air, the crust loses its crispiness and glassiness and becomes tough and leathery. Changes in the crumb are characterized by elevated firmness, loss of elasticity, and juiciness and it crumbles more easily. The deterioration of crumb quality is obviously not only due to drying but also to the retrogradation process of amylose and amylopectin (see Section 2.1.1). The latter process is reversible; therefore stale bread can be refreshed by heating up to at least 50–60 °C, the melting temperature of amylopectin. Bread firming can be retarded by certain amylases that cleave branched oligosaccharides from amylopectin and by emulsifiers that form complexes with amylose.

The level of aroma of freshly baked bread decreases rapidly on storage. It is an old experience: Fresh bread smells and tastes good and bread is eaten best within some hours after baking. During storage, aroma compounds either evaporate or, if they have reactive chemical structures, are degraded or entrapped within amylose helices.

Another possible way of deterioration of stored bread is microbial spoilage mainly by molds. Sliced, prepacked, and wrapped breads are most prone. Whereas the baking process eliminates most fungi, the following operations allow the contamination with fungal spores from the bakery environment and the outside atmosphere. Mold spoilage is caused by a number of different species such as *Penicillium*, *Aspergillus*, and *Cladosporium* strains. The prevention and control of contamination of stored bread with molds and their toxins is therefore indispensable. The most common method to control mold growth is the usage of antifungal agents. That can be done by the addition of food-grade preservatives like propionic acid, sorbic acid or acetic acid. Furthermore, sodium benzoate or propylparaben can be used. The admission of these preservatives is regulated by legislation in most countries. Alternatives for chemical preservation are natural antimicrobial agents like essential oils or the usage of modified atmosphere (e.g., carbon dioxide) for packaging.

4.2.1.7 Bread quality

The term "quality" cannot be simply defined and is related to different factors depending on the type of bread. While some quality parameters can be easily defined and measured (e.g., loaf volume and crumb porosity), many others are ephemeral and highly personal (e.g., appearance and eating quality). According to Stanley P. Cauvain, "bread quality may be described as being the sum of those quality traits that confer the sensory pleasure associated with smell, taste, and texture. In summary, when we taste and eat the product, we experience pleasure. Our individual perception will then depend on the combination of those parameters that characterize the mix of product appearance, texture, smell, and flavor that best suits each of us. We must also recognize that our perception of quality will change with the passage of time and personal circumstances, not least our health" [1].

Although the different bread types are variable in their overall characteristics, the judgment of bread quality by the consumers is generally influenced by the following parameters: external appearance (shape, size, and color), sensory properties (aroma, taste, and mouthfeel), state of crust (color, thickness, and crispness) and crumb (distribution and size of pores and elasticity), and shelf life. Moreover, it is important for the consumer that these properties, typical of each baked product, do not change from day-to-day. Fresh bread almost always has the best quality. Therefore most French people prefer eating baguettes within a few hours after baking. Most bread flaws are based on the following points: Shape and size are untypical, color is too light or too dark, taste is burnt, sour or bitter, flavor is weak or strange, crust and crumb are cracked, crust is tough, leather-like or spotted, crumb is hard and crumbly or damp and slippery, nonelastic and nonspreadable, unregularly pored with big hollows, bottom of the loaf is dark and burnt, and last but not least shelf life is short.

4.2.1.8 Bread types

The term "bread" includes a wide range of products that differ in the ingredients used, the breadmaking process applied, and the characteristics of the end product (e.g., shape, size, texture, color, crust and crumb properties, taste, and flavor). It is therefore quite difficult to classify bread into different types. In the following, breads are simply grouped into leavened and flat breads and the most important types within each group are shortly described.

Leavened bread

The major part of bread types mentioned below is leavened with yeast, sometimes in combination with chemical leavening agents. Some common products are shown in Fig. 4.3.

White (pan) bread is still the classical, predominant bread type worldwide. It is preferentially produced from flour with a low extraction rate (40–70%). The flours used usually need a relatively strong gluten. The basic recipe may include 100 parts of flour, 58–60 parts of water, 2 parts of salt (sodium chloride), and 1.5–2.5 parts of yeast. Frequently, small amounts of sugar (sucrose), fat, and ascorbic acid are used as ingredients. The straight-dough system is normally applied for dough preparation. Optimal water uptake and mixing time are pretested by instruments such as Mixographs or Farinographs (see Section 4.2.1.2). The optimal mixing time is mainly dependent on dough temperature, strength of gluten, and type of mixer. At a temperature of 22–30 °C, the typical mixing time is around 12–16 min, when a conventional mixer is used, and around 3–10 min, when a speed mixer is used. The conditions of proofing are dependent on dough properties and may vary in ranges of 20–50 min at 30–45 °C and 80–85% relative air humidity. The subsequent baking process for a 500 g pan bread is usually performed at 240 °C for 30–35 min. Breads of 500 g are the most popular bread sizes, but variants from 100 g to 1 kg are also on the

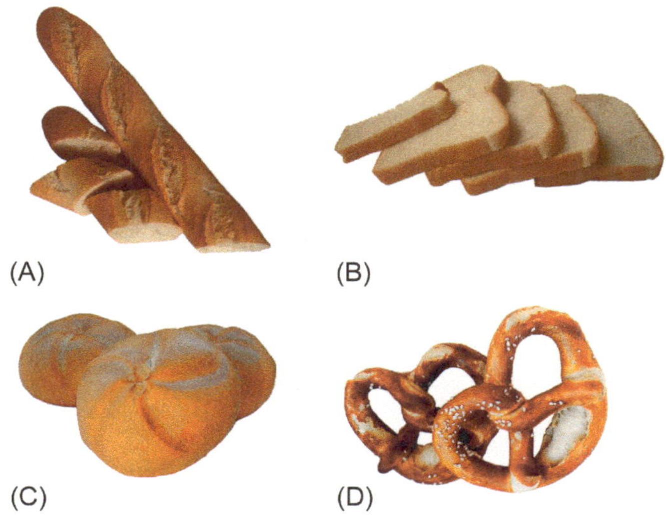

FIG. 4.3 Appearance of different leavened bread types: baguette (A), white pan bread (B), bread rolls (C), and pretzels (D).

market. Self-supported doughs allow the production of breads with different shapes, for example, round or oval. The percentage of packaged sliced bread has increased during the last decades. On average, white bread consists of 52% carbohydrates (including 39% starch and 3% dietary fiber), 37% water, 8% protein, 1.6% minerals, and 1.2% lipids [10] (Table 4.2). Predominant minerals are sodium, potassium, phosphorus, and calcium. Nicotinamide, pantothenic acid, and α-tocopherol are the most important vitamins. 100 g of white bread supply around 240 kcal energy.

Dark bread is made from flours with a high extraction rate (more than 80%) or from a mixture of wheat and rye flours. For baking, flours usually need the addition of vital gluten to reach the necessary dough strength. Due to higher contents of dietary fiber, minerals, and vitamins, the nutritional value of dark bread is much higher than that of white bread. Dark bread is especially popular in Central, Northern, and Eastern Europe.

Toast bread can be defined as sliced bread (each slice about 0.6–0.8 cm thick) that is browned by exposure to radiant heat. Typically, the recipe for dough preparation contains more fat and sugar as compared to conventional white bread and in addition milk or milk powder. Doughs are baked in pans of quadratic cross-section and rectangular longitudinal section. After baking, loaves are cut into slices and packaged. The crumb is characterized by small pores and the crust is thin and medium-brown with little crispness. In order to make the bread more palatable, slices are toasted. Toast bread is often served as accompaniment to soups and stews. Spread with butter or margarine and topped with jam or honey, it is a classical component of breakfast in many countries.

TABLE 4.2 Average chemical composition of white wheat bread [10].

Basic composition (g/100 g)		Minerals (mg/100 g)		Vitamins (μg/100 g)	
Carbohydrates	52.0	Sodium	540	Nicotinamide (B_3)	850
Starch	38.8	Potassium	132	Pantothenic acid (B_5)	690
Dietary fiber	3.2	Phosphorus	88	α-Tocopherol (E)	400
Water	36.9	Calcium	58	Thiamine (B_1)	86
Protein	7.6	Magnesium	24	Riboflavin (B_2)	60
Minerals	1.6	Zinc	0.7	Folic acid (B_9)	22
Lipids	1.2	Iron	0.7	Pyridoxine (B_6)	17

Baguette (French bread), considered to be one of the symbols of French culture, is now produced around the world. Traditional baguettes are made from white flour, salt, yeast, and water without any additives beside optional small amounts of bean flour, soy flour or wheat malt flour. Leavening of dough is achieved either by the direct addition of baker's yeast or the addition of a preferment to improve flavor ("artisan style"). Doughs are treated by a series of folding and rolling motions and slits are cut into the top surface to allow an expansion during baking. The shaped doughs are baked directly on a baking tray or in special perforated pans. A typical baguette loaf has a diameter of about 5–6 cm and a length of about 65 cm. Pieces of baguettes or small baguettes are often cut transversely, dressed with filling, and used for sandwiches or bread pizza.

Ciabatta, an Italian white bread, is made from wheat flour, water, salt, and yeast. The bread shape is somewhat elongated, broad, and quite flat. The crumb is soft and unregularly pored, whereas the crust is crispy and light to the touch. Many regions of Italy have their own variations of additives; for example, milk, herbs, or black olives are added to the dough. Ciabatta was created in 1982 by a baker in Verona and in 1999, the original recipe was licensed to bakeries in numerous countries by Cavallari's company. Today, ciabatta-like products are available in many parts of Europe, North America, and Australia.

Bread rolls differ from classical white bread only by their size, shape, and weight. They are made from white flour, salt, yeast, and water and are characterized by small, often round, or oblong loaves. Further additives used are regionally variable and may include sugar, emulsifiers, milk or milk powder, different spices, and seeds like sesame seeds. Sweet rolls (buns) have an increased sugar content. Bread rolls are common in Europe, especially in Germany and Austria, but many variants can be found in other countries such as Canada, Brazil, and Australia. The weight of the loaf may vary from 50 to 250 g. The brown-colored crust is crispy and tender, the crumb is rather uniform and its flavor is mild and bread-like. Sliced bread rolls are used for the preparation of hamburgers or cheeseburgers worldwide.

The basic recipe and dough preparation of *pretzels* are similar to those of bread rolls. However, dough strands are clipped, twisted, and set to a knot-shaped form ("pretzel loop"). The surface is dried with air, steeped in a warm solution of sodium hydroxide (2–4%), salted with coarse-grained salt, and baked. The crust hardens and caramelizes resulting in the typical flavor and the glaze and dark-brown color of the surface. Pretzels are most popular in German speaking countries and the United States but are also offered in many other countries around the world.

Flat bread

Flat breads are probably the oldest of all breads. Baking flat breads was discovered in the Neolithic period and remained common for a long period of time. By the Bronze and Iron ages however, flat bread was gradually replaced by voluminous breads leavened by lactic acid bacteria and yeast. At present, flat breads are still popular as specialties in many parts of the world. Though innumerable types of flat breads are produced, most of them have common characteristics [11]. They are made from soft wheat flour with a low protein content. The baked products have a low specific volume and a high crust to crumb ratio. The crumb is dense and the crust is light colored often with brown and dark spots. The most popular flat bread types are described in the following.

Chapatti is made from coarse flour of wheat or sometimes of other cereals mainly in India and Pakistan. Flour is mixed with water with or without the addition of salt and fat to a slack dough. The dough is allowed to relax and then divided, rounded, rested for some minutes, and sheeted with a rolling pin or by hand to a flat, round shape about 15 cm in diameter. The sheeted dough is transferred to a preheated hot plate and cooked for 70–80 s with turning two or three times.

Arabic flat bread is widely consumed in the Middle East and North Africa. White Arabic bread is made from flour of around 70% extraction rate and brown Arabic bread from flour of 90–95% extraction rate. Traditionally, flour is mixed with water (40–50%) and fermented by airborne bacteria and yeasts. In most automatic bakeries, salt (1%), yeast (1%), and sugar (0.5%) are additionally used as ingredients. After fermentation (up to 90 min), the dough is divided and shaped into balls (125–200 g), flattened to disks, proofed for 30 min, and baked at 500 °C for 40 s.

Tortilla, made from white wheat flour, is a flat bread originally derived from corn tortilla and a specialty of Mexico. It is made with an unleavened dough that is pressed and cooked like corn tortillas.

Crisp bread, a nutritious and tasty wafer-thin "cracker", has its origin in the Scandinavian countries and is now popular all over the world. Traditionally, the dough is made from rye flour (mostly whole grain flour), water, and salt.

Today however, blends of rye and wheat flours are frequently used as base material and milk, different spices, and seeds (e.g., sesame seed) as additives. The dough is either mechanically leavened by means of creating air bubbles during mixing or by yeast or sourdough fermentation. Another method is to mix the dough under pressure in an extruder. The rapid drop of pressure after leaving the extruder causes water to evaporate and bubbles to form in the dough. After mixing, the dough is rolled to a precise thickness (around 0.5 cm), dusted with flour, and cut to the desired dimension. The cut pieces are baked for 8–10 min at temperatures usually between 200 and 250 °C. Subsequent drying reduces the moisture content to a level of about 5%. Because of this low content, crisp bread has a longer crispness and shelf life as compared to most other bread types.

Pizza, originally from Italy, consists of a yeasted flat bread coated with different toppings and baked in an oven. Pizzas are produced and sold around the world and offered almost everywhere, for example in restaurants ("pizzeria"), kiosks, by home-services, and frozen in supermarkets. The recipe for the dough is rather simple and may consist of white flour, water, fresh yeast, and olive oil. The dough is usually kneaded by hand, fermented over 1.5–2 h, and rolled into plate-sized slices. Professional baking is performed in a special pizza oven. Temperature and baking time depend on the desired baked product, crispy ($\approx$200 °C for 15 min) or light and soft ($\approx$250 °C for 10 min). Countless different mixtures of toppings (e.g., cheese, ham, salami, mushrooms, and vegetables) determine the type of the end product.

4.2.2 Fine bakery goods

Previously, the manufacture of fine bakery goods was the domain of pastry cooks and cake shops. Today, they are produced and offered in many bakeries and the industrial production has grown substantially. The principle processes are similar to those of bread-making (selection of raw material and recipe, dough preparation, and baking) and are at least in parts automated [12].

4.2.2.1 Ingredients

Type and proportions of the raw materials used are highly variable and correspond to each specific product. In many cases, wheat flour (preferably from soft wheat in the American terminology) with a low extraction rate, milk and milk-derived products (cream, butter, curds, and cheese), sugar, fat, and eggs may be major constituents. Examples for further ingredients are different fruits, fruit-derived products (jelly and jam), special masses (marzipan, nut cream, and cocoa), spices (nutmeg, cinnamon, clove, and candied lemon peel), aroma compounds (vanilla, rum aroma, and fruit aroma), and acids (citric acid and tartaric acid). Due to the higher number of ingredients, the production of fine bakery goods is usually more complicated compared to regular bread-making.

Dough leavening is usually performed by means of yeast or chemical agents. The latter group comprises ammonium compounds that release ammonia and carbon dioxide (ammonium hydrogen carbonate, diammonium carbonate, and ammonium carbamate) as well as sodium compounds (disodium carbonate and sodium hydrogen carbonate) and dipotassium carbonate (potash), all of which release carbon dioxide. Baking powder is a ready mix of carbonate or bicarbonate salts with a weak acid (e.g., monocalcium phosphate, tartaric acid, adipic acid, or citric acid) combined with an inert filler (e.g., starch) to prevent both components from reacting with each other ahead of time. Baking powder works, when carbon dioxide is released in the wet dough by an acid-base reaction.

Thickening agents are added to increase the viscosity of the dough, to form gels, and to stabilize emulsions and suspensions. With the exception of the protein gelatin, they all belong to the class of polysaccharides (e.g., agar-agar, alginates, and carrageen) and are called hydrocolloids. To improve the appearance of fine baked goods, a number of food colors are available that adjust discoloration or color changes that happened during processing. Mostly used coloring agents are cocoa powder and caramel (brown), egg yolk, saffron, and carotene (yellow), and fruit and red beet juices (mostly red).

4.2.2.2 Selected products

It is in fact impossible to compile a comprehensive list of fine baked goods that are produced and consumed worldwide. Therefore only examples of the most important types are included in the following. Some of them are shown in Fig. 4.4.

Cakes are sweet, mainly chemically leavened wheat products that are very popular and consumed worldwide in numerous variations of recipes and processing. Well-known examples are flans, gateaus, pies, and Swiss rolls. Cakes are generally characterized by high contents of sugar and fat. A typical recipe includes soft wheat flour (100 parts), sugar (140), shortening (55), egg white (76), milk (95), baking powder (1.3), and salt (0.7). Common additional ingredients may include fresh or dried fruits, nuts, cocoa, and flavors. Compared to bread doughs, cake doughs contain more water and are batter-like. Countless cake recipes exist worldwide, but there is a lack of clear definitions of the different existing types of cakes. They may be best defined by how they are mixed. One of the oldest types is based on the multistage mixing procedure, in which sugar, shortening, milk, and eggs are mixed together first to produce a cream. In later stages, the other ingredients and finally flour are added. Another type is the so-called box cake corresponding to single-stage mixing. It has this name, because all ingredients are added into a box before mixing. The addition of emulsifiers is essential in order to increase the viscosity of the batter and thus to incorporate sufficient air bubbles during mixing. Box cakes are relatively easy to make and very tolerant to variations in recipes and processing. The third type includes the enhanced mechanical incorporation of air into the batter by using high-speed mixers. The conditions of cake baking depend on the recipe and the desired size and appearance of the final product. The aim is to manufacture a tender product with a pleasant taste and flavor. Formerly, the production of cake

FIG. 4.4 Appearance of different fine baked goods: ring cake (A), cream gateau (B), croissants (C), and pancake (D).

was the domain of housewives and confectionaries. Today, ready-to-use cake mix preparations enable almost everybody to bake a cake. Frosted prepackaged cakes are also staples of many supermarkets.

Puff pastry is a flaky light-weight product that consists of several layers alternatively composed of laminated dough and fat. The basic dough is made from white flour, water, fat, sugar, and salt. After intensive mixing and rounding, the dough is rolled out to obtain a thin sheet. The corresponding fatty layer consists of cooled butter or other solid fats. One layer is then placed onto the other layer. Preparing a classical puff pastry requires an envelope by placing the smaller fat layer inside the larger dough layer. The "inversed" puff pastry envelope places the smaller dough layer inside the larger fat layer. The prepared envelopes are rolled out and repeatedly folded and again rolled out in succession. The number of folds determines the number of dough/fat layers that can be more than a hundred with a thickness of less than one millimeter each. The material is cut into pieces of the desired size and baked. During baking, water evaporates and causes the layers to rise resulting in the typical flaky-like pastry.

Croissants are made from puff pastry leavened with baker's yeast. They have their origin in France and Austria ("Kipferl"). Croissants are offered without filling or are filled with savory or sweet fillings such as chocolate cream or apricot jam. Moreover, they are often covered with powdered sugar or other toppings.

Brioche is a highly enriched pastry of French origin. The dough is produced by mixing of flour, butter, eggs, liquid (water, milk, or cream), yeast (or sourdough), salt, and sometimes sugar. Common recipes suggest a flour to butter ratio of about 2:1. The dough is rolled into a ball or placed in a pan and fermented to about double of its volume, first at room temperature and then in the refrigerator. The dough is then shaped and placed in a container for a final proof, often brushed with an egg wash and baked at 230 °C until the crust is browning. Baked brioche is characterized by a tender, more or less fine crumb and a golden and flaky crust. Brioche has many local variations in ingredients, fillings, and toppings and is often served as a component of desserts and tea.

Pancake is a thin and round flat cake that is popular in many regions of the world. It is prepared from white flour, salt, and water. The recipe may also contain milk, eggs, and butter. The batter is often unleavened or leavened with baking powder or yeast. Cooking of the batter is performed on a hot surface like a frying pan or a griddle covered with butter or oil. The batter is evenly poured over the cooking surface in a thickness of about 0.5 cm, cooked for 1–2 min and then turned over to heat the other side. Structure and shape of pancakes can vary widely as well as toppings and fillings, e.g., jam, syrup, chocolate cream, or meat.

Crepe, typical of France, is a thin pancake cooked on one or two sides in a special pan ("crepe maker") and commercially served by creperies. Crepes are offered as sweet or savory variants. Sweet crepes are covered with sweet toppings such as nut cream, preserves, maple syrup or fruit spreads and mainly eaten as part of breakfast or as a dessert. Common savory crepes, often made with nonwheat flours like buckwheat, are topped with cheese, ham, eggs, and mushrooms and served for lunch or dinner. The fillings are put onto the center of the crepes and the edges are partially folded over the center. Crepes are particularly popular throughout France, but both sweet and savory variants are also produced and consumed in many other countries.

Donuts differ from other fine bakery goods in being heat-treated by deep-frying rather than baking. Apart from flour, salt, and water, other ingredients such as eggs, milk, sugar, fat, and flavors may be used and mixed to a dough or batter that is leavened by yeast or chemical agents. The dough can be formed into various shapes; most common types are ring doughnuts and ball- and finger-shaped donuts. Ring doughnuts are formed either by joining the ends of a dough strand into a ring or a piece is removed from the center of a dough ball. Ball-shaped doughnuts are usually filled by injection with jam, cream, custard or other sweet fillings. Yeast-leavened doughs are fried for about 150 s at 182–190 °C and cake doughnuts for about 90 s at 190–198 °C. After frying, the doughnuts are often covered with a glaze or powdered with sugar to improve their appearance and taste. Doughnuts are popular in many countries and offered in many various forms.

Gingerbread refers to a sweet, strongly spiced baked product with long shelf-life. The recipe contains flour, water, at least 50% (based on flour) sugar, and 2% spices. Traditionally, honey is used as source of sugar. The mixture of spices varies widely and is influenced by regional customs. Frequently used spices are anise, fennel, ginger, cardamom, nutmeg, clove, pepper, and cinnamon.

Candied fruits, dry fruits, nuts, and sweet fillings (jams, jelly, or marzipan) are further possible ingredients. Sometimes, eggs are added; in this case, water, fat, and milk are not used. The mixed heavy dough mass is usually leavened with a mixture of potash with baking powder or soda. Baking temperature and time are dependent on the type of product. Gingerbreads are often made in decorative forms (e.g., stars, hearts, or animals), decorated with sweets and icing, and specifically offered in many countries during Christmas time.

4.2.3 Dry baked products

The diversity of dry baked products, made from wheat flour, is immense. Their low water content (< 10%) and thus their long shelf life are typical. Doughs are mainly made from soft wheat flour with low particle size and the recipe is characterized by high levels of fat or sugar (or both) [12]. Some economically important dry baked goods (examples are shown in Fig. 4.5) are described in the following.

Cookies (called "biscuits" in Europe) are small, flat, dry, and crisp baked products with high contents of sugar and fat (ca. 20–25% each) and a low content of water (< 10%). They do not have a developed gluten network and therefore break easily. The diversity of cookies, for example in appearance and taste,

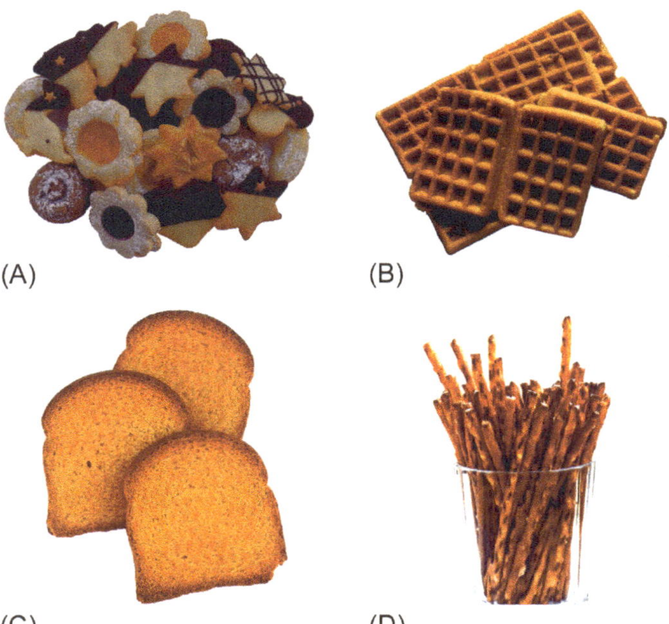

FIG. 4.5 Appearance of selected dry baked products: cookies (A), waffles (B), zwieback (C), and salted sticks (D).

is wide and mainly dependent on recipe and processing. Soft wheat flour with a low protein content is usually used for production, because hard wheat flour provokes an undesirable hard texture of cookies. Ingredients besides flour may include milk, eggs, butter, sugar, fat, baking powder, and salt as base materials. They are mixed with water to a batter-like dough with limited gluten development. The optimal consistency of the dough depends on the method to form dough pieces before baking. Dry and crumbly doughs are processed on a rotary-mold machine. The dough is made into a continuous sheet and pieces are cut out from. Softer doughs are formed by an extrusion process and pieces are cut to size by a reciprocating wire. The dough pieces obtained are then baked on a solid steel band in long tunnel ovens using a baking temperature and time that are optimal for each desired product.

Crackers are small baked products with a light and flaky texture. They are produced in many shapes and sizes, round, square or triangular, for instance. As a salty product with 3–4% salt in many cases, crackers are distinguished from cookies by a low or zero sugar content. The dough contains low levels of water (20–30%) and its leavening is accomplished by yeast or chemical agents. Depending on the product type, proofing time may vary from 3 to 18 h and baking time is around 2.5 min at 230 °C. After baking, crackers contain only around 2% water. Crackers can be subdivided into three types: saltine crackers, crackers sprayed with shortening, and savory crackers with cheese. Different flavorings or seasonings, for example, herbs, seeds, or cheese added to the dough before baking, allow the production of numerous cracker variants. In recent years, the type of crackers has widely expanded by using further additives such as cracked wheat or vegetables. So-called snack crackers are traditionally flavored with cheese.

Wafers are small, flat, thin, and crispy baked products. They are made from a fluid batter consisting of soft wheat flour (35–40%), oil or emulsifier (2–3%), salt (0.25%), and baking powder (0.3%). Special sweet variants contain sugar in amounts up to 60% of flour weight. Small amounts of other ingredients may be added to influence flavor, color, or texture of the final product. The batter is baked between pairs of heated metal plates that are engraved with crisscross lines or other patterns. Wafers are frequently served and eaten with ice cream and used for the production of chocolate bars. Special round wafers serve as Christian communion wafers.

In comparison to wafers, the batter, used for the production of *waffles*, is much richer in fat (25–30% of flour weight) and sugar (10–25%) and eggs and milk solids may be additionally added. The waffle batter is baked between heated metal plates like wafers. The baked products are thicker than wafers and have a crisp surface and a soft, chewy interior. Waffles are usually eaten immediately after baking while still hot or after heating of deep-frozen products.

Zwieback (Melba toast) is a yeast-leavened dry baked product that is sliced and baked a second time. Basic ingredients are flour, milk, sugar, fat (butter), yeast, and salt. The additional usage of eggs or emulsifiers improves crumb properties. After intense mixing, the dough is proofed at room temperature until reaching about double of its volume. Proofed doughs are placed into a pan and baked at 220–230 °C until browning of the surface. Afterwards, the material is cut into pieces of about 0.6 cm thickness and again baked at 150–160 °C. After cooling, the slices are packed airtight and due to the low water content (< 10%), they have a long shelf life. The fine-pored, crisp, and brittle slices are often used to feed pediatric and adult patients with an upset stomach and other digestive problems.

Salted pretzels and *salt sticks* are made from doughs that are quite similar to those of crackers. Doughs are extruded, cut into strips, and formed corresponding to the desired product. The uniqueness of the process is the alkaline dipping step: The formed dough pieces quickly pass through a caustic bath (e.g., 1.25% sodium hydroxide at 85–90 °C). Afterwards, the surface is salted with about 2% coarse-grained salt and baked for 5 min at about 240 °C. The baked products are further dried from about 15% to 2–4% water content. Upon baking and drying, the material hardens and the surface caramelizes due to the alkaline milieu, which forms the typical surface glaze, the brown color, and the unique taste and flavor.

4.3 Paste products

Paste products made from wheat comprise a large number of staple foods, in which paste is referred to as any shaped dough prepared from semolina, farina or flour or a mixture of these with water, salt, with or without eggs or other ingredients [13]. After mixing and kneading, the nonfermented dough is shaped, cut into pieces, and either dried and packaged or used freshly. Both dried and fresh products are usually consumed after boiling in water. Paste products have become an important part of the diet around the world. Most of them are produced in factories and only a small portion is home-made. Paste products have lost their "cheap and filling" image and have gained positive attributes of convenience, economics, taste, and nutrition. Moreover, they are widely versatile in that they can be served as main dish, side dish, soup or salad. Pasta and noodles are the most common types of paste products. The term "pasta" is used to describe paste products fitting the "Italian" style of extruded foods majorly made from durum wheat and is distinguished from the "Oriental" style of sheeted and cut foods termed "noodles" that are predominantly made from common wheat. The following sections describe the preparation and characteristics of different pasta and noodle types and some other paste products.

4.3.1 Pasta

In this section, the term "pasta" is used for those extruded products that are majorly made of semolina or farina and water and may be supplemented by other ingredients. Italy, the classical pasta-making country, is very restrictive about the ingredients that may be used to make pasta: Only durum wheat semolina, salt, and water are admitted for "pasta secca" (e.g., spaghetti and macaroni); the usage of durum wheat flour or common wheat farina is not allowed. Although not shaped by extrusion, flat-shaped lasagne is usually assigned to pasta products. For "pasta fresca" (e.g., fettucine and tagliatelle), the usage of common wheat farina and eggs is allowed. The following subsections include a short characterization of durum wheat and semolina because of their important roles in pasta production followed by the description of pasta preparation and pasta types.

4.3.1.1 Durum wheat

The bulk of durum wheat, going into food production, is used for pasta products. Durum wheat is a free-threshing tetraploid *Triticum* species having the genome AABB and 28 chromosomes (see Section 1.1). Accounting for around 5% of the world wheat harvest, durum wheat is the only economically important wheat species apart from common (bread) wheat. It is best grown in warm areas with low precipitation. Therefore major producing regions are the Middle East, North America, Southern Europe, and North Africa. Durum wheats of North America are mostly spring wheats; in other regions, winter wheat is preferred. Durum wheat grains differ significantly from those of other wheat species by two characteristics: the hardness and the yellow color of the endosperm. Hard vitreous grains are synonymous with durum wheat (Latin "durum" = hard) and are not found in common wheat and other wheat species. The softness of common wheat endosperm is based on genetically controlled chemical linkages between starch and puroindolines a and b (see Section 2.2.5). These linkages mediate softness of the endosperm. Soft common wheat varieties possess both puroindoline types and hard common wheat varieties only one. In durum wheat, puroindolines a and b are missing resulting in grain hardness. This makes durum wheat most suitable as a raw material for use in the manufacture of pasta. Moreover, durum wheat endosperm is rich in yellow pigments colored by carotenoids (e.g., β-carotene and lutein) with contents in a range of 4–8 mg/kg. The level of pigmentation is a sign for durum wheat quality and an important criterion for consumer's acceptance of durum wheat products. The color can be intensified by selecting varieties with a high carotenoid content.

It is not surprising that most durum food products have their origins in durum producing regions, especially in Italy. Pasta, the most popular product, is mainly made from durum wheat semolina because of better cooking and biting strength and higher levels of carotenoids as compared to pasta from common wheat farina.

4.3.1.2 Semolina

The basic material for the manufacture of most durum products is semolina that satisfies the requirements of the customer, usually the pasta producer. It is predominately produced in special durum mills. The first steps of the miller's handling correspond to those of common wheat milling: selection of suitable durum wheat varieties and cleaning the grains (see Section 3.3). Selected durum wheat should have an appealingly yellow-colored endosperm and a high level (> 11%) of protein. The typical process of cleaning includes gravity tables, destoners, disc separators, cockle separators, and optical sorters. The hard grains must be moistened to a relatively high water content (16–17%) before grinding. This makes it easier to separate the bran from the endosperm. The equilibration time should be short (4 h or less), because only the outer layers and not the endosperm has to be softened. Durum milling differs decisively from common wheat milling. Durum mills have more break passages and the profile of the corrugations is generally sharp to minimize the amount of fine material in favor of coarse semolina. After sifting and purifying, coarse semolina is "sized down", i.e., it is gradually reduced by corrugated sizing rolls and cleaned by sizing purifiers.

Several common factors have to be considered when judging semolina quality, which is especially important for pasta manufacturers. Optimum preconditions for high quality are as follows: Semolina is made from grains with high durum wheat portions (98–100%). The optimum particle size of semolina is in a range from 150 to 350 µm. The range should be as narrow as possible, because water absorption of semolina is dependent on the particle size and has to be homogenous and relatively low. The grit content of semolina should be below 0.005%. Desirable yellow color units should be high and speck counts low. Regarding the chemical composition, semolina of high quality contains 13.5–14.5% moisture, 11% protein at a minimum, and 0.8% ash at a maximum. The functional properties of gluten (e.g., cohesivity and viscoelasticity) are especially important for semolina quality. They can be determined by means of a Mixograph or a Farinograph (see Section 4.2.1.2).

Semolina is widely used for the manufacture of paste products; pasta (e.g., spaghetti and macaroni) is the most popular common product. Moreover, semolina can be used for the production of noodles, couscous, bulgur, soups, gruels, porridge, pudding, and pastries.

4.3.1.3 Pasta preparation

For centuries, pasta-making was more of an art than a craft without any mechanization. Crude forms of mechanization were introduced in the 18th century and a simple extruder was developed in the 1930s. The first production line that converted semolina into dry pasta in one continuous process was unveiled in 1946. Basic operation steps of pasta production include mixing and hydration of the ingredients, further kneading, and converting unleavened dough into pieces of desired size and shape. Fresh pasta is traditionally hand-made at home often with the aid of a kitchen pasta machine for extrusion. Today, fresh pasta is also commercially produced by large-scale machines and available in shops and supermarkets. Dry pasta is usually made in factories. The core of industrial pasta production is an extruder, a device that mixes, transports, shears, stretches, and shapes material under elevated pressure and temperature. Most industrial process lines for pasta production have capacities in the range of 200–1200 kg pasta per hour.

For mixing and hydration, the raw material (semolina, farina or flour) and water are put into a mixing chamber in a predetermined ratio. When whole eggs (2–4 eggs/kg flour) are included in the recipe, the addition of water is not necessary in most cases. The water content of the dough should be about 30% corresponding to around one-half of that in a bread dough. At this point, other additives can be added such as minerals, heat-stable vitamins, and colors. Some dough-improving agents may also be included. For example, cysteine hydrochloride (≈0.01%) can be added to reduce the mixing time of strong doughs. Ascorbic acid (≈0.01%) prevents carotenoid bleaching by lipoxygenase and the cooking quality can be improved using emulsifiers like glyceryl monostearate or propyleneglycol monostearate.

The mixing chamber of the extruder is either airtight or operated under vacuum to exclude air in order to minimize pigment bleaching and to prevent air bubble formation that would cause points of weakness in the dried pasta. The crumbly dough formed by mixing is introduced into the extrusion chamber that is equipped with a screw, powered by an electric motor or driven by hydraulic pressure, and a so-called die. The dough is kneaded under pressure exerted by the screw (15–20 MPa), until a uniform and smooth paste without air bubbles is formed. The chamber has to be cooled with circulating water to keep the temperature below 50 °C. Thus "cold" extrusion is used for the manufacture of pasta and related products. Higher temperatures are applied for the production of ready-to-eat snacks, for instance. The material is then pressed through the extruder head die. Mostly, direct extrusion is used with the screw moving towards a stationary die. The design of the die creates fixed cross-sectional profiles, which allows the production of pieces with different shapes and sizes. Formerly, dies were made from bronze alloy and they are still in use in some factories. On the one hand, bronze dies allow the production of excellent products but on the other hand, they wear out rapidly and have to be cleaned carefully. Therefore they are often replaced by stainless steel or tetrafluoroethylene (Teflon)-coated dies. These "new" materials have a distinct influence on the quality of pasta products that are smoother, cook more slowly, and tend to get mushy.

The conditions of the subsequent drying step in drying chambers are particularly critical for end product quality. First, freshly extruded material is rapidly dried from the outside, until it is no longer sticky, while the inside remains moist. Drying is then continued at 45–60 °C, either very slowly or stepwise, so that the moisture content of the total material becomes more uniform (20–24%). The following final drying results in a moisture content of about 12%. Major quality factors of the end product comprise appearance (e.g., color, uniformity, clarity, and surface texture), mechanical strength (e.g., fragility and hardness), and cooking properties (e.g., nonsticky and firm to the bite).

4.3.1.4 Pasta types

The number of pasta types, existing all over the world, is virtually unlimited. In particular, Italy has hundreds of pasta types differing in recipe, shape, size, and thickness. Pasta may be divided into two main categories, dry ("pasta secca") and fresh ("pasta fresca"). Classical representatives of pasta secca are spaghetti and macaroni and those of pasta fresca are fettuccine and tagliatelle.

Spaghetti belong to the most popular pasta products worldwide. Their typical appearance may be described as thin, cylindrical, dry sticks that usually have a length of 25–30cm (Fig. 4.6). The thickness may vary, thicker variants are termed "spaghettoni" and thinner variants "spaghettini"; "capellini" are very thin variants ("angel hair spaghetti"). Spaghetti have their origin in Southern Italy of the 12th century and Italy has remained the classical spaghetti producer (≈3.5million tons per year) and consumer (≈28kg per capita and year) up to this day. Spaghetti are staple foods in Italy and usually eaten every day, often served as a first course in small portion sizes. Many Italian families have their own pasta machine at home and eat freshly prepared spaghetti immediately. In the course of time, spaghetti production as well as consumption have spread over the whole globe. Factors that contributed to the growing popularity were particularly the low price, ease of cooking, and variability of usage.

Mass production of dried spaghetti is done in factories using auger extruders mostly with capacities in a range of 200–600kg/h. After extrusion, the strands delivered from the dies are picked up on rods in form of a curtain and subjected to a blast of air. This surface drying prevents the strands from stretching. The curtain is then conveyed into several drying chambers and cut to the required strand length.

Pasta quality is important for both producers and consumers. The finished product should be appealing regarding appearance (translucent bright yellow free from specks and cracks), stability (resistant to breakage during packaging, shipping, and handling), cooking quality (not sticky or mushy), flavor, mouthfeel, and firm to the bite. Dry packaged pasta represents the largest segment of the market. Packaging, mostly in 500g sizes, is done using paper wrapping, plastic bags, or boxes. Medium-moisture (28–40%) products are marketed canned or frozen. Canned pasta is combined with stuffed products, often tomato sauce. The advantage of frozen products is that they can be heated in a microwave oven.

(A) (B) (C)

FIG. 4.6 Appearance of selected pasta products: spaghetti (A), macaroni (B), and lasagna (C).

Before consumption, dry spaghetti have to be cooked in salted boiling water and then drained in a colander. In Italy, spaghetti are cooked "al dente", i.e., firm to the bite, whereas they are often cooked to a softer consistency in other countries. Spaghetti are served in a number of different dishes. Most common is *pasta asciutta* served on a plate with a complementary sauce, frequently tomato sauce containing oregano and basil. Other well-known dishes are *all'Bolognese* (with hash), *all'carbonara* (with bacon bits and cream), *all'aglio* e oleo (with garlic and olive oil), and *all'vongole* (with clams). Grated hard cheeses like parmesan are usually sprinkled on top of cooked spaghetti. Further categories of dishes are *pasta in brodo*, a soup containing pasta, and *pasta al forno*, a baked pasta-containing dish.

Macaroni are dry pasta products that are shaped into narrow tubes, mainly thick with short length (Fig. 4.6), but also variants exist that are long and thin like spaghetti. The hollow cavities of macaroni are created by special extruder dies. Curved macaroni are termed *elbow macaroni*. The curve is formed by different speeds applied at the opposite ends of macaroni, when they leave the extruder. Macaroni are commercially produced in large-scale factories and served as dishes like spaghetti.

Fettuccine and *tagliatelle* are traditional pasta types that are made from flour and eggs (usually one egg for 100g of flour). Both types are shaped as long flat ribbons but are different in wideness (3–5mm vs. 4–10mm). They are offered fresh or dried and eaten with a variety of sauces, in particular meat sauces.

Lasagne is probably one of the oldest pasta types. It was created in Southern Italy in the Middle Ages and became a traditional dish that is now served in many parts of the world. Traditional base components of lasagne are dry paste sheets made from semolina and water (Fig. 4.6). Several layers of these sheets are interspersed with layers containing different fillings. Many variants of fillings exist and may consist of cheese and tomato sauces, meats or vegetables and flavored with wine, garlic or oregano, for instance. The product is usually baked at around 180°C for 30–40min, until the crust is golden brown.

4.3.2 Other durum products

Apart from pasta, a number of other durum wheat-based foods exist. They often are niche-products but are regionally important. *Couscous*, one of the staple foods in North African and Arabian countries, is simply made from semolina and water. Semolina particles are sprinkled with water and rolled to form agglomerated small pellets ($\approx$2mm in diameter). Pellets, too small for couscous production, are separated by sieving, sprinkled with dry semolina and again rolled into pellets of suitable size. The material is then steam-treated and can be eaten directly or it is dried and stored or packaged. Couscous is traditionally produced by hand or commercially by largely automated processes. Most couscous sold in supermarkets has been presteamed and dried ("instant couscous"). After short cooking, instant couscous is ready for consumption.

Cooked couscous of high quality is characterized by a light and fluffy appearance, pale color, a natural bland flavor, uniform particle size, absence of particles that stick to one another, and the ability to absorb sauce. Like pasta, couscous is a versatile food that is served in many variations and in different combinations with sauces and other foods.

Bulgur is a traditional food in Mediterranean regions, the Middle East, and North India but also available and consumed in many other countries. Bulgur is preferably made from durum wheat, but common wheat can also be used. Durum wheat gives yellow color and firm texture to typical bulgur products. Bulgur is usually made by soaking the cleaned grains in water and cooking, until the starch gelatinizes. Stirring during heating causes light foreign material to float to the surface for removal. After cooking, the grains are cooled and dried, cleaned by winnowing, and milled. Sometimes, dried bulgur is peeled to remove the bran. By milling, bulgur is classified into up to four classes: coarse, fine, very fine, and flour. Within each class, particle sizes should be uniform, which allows a more consistent cooking time. The preparation for consumption depends on the particle size. Coarse bulgur is usually cooked or steamed in combination with other foods such as meat or vegetables. Moreover, it is frequently used as a side dish like rice and for soups. Fine and very fine bulgur are used in dishes, where coarse granulation is not accepted (e.g., in salads). Flour and bran are utilized for animal feed.

Frekeeh is a very special product consumed in several countries of the Middle East and North Africa. It is preferably made from durum wheat because of the harder and larger grains compared to common wheat. Immature plants with soft green to yellow grains are cut at the stem, dried in the sun, bunched together, and scorched by flames, so that only awns and leaves are burnt not the seeds. Parching gives frekeeh its characteristic flavor. The parched wheat heads are dried in the sun and separated from ash, charred awns, and leaves by winnowing. The roasted grains are threshed either by hand or mechanically, separated from the chaff, and ground or chopped into small pieces that are usually boiled or steamed before consumption. Typically, frekeeh is served with meat of lamb, sheep or poultry or in soups prepared with onion and tomatoes, for instance.

Apart from the products described above, durum wheat is used to make a number of other foodstuffs that are popular in some regions. For example, puffed grains are consumed as breakfast cereals and semolina is used for hot porridges. In the Middle East and Southern Italy, flat bread is manufactured from durum flour or blends with common wheat flour. Appetizers and desserts can be made by boiling semolina in vegetable oil, fruit juices or sugar syrups combined with spices or nuts.

4.3.3 Noodles

Noodles are staple foods in many regions of the world. Their most likely origin is China, where 4000-year-old noodles were found at the Lajia archeological site. Noodles can be made from various raw materials, for example, from cereals and pseudocereals with wheat and rice being predominant. In Asia, wheat noodles are so popular that around 40% of wheat consumption are ascribed to noodles. Most noodle types share the common processing steps of mixing ingredients, kneading, rolling or sheeting the dough, and cutting into pieces. There are many types of noodles. They vary in their ingredients, type of processing, size and shape, cooking properties, and end-use quality. Noodles are usually consumed in a wet, boiled, steamed or fried form.

Wheat noodles are mainly made from common wheat flour and their appearance and texture strongly depend on flour characteristics. The protein content of flour should be in a range of 10–14%, which provides an elastic texture and a strong bite of noodles. Flour with protein levels below 10% results in products with soft bite and poor cooking tolerance. Starch properties are also important. For example, its swelling and gelatinizing characteristics determine whether the surface of noodles is sticky after cooking. Further primary ingredients for noodle production are water and salt. The level of water, added to the flour, is in a range of 30–38% and the level of salt (sodium chloride) is between 1% and 3% of flour weight. Salt strengthens the dough structure, improves sheeting properties, makes the noodles more elastic, and offers the basic salty taste. In Asian countries, kansui, a mixture of sodium and potassium carbonates, is often used instead of sodium chloride. These alkaline reagents confer firmness and a bright yellow color to noodle products. Further additives may improve process performance and noodle quality. For instance, whole eggs or egg powder enhance cooking strength and color of noodles. The major part of Asian noodles contain no eggs. In contrast, US legislation states that noodles must be made from wheat dough containing more than 5.5% eggs.

The preparation of noodles is relatively simple. Flour, salt, and other ingredients are mixed with water for 5–10 min. The crumbly dough is then allowed to rest for 10–15 min, so that soluble ingredients are evenly distributed and particles are completely hydrated. Afterwards, the dough is pressed between two rotating rolls to form a single noodle sheet. This process is repeated several times ("lamination") and by adjusting the gap of the rolls, the sheet is made to the desired uniform thickness. Sheeting speed and final thickness are important factors for noodle quality. Finally, the dough sheet is passed through the slitter and different cutting rolls resulting in noodles of required size.

A large variety of shapes contributes to the diversity of noodles. Most common are long thin strips, strings, waves, tubes, helices, and shells. Noncooked fresh noodles contain around 35% moisture and deteriorate rapidly at room temperature. The shelf life under refrigeration is about two to three days. In contrast, dried noodles have a shelf life of up to 1 year. Hot air drying is the main drying method in Western countries and frying is favored in Asian countries. In any type, freshly prepared and shaped noodles are steamed for 1–5 min at around 100 °C, so the texture is fixed by gelatinized starch and noodles are less fragile. Hot air drying is performed for 30–40 min at 70–90 °C resulting in a moisture content of 8–10%. To obtain fried noodles, freshly shaped and steamed noodles are fried in plant oil, mostly palm oil, at 140–150 °C for 0.5–2 min. The moisture content rapidly decreases to 5–8%. Both fresh and dried noodles have to be cooked for about 10–15 min in boiling salty water and drained, before they are served as dishes. Instant (precooked) noodles are produced by steaming of fresh noodles and drying (traditional instant) or by steaming and frying (ramen style). Instant fried noodles are characterized by a firm nonsticky surface and a relatively strong bite.

Ingredients, preparation processes, sizes, and shapes of noodles vary widely, both among and within countries and regions and requirements in noodle properties may be strongly different all over the world. For example, soft noodle texture usually is preferred in Japan, while noodles with chewy, more elastic texture are favored in other Asian countries. The most popular Asian noodles are *ganmien* (China), *ramen* and *udon* (Japan), *kooksoo* (Korea), *mi chay* (Vietnam), and *miswa* (the Philippines).

Noodles are widespread components of various dishes. They are often served with an accompanying sauce or in a soup. Fried noodles are combined with meat, seafood, dairy products or vegetables and cold noodles can be used as additives for salads. Instant noodles are usually sold with flavoring powder and/or seasoning oil and are served after 1–2 min cooking in boiling water or soaked in hot water for 3–4 min.

4.3.4 Miscellaneous products

Tortellini are small (≈ 2 g), ring-shaped products of Italian origin; larger tortellini (≈ 5 g) are called *tortelloni.* Both are typically made from wheat flour, water, eggs, olive oil, and salt that are mixed and kneaded. The resulting dough is rolled out, outrivaled, formed into pockets, and stuffed with meat (veal, chicken or pork), cheese (mortadella or parmesan) or vegetables (mostly spinach). Closed pockets are cooked in boiling salted water for about 3 min. Tortellini are usually served in broth made from beef or chicken or in sauces made from cream and ham or ground meat. Industry produces large amounts of packed tortellini that are sold mainly refrigerated or frozen in many parts of the world.

Ravioli are produced by the same procedure as tortellini but typically have a square form. They have their origin in Italy and were first mentioned there in the 14th century. The filling varies, most popular are beef or ricotta cheese. Ravioli are mainly served in tomato sauce often supplemented with meat or cheese. In many other cultures, products similar to ravioli are wide-spread dishes, for example, *maultaschen* in Southern Germany, *kreplach* in Israel, *manti* in Turkey, *jiaozi* or *wonton* in China, and *gujiya* in India. Toasted ravioli are popular snack foods in the USA. Today, ravioli are industrial mass products. In particular, canned ravioli, filled with meat or cheese and immersed in tomato sauce, are favorable products around the globe.

4.4 Breakfast cereals

Cereals are important components of the breakfast in many parts of the world. Breakfast cereals are consumed not only at home, but they also are obligatorily served in cafes and hotels. Combined with milk, yogurt, or fruits, breakfast cereals can contribute to a nutrition of very high value. They are made from a number of different cereal sources such as wheat, oats, corn or rice [14]. The main products made from wheat are ready-to-eat materials, suitable for consumption without further cooking like flakes or popped and extruded materials. Other products such as porridge require cooking, before they are ready for consumption or they need the addition of hot water or hot milk in the bowl.

For the preparation of porridge, one of the oldest wheat-derived dishes, farina from common wheat is typically used. Farina is the fraction of middlings, obtained from grains by roller milling (see Section 3.3), with a particle size range from 350 to 800 μm. Before eating, farina must be heated in water for several minutes, so that the particles are penetrated with water and the starch gelatinizes. Commercial farina, offered for making porridge, is often enriched with minerals and vitamins and flavored with malt or cacao. Farina with short heating time ("instant" farina) is usually supplemented with hydrolyzing enzymes that enable water to penetrate the particles faster.

To make flakes, whole wheat grains are first broken ("bumping"), slightly steamed and then distorted by passing special rolls. Mixed with water, fine-granulated sugar, salt, and malt syrup, they are cooked under pressure (about 90 min at 140 MPa) resulting in tightly stuck-together balls, the size of which is reduced to around 1 cm in diameter by lump breaking followed by drying to around 20% moisture. After cooling, the material is tempered and rapidly heated to around 90 °C. It is then immediately flattened by flaking rolls; the selected distance between the rolls produces flakes of the desired thickness. Thereafter, they are toasted in an oven to obtain crispness and color. Rotary toasting ovens are typically used that heat all surfaces evenly and provide uniform browning. Finally, flakes are dried to less than 3% moisture and packaged. A modification of the traditional process is extrusion before flaking. The mixed ingredients are put into an extruder, formed by the die, cut into the desired size, and dried. Flaking and toasting are then performed as described above.

Durum wheat is best for puffing, because the grains are large and make large puffs. The principle of puffing is the sudden change of temperature and/or pressure upon wet material, which allows water to evaporate very quickly. Puffed material can be produced by various techniques. Pressure-puffed (so-called gun-puffed) products can be made from steamed pearled or whole grains as well as from dough pieces. They are loaded into the guns that afterwards are sealed, rotated, and heated to around 400 °C. The pressure increases to around 1.5 MPa. The trip valve is then opened and the material blows out. The final steps consist of removing unpuffed or broken grains, bran, and dust particles and drying the material to 1–3% moisture.

Extrusion cooking is another technique for expanding. Flour or farina is moistened by steam or hot water and mixed with water and other ingredients. The dough is first loaded into a cooking extruder and then into a forming extruder. At a temperature of $\approx 175\,^{\circ}C$ and a pressure of $\approx 3.5\,MPa$, the material forms a plastic mass that readily adapts to the die configuration. When the mass exits the die, pressure is released and moisture is flashed off. The extruded product is then dried to a moisture of less than 3%.

Flakes and puffed or extruded products are often sugar-coated, which protects the products from moisture and provides a sweet taste. For that purpose, molten starch syrup is dripped onto the hot material and the syrup hardens after cooling. Another type of supplementation is the fortification with minerals and vitamins. The most common principle of fortifying ready-to-eat cereals is to add minerals and stable vitamins (e.g., nicotinamide and riboflavin) to the basic recipe mix or to spray the labile vitamins (e.g., vitamins A, B_1, C, and D) on the end product. A second approach is to spray all the vitamins in form of a premix on the end product. Flavor and color can also be applied by the spraying technique. The package has to protect the contents from breaking and from the entry of moisture and exogenous odor. The materials used to package breakfast cereals typically include a highly sealed plastic bag inside printed paperboard cartons.

4.5 Beverages

The production of cereal-derived beverages is almost exclusively restricted to alcoholic beverages and is based on the enzymatic conversion of starch to sugar and then to ethanol by yeast. In principle, grains of all cereals are suitable for producing this kind of beverages, but traditionally, the most commonly used cereal is barley. Wheat is utilized to a much smaller extent; one of the reasons, at least in the past, has been that wheat should be reserved for making bread. Thus beverages made from wheat are more or less niche products and only of regional importance. Examples are wheat beer and distilled products.

Beer brewing usually involves the usage of barley malt, hops, yeast, and water. In addition to barley malt, the recipe can include wheat malt providing a special beer, called "wheat beer". The most popular wheat beers have their origins in Germany ("white beer") and Belgium ("witbier"). Classical white beer (white has the same etymological root as wheat in the German language) is made with a mixture of barley and wheat malts and top-fermenting yeast. In comparison to traditional barley beer, white beer has lower hop bitterness and much more carbon dioxide [15]. Depending on the yeast type used, the predominant flavor notes may be "banana", "clove", or "vanilla". Wheat beer is produced in a number of different variants, for example, "yeast white beer" (unfiltered beer), "crystal white beer" (filtered beer), "dark white beer" (made from dark malt), or "strong white beer" (higher ethanol content). Additionally, alcohol-reduced and alcohol-free wheat beers are on the market. Sales of white beer have significantly increased, not only in German speaking countries but also in other European and

North American countries. Witbier, the Belgium variant, is characterized by its pale and cloudy appearance due to being unfiltered and its special flavor produced by spices such as coriander, orange peel or herbs.

The production of beverages with high alcohol contents ("spirits") is accomplished by distillation of fermented, ethanol-rich mashes. Grains are ground, mixed with water, and cooked to gelatinize the starch and to sterilize the mash. Saccharification of the starch is achieved by addition of enzyme-rich malt and constant stirring at elevated temperature (50–60 °C). The mash is then heated to inactivate enzymes and rapidly cooled to around 20 °C. After addition of yeast, the mash is fermented for several days. When ethanol production is completed, the product is distilled, either in simple stills used by small plants or in sophisticated rectification columns in industrial production. For aging, the distillate is stored in special barrels. Examples of wheat-based spirits of importance are special American whiskeys and "wheat korn" from Northern Germany. According to US federal laws for wheat whiskey, a mash of at least 51% wheat has to be used and the distillate has to be aged in new charred white-oak barrels. The production is similar to that of bourbon whiskey (51–80% corn). Wheat korn is produced like wodka (based on rye) but with lower alcohol level and less rigorous filtration resulting in more cereal flavor. Wheat korn is one of the cheapest spirits and very popular among the working class.

4.6 Animal feed

Wheat is also used as animal feed worldwide [16]. For economic reasons, feed wheat usually comes from those varieties of common wheat that do not comply with the requirements for bread-making. Triticale, a man-made hybrid of durum wheat and rye, is another source of feed material. Additionally, wheat harvests with inferior grain quality, for instance shriveled or sprouted, are in use for animal feed production. It is important that grains are not contaminated with mycotoxins. The general suitability of wheat grains for animal feed is based on their high energy value, resulting from the high starch content, and their nutrient digestibility. However, in many cases and depending on the animal type, feed based on wheat has to be complemented with minerals, vitamins, and essential amino acids. Because some animals (e.g., poultry) lack intestinal enzymes capable of digesting dietary fiber, wheat-based feed may be enriched with xylanases, cellulases, and β-glucanases to improve its digestibility.

Many various forms of feed, targeting the need of single animal types, are on the market. They may include whole grains or different milled products, may be mixed with other sources for animal feed and are offered either as loose material, pelleted or canned. Extrusion cooking has long been utilized as part of pelleting. It improves digestibility, nutrient availability, and palatability of the products. The usage of whole wheat grains is a century-old tradition in feeding poultry and is considered to be more natural and to have a positive effect on bird health. Bran, shorts, and other milled products are useful component feeds for livestock such as horses, cattle, pigs, and rabbits. Pet food, specifically that for dogs and cats, often contain meat products that are blended with wheat-derived products. For thickening and binding, gluten or starch derivatives may be used. Fish needs feed that has to fulfill special criteria, for example, slow water absorption and floating on the water surface. This can be accomplished by extrusion cooking of grains resulting in a glassy texture of the material. Crustaceans usually are slow eaters and feed pieces should stay intact for longer time.

4.7 Economics

The global economic importance of wheat is enormous. Around 700 million tons of wheat grains are harvested worldwide each year, which corresponds to a value of around 100 billion US dollars. Therefore wheat is an essential component of the world's trading network. Many millions of producers and employees receive their profits and salaries in relation to wheat, for example, farmers, millers, bakers, handlers, and persons engaged in the corresponding industry and logistics. Farming is the first constituent of the economical chain of wheat utilization. A statistic, presented by the Washington Grain Commission for the years 2015/2016, may exemplarily demonstrate the economic importance of wheat farming. In the Eastern Washington State region, 3715 wheat farmers support 4013 additional jobs sustained by their families and employees. Another 3406 off-farm jobs are provided by businesses that support farmers. In summary, each wheat farming dollar generates 98 additional cents of off-farm economic activity, most of it in rural areas, where agriculture serves as a crucial factor for regional economies. Wheat grain processing usually starts in mills that are distributed over all the regions, where wheat is cultivated. They comprise both small handcraft businesses and big factories that provide countless jobs mostly in rural regions and small towns. Altogether, farming and milling of wheat is an enormously important factor to keep economic potential in rural regions that are often underdeveloped. The transportation of many million tons of grains and flours is an additional important factor for the regional economy.

The manufacture and distribution of wheat products, in particular baked products, pasta, and noodles, are certainly the most important economic factor concerning wheat. The processing activities comprise small and medium-sized enterprises as well as big factories located in both rural and urban regions. Similarly, selling activities occur almost everywhere, in small specialized shops as well as in big supermarkets. Widespread gastronomy and luncheonettes additionally benefit from the disposal of wheat products. Apart from employees, countless ancillary companies contribute to the economic activities of production and trade. Moreover, companies that produce special goods such as starch, gluten, and animal feed are of increasing economic importance. Last but not least, around 700 million tons of grains and several billion tons of wheat products have to be transported each year by truck, train or ship. Thus logistics is another substantial economic factor concerning wheat.

Apart from earning money, the advantages for consumers is another aspect of wheat economics. Wheat products, especially bread and paste products, are staple foods and belong to the cheapest foodstuffs worldwide. People with wheat hypersensitivities such as celiac disease are highly experienced with dietetic wheat/gluten-free foods and their high prices. For example, dietetic gluten-free products, purchased across five geographic US regions and four venues, were 139% more expensive than their wheat-based counterparts [17]. Within the UK "Bread and Cereal" category, retail price data showed that the average price of products labeled gluten-free was £ 1.12 per 100 g relative to £ 0.59 per 100 g of wheat-containing products and consumers with CD had to pay an extra £ 10 per week to maintain their diet [18]. Packaged gluten-free food products, obtained from different Austrian supermarkets, were substantially higher in cost, ranging from +205% (cereals) to +267% (bread and bakery products) compared to similar wheat-containing products [19].

The mechanization of wheat farming and milling as well as food processing has enabled cost-efficient production lines in many countries of the world. Table 4.3 exemplarily shows the comparison of costs of basic wheat products (white flour, white bread, whole grain bread, and white spaghetti) with other staple foods purchased in a German supermarket (July 2018). Relative price units (PU), enlisted in Table 4.3, are based on the price of white wheat bread set to 1.00 PU/kg. Surely, prices fluctuate and depend on countries, regions, seasons, and stores, but the values may be roughly representative for many regions. White wheat flour, used as starting material for countless products, is distinctly cheap (0.54 PU, special offers are below 0.40 PU) and only milk (0.38 PU) and potatoes (0.52 PU) are comparably inexpensive. Both white bread and whole grain bread, main products of wheat flour, are also in a low range (1.00 PU) and comparable with rice (1.12 PU), apples (0.97 PU), and tomatoes (0.97 PU). In contrast, meat, cheese, and fish are extremely expensive (2.99–7.91 PU).

However, not only the price but also the valuable constituents are important for consumers and their expenses. Wheat products are leading in the supply with energy and particularly in the contents of dietary fiber (DF), B vitamins, and vitamin E, when whole grain flour is used for production [20] (see Chapter 6). Regarding energy supply, wheat bread and spaghetti together with rice provide the cheapest energy ($\approx$0.40 PU/1000 kcal). Energy delivered by meat, for example, has a sevenfold higher price ($\approx$2.76 PU) compared to bread. Tomatoes, boiled ham, and codfish are entirely inappropriate as a cheap energy supplier (4–10 PU). Regarding the price (PU/100 g) of dry matter (DM) that in fact decisively contributes to nutrition, bread and spaghetti are cheapest ($\approx$0.15 PU) among staple foods except pulses like dry beans (0.14 PU). The costs of DM of vegetables and fruits depend on their moisture content: DM of vegetables like beans (0.14 PU) is cheap, bananas are in a medium range (0.29 PU), whereas products with high moisture contents (>90%) like tomatoes (1.66 PU) are quite expensive. On average, DM of animal products (except milk) are most expensive with codfish at the top (4.05 PU). Consumers may be surprised, when they

TABLE 4.3 Price units (PU) for staple foods and their energy, dry matter (DM), water (H_2O), dietary fiber (DF), B vitamins (B), and vitamin E (E).

Food	Price[a] (PU/kg)	H_2O[b] (%)	Energy[b] (kcal/kg)	PU 10^3 kcal	PU 100 g DM	PU H_2O/kg	PU 25 g DF	PU 25 mg B	PU 13 mg E
White wheat flour	0.54	13.0	3350	0.16	0.06	0.07	0.34	1.14	2.33
White wheat bread	1.00	36.9	2380	0.42	0.16	0.37	0.78	1.54	2.22
Whole grain wheat bread	1.00	42.6	1990	0.50	0.17	0.43	0.34	0.56	1.59
Spaghetti (white)	1.23	10.7	3540	0.35	0.14	0.13	0.91	1.22	7.63
Rice (brown)	1.12	13.1	3450	0.33	0.13	0.15	1.27	0.36	1.98
Potatoes[c]	0.52	77.8	700	0.74	0.23	0.40	0.61	0.62	12.64
Beans (dry)	1.23	10.3	237	0.52	0.14	0.13	0.13	0.74	7.63
Tomatoes[c]	0.97	94.2	170	5.68	1.66	0.91	2.41	2.36	1.55
Apples[c]	0.97	84.9	540	1.79	0.64	0.82	1.21	4.16	2.56
Bananas[c]	0.77	73.9	880	0.88	0.29	0.57	1.07	1.42	3.72
Milk (whole)	0.38	87.4	650	0.59	0.30	0.33	–	1.34	7.06
Cheese (Emmentaler)	4.42	35.0	400	1.10	0.68	1.55	–	10.62	10.76
Eggs[c]	1.82	74.7	1550	1.17	0.83	1.36	–	2.01	1.18
Beef (shoulder)	3.48	73.1	1290	2.70	1.29	2.55	–	1.37	–
Pork (escalope)	2.99	74.8	1060	2.82	1.19	2.24	–	1.17	–
Salami	7.98	37.3	4020	1.98	1.28	2.98	–	?	–
Boiled ham	5.78	70.3	1250	4.62	1.95	4.06	–	2.64	–
Codfish (filet)	7.91	80.5	770	10.27	4.05	6.33	–	8.24	10.28
Salmon (filet)	6.67	65.5	2020	3.30	1.93	4.37	–	2.12	3.94

[a] *Regular price (special offers are not included).*
[b] *Taken from Ref. [10].*
[c] *Edible portion: –, not present; ?, value not available.*

find out the prices of water present in foods. When they buy 1 kg of codfish or boiled ham, for example, they have to pay 6.33 PU and 4.06 PU for water, respectively, whereas water of 1 kg spaghetti costs 0.13 PU and that of 1 kg bread around 0.40 PU. Only water from beans and rice is similarly cheap.

Aside from the entire DM, DF, B vitamins, and vitamin E are important nutritive ingredients of wheat grains. Especially, whole grain bread is rich in these components (approximately 7–8 g DF, 4–5 mg B vitamins, and ≈0.8 mg vitamin E per 100 g) and thus much more valuable than white bread. Based on the purchase price (1.00 PU/kg), 25 g of DF (the daily requirement of adults [20]) in whole grain bread cost 0.34 PU and only dry beans (0.13 PU) underprice this value. DF of fruits like apples (1.21 PU) and salads like tomatoes (2.41 PU), often acclaimed as important DF suppliers, is much more expensive. 25 mg of B vitamins (the daily requirement of adults [21]) are cheapest in brown rice (0.36 PU), whole grain bread (0.56 PU), and potatoes (0.62 PU). B vitamins of meat, fish, and most vegetables and fruits are much more expensive. The intake of 13 mg of vitamin E (the daily requirement of adults [21]) is cheapest in eggs (1.18 PU) followed by tomatoes (1.55 PU) and whole grain bread (1.59 PU). In parts, 13 mg of vitamin E cost 10 PU and more (e.g., in potatoes, cheese, and codfish).

In summary, wheat-based staple foods belong to the cheapest foodstuffs concerning energy and dry matter supply. When whole grain products are considered, they are leading in low prices for DF, B vitamins, and vitamin E. Altogether, the comparison of staple foods clearly demonstrates that poor people, a considerable portion of the global population, are widely dependent on cheap wheat products to achieve an adequate nutrition.

References

[1] Cauvain SP. Breadmaking—improving quality. 2nd ed. Cambridge: Woodhead Publishing Limited; 2012.

[2] Pomeranz Y. Composition and functionality of wheat flour components. In: Pomeranz Y, editor. Wheat—chemistry and technology, vol. 2. St. Paul: AACC; 1988. p. 219–370.

[3] Goesaert H, Brijs K, Veraverbeke WS, Courtin CM, Gebruers K, Delcour JA. Wheat flour constituent: how they impact bread quality, and how to impact their functionality. Trends Food Sci Technol 2005;16:12–30.

[4] International Association for Cereal Science and Technology: ICC Standard Methods.

[5] Wieser H, Kieffer R. Correlations of the amount of gluten protein types to the technological properties of wheat flours determined on a micro-scale. J Cereal Sci 2001;34:19–27.

[6] Wieser H, Zimmermann G. Importance of amounts and proportions of high molecular weight subunits of glutenin for wheat quality. Eur Food Res Technol 2000;210:324–30.

[7] Bloksma AH, Bushuk W. Rheology and chemistry of dough. In: Pomeranz Y, editor. Wheat—chemistry and technology, vol. 2. St. Paul: AACC; 1988. p. 1–217.

[8] Kulp K. Bread industry and processing. In: Pomeranz Y, editor. Wheat—chemistry and technology, vol. 2. St. Paul: AACC; 1988. p. 371–406.

[9] Prost C, Poinot P, Rannou C, Arvisenet G. Bread aroma. In: Cuavin SP, editor. Breadmaking—improving quality. 2nd ed. Cambridge: Woodhead Publishing Limited; 2012. p. 523–61.

[10] Souci SW, Fachmann W, Kraut H. Lebensmitteltabelle für die Praxis. 5th ed Stuttgart: Wissenschaftliche Verlagsgesellschaft; 2011, p. 242.

[11] Faridi H. Flad breads. In: Pomeranz Y, editor. Wheat—chemistry and technology, vol. 2. St. Paul: AACC; 1988. p. 457–506.

[12] Hoseney RC, Wade P, Finley JW. Soft wheat products. In: Pomeranz Y, editor. Wheat—chemistry and technology, vol. 2. St. Paul: AACC; 1988. p. 407–56.

[13] Dick JW, Matsuo RR. Durum wheat and pasta quality. In: Pomeranz Y, editor. Wheat—chemistry and technology, vol. 2. St. Paul: AACC; 1988. p. 507–47.

[14] Fast RB, Caldwell EF. Breakfast cereals and how they are made. AACC: St. Paul; 1990.

[15] Faltermaier A, Waters D, Becker T, Arendt E, Gastl M. Common wheat (*Triticum aestivum* L.) and its use as a brewing cereal—a review. J Inst Brew 2014;120:1–15.

[16] Yang W, Shen Y. Quality assessment of feed wheat in ruminant diets. In: Fahad S, Basir A, Adnan M, editors. Global wheat production. London: InTechOpen; 2018. p. 115–32.

[17] Lee AR, Wolf RL, Lebwohl B, Ciaccio EJ, Green PH. Persistent economic burden of the gluten free diet. Nutrients 2019;11. pii: E399.

[18] Capacci S, Mazzocchi M, Lucci AC. There is no such thing as a (gluten-) free lunch: higher food prices and the cost for coeliac consumers. Econ Human Biol 2018;30:84–91.

[19] Missbach B, Schwingshackl L, Billmann A, Mystek A, Hickelsberger M, Bauer G, et al. Gluten-free food database: the nutritional quality and cost of packed gluten-free foods. Peer J 2015;3:e1337.

[20] EFSA. Scientific opinion on dietary reference values for carbohydrates and dietary fibre. EFSA J 2010;8:1462.

[21] Deutsche Gesellschaft für Ernährung (DGE). D-A-CH Referenzwerte für die Nährstoffzufuhr, 2. Ergänzungslieferung; 2017.

Chapter 5

Wheat-based raw materials

For a long time, wheat has been recognized as carrier of constituents that are important raw materials for food and nonfood applications. Single compartments of the grains serve as valuable sources: the starchy endosperm for white flour, starch, and gluten production, the germ for the recovery of edible oil, and the outer layers for the isolation of dietary fiber. Straw is another usable part of the plant. Chemists and engineers have developed numerous techniques that enable the conversion of wheat constituents into materials with a wide range of properties. Starch and gluten are the economically most valuable constituents of wheat grains. Both fractions are available in native and modified forms and are surprisingly versatile commodities for food and nonfood applications. Although the germ and outer layers are byproducts of white flour milling and make up only a small portion of grains, they are widely used for the preparation of plant oil and dietary fiber, respectively, products that are closely linked to healthy nutrition. Straw, the second predominant part of wheat plants apart from grains, has also found a number of variable utilizations. The following sections will demonstrate that the wheat plant constituents are widely used as ingredients for both diversified diets and many nonfood applications.

5.1 Starch

Starch is the most abundant carbohydrate component in wheat grains and flour (see Table 2.1). It is exclusively present as discrete granules within the cells of the starchy endosperm. Wheat starch closely resembles the starches of rye and barley, but it is quite different from the starches of other cereals like corn and rice, legumes, roots, and tubers. The starch properties of the different wheat species and varieties are comparatively uniform and show only small variations mainly attributable to growing conditions and the milling process. The external appearance of dried starch powder is bright and white in color without any significant flavor. Starch has two populations of granules. The large A-granules are lenticular and have an average diameter of 25 μm. The small B-granules appear spherical and are less than 15 μm in diameter. A-granules contribute only 3–4% to the number of granules but up to 75% of the total weight of starch. Both granule types are composed of two major polysaccharides, (amylose and amylopectin), small amounts of so-called starch lipids (mainly lysophospholipids), and proteins (mainly granule-bound surface proteins). A-granules have a

Wheat—An Exceptional Crop. https://doi.org/10.1016/B978-0-12-821715-3.00005-8

higher amylose content and less amylopectin, lipid, and protein contents than B-granules.

The numerous applications of wheat starch arise from its unique physical and chemical properties, particularly its gel-forming and water-holding ability [1]. Starch granules swell when heated in water suspensions. At the end of swelling, they lose their shape and the crystalline structure of amylose and amylopectin gelatinizes and forms a gel. Upon subsequent cooling, the amorphous system reassociates to a more ordered crystalline state, the so-called retrogradation state (see Section 2.1.1). Due to these specific physico-chemical features, both food and nonfood industries make wide use of starch and starch derivatives, for example, in form of intact native granules, the swollen granular state, the dispersed form or as a film dried from a dispersion. Moreover, starch properties can be adapted to the required application by diverse physical or chemical modifications, the latter may be nondegradative or degradative. Last but not least, starch is a major source of sweeteners for the food and beverage industry, because it can be easily converted to oligosaccharides and glucose.

5.1.1 Starch production

The production of wheat starch continuously increased during the last decades and reached a worldwide annual capacity of around eight million metric tons in 2016. The principle of wheat starch (and gluten) production consists of the preparation of a dough or batter by mixing wheat flour with water, separation of agglomerated gluten from starch and soluble compounds, and removal of soluble matter from starch. Finally, starch is refined and dried. The unique cohesive property of gluten is the basic precondition of starch production, because the special agglomeration behavior of gluten makes the process economically valuable. Usually, the starting material is high-extraction flour from common wheat varieties (e.g., 78% extraction rate and 0.8% ash) with a high protein content. Wheat varieties with a high proportion of large granules are preferable, because a high yield of A-starch is required. The use of whole grain flour obtained by wet-milling has a number of disadvantages (e.g., poor quality of starch and gluten products caused by relatively high contents of fiber, lipids, and ash) as compared to white flour. The content of damaged starch granules should be as low as possible, because they reduce the yield of intact starch, disturb gluten agglomeration, and restrict starch quality in certain applications. Wheat varieties are quite variable regarding their suitability for starch production. Simple laboratory tests that include mixing of small amounts of flour with water, coagulation of gluten, and sieving with different sieves can provide information on yield and quality of starch and gluten.

The technology of starch production has advanced during the last decades and capacities today reach the processing of 40–50 metric tons of flour per hour. Two main types of processes that usually include a lot of technical variations have been developed for the simultaneous production of starch and gluten: the dough process (known as Martin process) and the batter process [2]. The traditional Martin process has been used since more than hundred years. Flour is mixed with water to make a dough, whereby gluten proteins agglomerate by kneading. The dough is then washed with large amounts of water to separate starch granules and soluble matter from the mesh-like gluten agglomerate. Starch granules are then recovered from the separated milky liquor by centrifugal methods using hydrocyclones or decanters for instance. The major problem with this system is the large amount of process water used for dough washing (more than 15 parts of water per one part of flour).

With the aim to maximize yield and product purity while minimizing water usage and energy, technologies have advanced with the advent of the batter process in the 1940s. The modern batter process differs from the Martin process by preparing a batter rather than a dough. Flour is mixed with water in a continuous mixer and the suspension is pumped to the homogenizer, where the agglomeration of gluten takes place. The dispersion obtained is diluted with water prior to the centrifugal separation of starch, gluten, and soluble matter. Modern centrifugation technologies that use hydrocyclones or decanter centrifuges enable a separation of the suspension into process water containing soluble material and three solid-containing fractions: A-(primary) starch, B-(secondary) starch with pentosans, and the gluten fraction with residual B-starch and pentosans. The advantage of centrifugal separation is that A-starch is separated out in a matter of seconds as compared to the washing process of the Martin process that takes up to half an hour. Refining of the obtained A-starch fraction with fresh clear water allows the preparation of pure A-starch that is then dewatered and dried in a flash dryer. Typical yields of A-starch, containing less than 0.4% protein, are between 50% and 54% based on flour. The B-starch fraction is handled similarly to A-starch. Yields of purified B-starch with less than 1.2% protein are in a range from 6% to 10%. B-starch has always been a product with few applications: native as animal feed, as core binder in the foundry industry, and gelatinized as milk replacer for calf feed and core binder for instance. The gluten fraction is refined with water and dried by a so-called ring dryer (see Section 5.2.1). Process water can be used either for the dilution of the batter (see above) or it is put through a multistage evaporator and concentrated to a thick syrup for sale as animal feed.

5.1.2 Native starch

Native wheat starch, especially A-starch, is a surprisingly versatile raw material used in many forms of food and nonfood applications. One of its most important features is the ability to swell and gelatinize upon heating with water combined with the formation of pastes or gels. Thus starch has become a well-known thickening and binding agent. Due to these properties, it is widely used as additive in food production, for example in puddings, custards, soups, sauces, salad dressings, mayonnaises, diet food for infants, and pasty fillings [3].

Starch concentrations of 4–8% are usually suitable for most purposes. Lower concentrations produce softer and higher concentrations firmer gel-like products. Another use in the food industry is molding starch in the confectionery industry, for example as a filler, where the ability of starch to retain the more volatile flavor components is used. Moreover, heating of starch suspensions followed by drying provides material that is soluble in cold water and forms gels without heating. This pregelatinized starch is mainly used in the production of instant products (e.g., puddings and baby food). Similarly, cold-water swelling is enabled by heating native starch in an alcohol-water mixture, so that foods can be thickened without heating.

Native starch is often included as an additive into the recipe for baked products; examples are cakes and biscuit fillings. It is further used for coating or dusting of products such as donuts and cakes. Wheat starch plays an important role in the production of gluten-free baked goods, in particular bread, provided that it is gluten-free according to the Codex Alimentarius regulations (see Section 12.2). Correspondingly, the gluten content of starch may not exceed 20 mg/kg. Commercial wheat starch usually contains around 2000–5000 mg gluten/kg. However, special washing steps or treatment with gluten-degrading enzymes allow the production of gluten-free material. Wheat starch is highly appreciated for baking gluten-free products because of its unique properties that include high water absorption leading to high dough density and product volume. Corn and rice starch or food-grade hydrocolloids (e.g., carboxymethyl cellulose or guar gum), used as substitutes for wheat starch, cannot replace the technological functions of wheat starch satisfactorily.

In the field of nonfood applications, the main outlets for wheat starch are the paper, adhesive, and textile industries. It is well known that Egyptian papyrus dating to 4000 BC already consisted of starch as an additive. Today, printing and writing papers usually contain 4% starch and paper board and domestic and industrial papers contain 2% starch. The growing usage of recycled paper requires higher starch quantities. Starch is known to improve the printability and writing properties of paper. Depending on the properties required for the finished paper, both native and modified starches are used. For beater and tub sizing, starch (≈2% of dry pulp) is added to the fibers from the pulping of wood and is evenly distributed before sheet formation. Thereby, paper strength and binding of fillers are increased. After a continuous fiber network is formed, surface sizing is carried out by application of additional starch. It prevents sticking of short fibers to the rollers and provides a smooth, printable paper surface.

Apart from the paper industry, the adhesive industry makes use of many starch applications, often in the form of a mixture of native and pregelatinized starch. For example, this material is used in paper bags, tissue and package papers, corrugating board, stationery, and postage stamps. The production of starch-based adhesives is quite simple as shown in the following example. Starch is pregelatinized by cooking in water for 10 min and cooled to 50 °C. Thereafter, native starch and water are added, stirred, and mixed with a sodium hydroxide solution at temperatures lowered to less than 38 °C. Sodium tetraborate and fungicides are finally added in small amounts. The liquid product is now ready for use.

One of the oldest industrial uses of starch is to size and stiffen textiles. It is added to individual yarns to increase mechanical strength and resistance to friction wear, helps to resist penetration of moisture, and serves as stabilizer and filler of colored inks, when fabrics are overprinted.

The pharmaceutical and cosmetic industries make use of starch as additive for toothpaste, tablets, liquid medicine, emulsions, lotions, and creams. The lack of odor and taste makes starch ideal for these products. In some applications, starch powder has to flow more readily, for instance when used in a tableting machine. A significant improvement in flowability can be achieved by admixture (0.2–0.5%) of substances like tricalcium phosphate. In addition, starch plays a role in the production of antibiotics, vitamins, dialysis solutions, enteral nutrition, and even as a substitute for blood plasma.

Other industrial usages of wheat starch are the production of biodegradable detergents, amino acids, enzymes, and yeast by means of fermentation processes and the production of surfactants, resins, and biodegradable plastics. In principle, it is possible to replace a number of petrochemicals by starch-based products. Starch is further used in oil drilling and mineral and metal processes.

5.1.3 Amylose, amylopectin

The separation of amylose and amylopectin is rather difficult, because the molecules are highly associated in the starch granules. They can be roughly separated by heating the starch suspension (e.g., 2% solids) to 80–90 °C, upon which amylose leaches from the starch, whereas amylopectin remains insoluble. However, the extracted amylose is contaminated with amylopectin to a considerable extent. A better method is based on the precipitation of amylose complexed with butanol. Starch has to be dispersed first in dimethyl sulfoxide or even better in a potassium thiocyanate solution at increased temperature (≈60 °C) followed by addition of butanol. The amylose/butanol complex precipitates and can be removed after slow cooling of the dispersion, refined, and dried. The remaining amylopectin is precipitated by the addition of ethanol and then filtered and dried.

Amylose forms clear films and thus can be used as protecting layer for fruits and candied fruits, where it prevents sticking together, and for French fries, where it decreases oxidation. Amylose films are further applied as edible food wrapping (e.g., for instant coffee and tea). Amylopectin is widely used as thickener or stabilizer in different food products (e.g., sterilized canned or frozen food and snack food) and as adhesive and binding agent, for example in the paper industry. Amylopectin is further utilized in the textile industry to size yarns of cotton, wool, and rayon. In many cases, blends of amylopectin and native starch are successful as well.

5.1.4 Chemically modified starch

The properties of native starch do not fit to a number of applications, particularly because of the tendency of its gels to retrograde and their rubber-like nature. Modifications by chemical agents are therefore often used to change the properties and functionalities of starch in order to adjust the material to special applications. Modification may be achieved by treatment with acids or by derivatization of functional groups with specific reagents [4].

The oldest method to modify starch is the treatment with an acid under mild conditions. Usually, starch slurry is heated with 1–3% hydrochloric acid at about 50 °C for 12–14 h. Afterwards, the slurry is neutralized, filtered, and dried. Whereas the covalent structures of amylose and amylopectin remain intact, noncovalent binding forces like hydrogen bonds are changed. In consequence, the modified starch paste is much less viscous than that of native starch after cooking in water and forms a rigid gel upon cooling. Examples of food applications are jelly beans and other gum candies. Harsher conditions such as heating at 200–250 °C provoke limited hydrolysis of glycosidic bonds and degradation to lower molecular weights, which leads to products with strong adhesive properties.

Most chemical modifications of starch are focused upon the derivatization of functional groups. Because amylose and amylopectin contain reactive groups exclusively in form of hydroxyl groups (two to three groups per glucose unit), only few reaction types are possible and used for the following modifications: oxidation to carboxyl groups, ring-opening of glucose units, esterification with acids, and conversion to ether derivatives. Cross-linking by ester and ether bonds complements the possibility to modify starch. In most cases of modification, the percentage of derivatized hydroxyl groups is much lower than 50%. The reactions are complex and yield diverse mixtures of products. The main reactions are shown in Fig. 5.1.

The oxidation of the hydroxyl group of the C6-atom results in various properties that mainly depend on the type of oxidant and the extent of oxidation. Hydrogen peroxide (reaction I a), sodium hypochlorite, and sodium perborate are most frequently used as oxidants. On average, one carboxylic group per 25–50 glucose units is obtained. Oxidized starch has lower viscosity, clearer gels, and increased adhesiveness as compared to native starch. Main applications are low-viscosity fillers of salad dressing and mayonnaise and sizing of paper, yarn, and textiles. Ring-opening of glucose units and the formation of dialdehydes occur, when starch is oxidized with periodic acid (reaction I b). The product has applications in the paper, oil drilling, and leather tanning industries.

Starch ethers can be obtained by reaction with ethylene oxide, 2-chloroethanol or propylene oxide in an alkaline medium, which results in hydroxyethyl- and hydroxypropyl-derivatives (reactions II a, b). The introduction of these groups, together with a small extent of cross-linking (see below), improves starch swelling power and stability, lowers gelatinization temperature, increases freeze-thaw

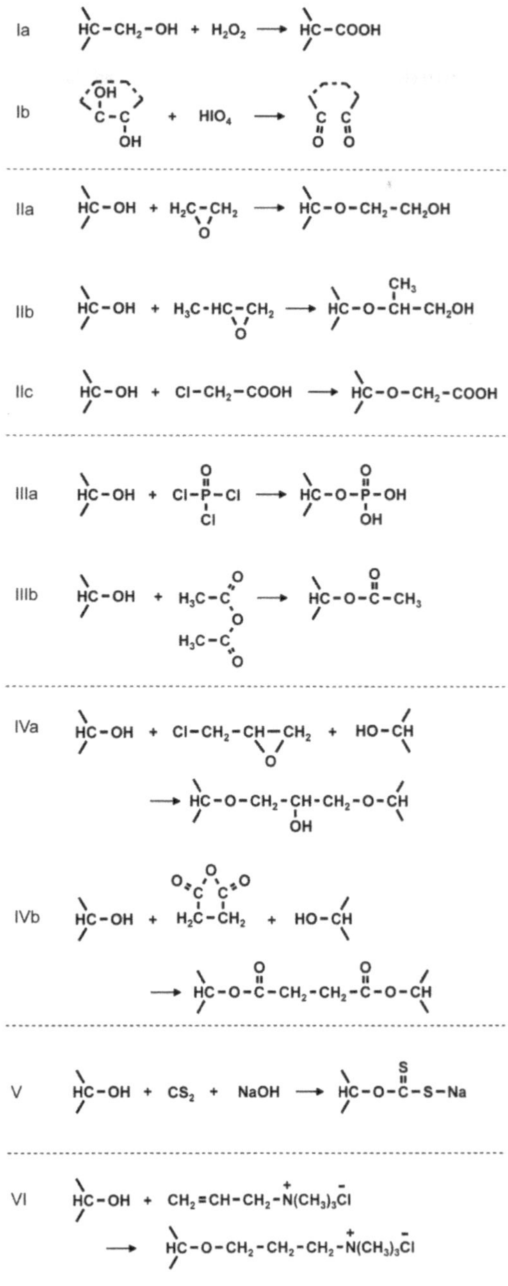

FIG. 5.1 Chemical modifications of wheat starch.

stability, and paste clarity. Therefore these derivatives are frequently used as thickeners in refrigerated products, fruit pie fillings, and heat-sterilized canned food. Ether bonds with carboxymethyl groups, formed by reaction with chloroacetic acid (reaction II c), provide starches that swell instantly even in cold water and are used as thickeners and gel-forming agents and as sizes in textile products.

Most common starch esters are derivatives of phosphate and acetate. Other ester types such as succinate and adipate starches are of less importance. Starch phosphates can be produced by reactions with phosphoric oxychloride (reaction III a) or sodium tripolyphosphate. Starch acetate is easily prepared by reaction with acetic acid anhydride (reaction III b). Starch esters improve thickening and paste clarity of native starch. Moreover, they show good freeze-thaw stability, resistance to acid pH, and increased tendency to swell in cold water. Starch esters are mainly used in the food industry for improving bakery products, soup powder, sauces, pie fillings, refrigerated foods, heat-sterilized canned food, and coating of dried food.

Cross-linked starches (covalent bonds between amylose and amylopectin molecules) are accomplished by reactions of hydroxyl groups with bi- or polyvalent reagents such as epichlorohydrine or succinic acid anhydride (reactions IV a, b). High levels of cross-linking increase the gelatinization temperature and delay crystallization, retrogradation, and the time, when starch becomes opaque. Moreover, viscosity does not increase, when the starch is boiled in water or sterilized in an autoclave. In contrast to the native state, starch with sufficient cross-links retains the desired properties under acid conditions; an example is thickening a cherry pie filling. Thus cross-linked starches are mainly used, when high stability of the starch paste is needed. Another characteristic feature of cross-linked starch is that the texture of the paste is modified. For example, cross-linking makes the paste much shorter and thus "spoon-well".

Starch xanthate is made by reaction with carbon disulfide in aqueous alkali (reaction V). It is known to strengthen paper and therefore is used in the paper industry for the production of corrugated cardboard boxes. Similarly, cationic starches have their application in the paper industry. They increase fiber and filler binding and are used in paper-coating formulations for instance. The synthesis of a typical cationic starch, made from an allyl derivative substituted by trimethylamine, is shown in reaction VI.

5.1.5 Starch-derived sweeteners

Aside from corn, rice, and potato starches, wheat starch is the most important source for the production of natural sweeteners including different starch syrups, dextrins, and sugars. A considerable portion of starch production is used to make these products. The principle of the process is the conversion of tasteless, water-insoluble starch molecules (amylose and amylopectin) into sweet, water-soluble oligosaccharides (dextrins and maltotriose), disaccharides (maltose), and monosaccharides (glucose and fructose). The relative sweetness (related to sucrose = 100) of the sugars (maltose = 30, glucose = 70, and fructose = 180) and their concentrations determine the sweetness of the syrups produced from starch. The underlying chemical reaction of the process is the partial or total cleavage of the glycosidic bonds between glucose units, which can be performed with acid and/or amylases. Conversion of glucose into fructose can be achieved by the enzyme glucose isomerase. The degree of starch degradation is an important parameter for the properties, in particular the sweetness, of the end product and is usually expressed as so-called dextrose (= glucose) equivalent (DE). DE represents the percentage of glycosidic bonds that are cleaved during the process (DE starch = 0, DE glucose = 100) and can be determined by measuring the reducing potential of the liberated glucose units, for example by reduction of Cu(II) to Cu(I) combined with a color reaction.

Starch syrups are the main products of starch saccharification. The quality of starch, used for the process, should be of high grade, since impurities like proteins and lipids and their resulting byproducts are difficult to remove after hydrolysis. Before the hydrolysis starts, starch granules have to be suspended in water (up to 35%) and gelatinized by acid and/or heat. This makes the starch slurry pumpable and more sensitive to the attack of hydrolyzing agents. The quantitative composition of saccharides in the syrups obtained depends on the conditions of hydrolysis and is closely related to the DE (Table 5.1) [5]. High DE values correlate with high glucose contents and thus with a high sweetness intensity.

TABLE 5.1 Average composition of sugars (%) in different starch syrups [5].

DE[a]	Glucose	Maltose	Maltotriose	Higher oligosaccharides
Acidic hydrolysis				
30	10	9	9	72
40	17	13	11	59
60	36	20	13	31
Enzymatic hydrolysis[b]				
20	1	5	6	88
45	5	50	20	25
65	39	35	11	15
97	96	2	0	2

[a] *Dextrose equivalent.*
[b] *Partly combined with acidic hydrolysis.*

There are various strategies to produce starch syrups: (1) acidic hydrolysis, (2) acidic hydrolysis followed by enzymatic treatment, and (3) enzymatic treatment alone. Acidic hydrolysis is usually performed with hydrochloric acid (pH ≈ 2) at temperatures of 130–170 °C. The heating time depends on the temperature and the desired DE of the end product. When DE of 40 is exceeded, a number of side reactions occur (e.g., transformation of glucose and Maillard reaction) and dark-colored undesirable syrups are obtained. Therefore most commercial so-called dextrin syrups, produced by acidic hydrolysis, have DE values from 10 to 35. After hydrolysis, partial neutralization with sodium hydroxide and purification on activated carbon for instance are necessary. Dual conversion syrups are made by combining acidic and enzymatic treatments. Starch is first gelatinized and partially hydrolyzed with acid, until a syrup of ≈ 30 DE is obtained, followed by pH adjustment and cooling. Afterwards, several enzymes (see below) can be used to produce syrups with desired properties. Typical dual conversion syrups have a DE of ≈ 65. Glucose and maltose are the major saccharides present in similar amounts (35–40% each).

Syrups prepared by enzymatic treatments alone are the most common products nowadays. There are several types of enzymes that cleave glycosidic bonds in amylose and amylopectin. Most enzymes are of bacterial or fungal origin. Heat-stable α-amylase is normally used first. It has a temperature optimum of 70–90 °C at pH 6.5; thus starch granules are simultaneously gelatinized and hydrolyzed. After deactivation of α-amylase at 120 °C and cooling, β-amylase can be used for further hydrolysis. The successive treatment of starch with both enzymes results in syrups with DE values of about 70. For the production of high-maltose syrup (up to 95% maltose on a dry matter base), the enzyme pullulanase has to be applied together with β-amylase. Pullulanase is a debranching enzyme that hydrolyzes α-1,6-glycosidic bonds of amylopectin.

High-glucose syrup can be made by using glucoamylase. It hydrolyzes α-1,4 and α-1,6 glycosidic bonds from the nonreducing ends of amylose and amylopectin chains. Starch is first pretreated with acid and/or α-amylase and then incubated with glucoamylase, which allows the production of glucose syrup with a DE value of 97–98. To obtain higher sweetness, glucose has to be partially converted to fructose. This can be achieved with the enzyme glucose isomerase that transfers high-glucose syrup into a product with around 40% fructose and a sweetness just as that of sucrose. Syrups with higher levels of fructose (up to 90%) can be produced by different enrichment techniques like ion exchange.

Modern technology often uses enzymes that are attached to a support (so-called immobilized enzymes) and filled into a column. In a continuous process, the material to be hydrolyzed is pumped through the column, where the enzymatic reaction takes place. Thus there is no need to deactivate and remove the enzyme and the process becomes faster and more economic.

Raw syrups can be purified by different techniques, for example by cross-flow membrane filtration, treatment with activated carbon or ion exchange. Evaporation of the refined syrup is finally performed to yield products in the range of 71–85% solid contents. Syrups with a lower water content (3–4%) are produced by spray drying. The sweet, white or slightly yellowish powder is readily soluble in water or aqueous alcohols. Variable processing of syrups allows the adjustment of widely different properties that fit to diversified fields of food application. Nonfood applications represent a minor proportion of the total market. Most starch syrups are sold as sweeteners in admixture to sucrose in the confectionery industry. Apart from sweetness, they have the following favorable features: they decrease the water activity, retard or prevent the crystallization of sucrose, retain moisture in the product, act as softening agent, and add body and texture to the product. Well-known products are hard candies, caramels, fondants, gummy bears, jelly beans, and marshmallows. Besides confectionery, a number of other sweet products like jam, ice cream, and soft drinks are produced using starch syrups as additives. Furthermore, cane sugar has been partially replaced by starch syrup in the production of brewed beverages and yeast, the processing and canning of fruits, and in the baking industry. Dried starch syrup is used in sausage production as a red color enhancer.

Glucose syrup is the starting material for the production of glucose. The syrup is evaporated, upon which glucose crystallizes as monohydrate. Water-free glucose can be obtained from the monohydrate by drying in a stream of warm air or by crystallization from ethanol or glacial acetic acid. Aside from its usage as natural sweetener, glucose serves as an invigorating and strengthening agent in many nourishing formulations and medicines. Industrial fermentation of glucose enables the production of numerous chemicals such as alcohols, polyols, ketones, and organic acids. Ethanol is the compound that is produced in greatest abundance from glucose. Hydrogenation of glucose can be used for the production of the sugar alcohols sorbitol and mannitol. Sorbitol is extensively used for the production of surfactants and emulsifiers especially for food applications. In smaller amounts, it serves for making polyethers used in urethane foam production.

5.2 Gluten

Gluten has been classically defined as the rubber-like proteinaceous mass that remains when wheat dough is washed with water or salt solution to remove soluble constituents and starch granules. This primary definition is different from that applied in the field of wheat hypersensitivities, where gluten is a collective term for the storage proteins (prolamins and glutelins) of wheat, rye, barley, and oats (see Section 2.2.2 and Chapters 8–11). Gluten is a byproduct of wheat starch production and industrially produced in amounts of around one million metric tons per year. Gluten mainly consists of proteins that are unique regarding their chemical and physical properties. The combination of different physico-chemical characteristics of proteins (e.g., cohesivity, viscosity, elasticity, and emulsifying and film-forming capabilities) makes gluten particularly interesting for many food and nonfood applications. The modification of gluten by chemical and physical treatments offers additional possibilities to create various valuable products.

5.2.1 Gluten production

In 1745, the Italian biochemist Jacobo Beccari was the first, who prepared gluten by washing out the starch and soluble compounds from wheat dough. In his article "De Frumento" (concerning grain), he described gluten as the first protein of plant origin. Before that time, it was believed that proteins were present only in materials of animal origin. Beccari's procedure is extremely simple and everybody can prepare gluten on a small scale by mixing flour with an appropriate amount of water into a dough and manually washing out the starch and soluble matter from the remaining gluten network in a stream of tap water. This small-scale process can be automatized by means of instruments like the Glutomatic, by which dough mixing, washing out, and centrifugation are performed under standardized conditions (see Section 4.2.1.1.1). The remaining "wet gluten" can then be dried and ground into "dry gluten" known in commerce as "vital gluten".

The world production of gluten increased greatly in the past decades. In 1980, the annual production was estimated at 80,000 metric tons and in 1986 at 253,000 tons; by 2016 it had grown to around 1.2 million tons. Major producers are the European Union, USA, and Canada. Industrial production is performed in starch plants, where gluten is separated from starch by centrifugation and refined by washing with water (see Section 5.1.1). The resulting wet gluten forms compact lumps. Before drying, these lumps must be chopped up by means of a disintegrator to increase the surface area. For drying, ring dryers are now mainly used. They have the advantage of higher drying capacity as compared to other systems like dryer mills.

The design of a ring dryer is schematically illustrated in Fig. 5.2. The principle of a ring dryer is to evaporate moisture of the wet gluten particles ($\approx$70% moisture content) in a circulating hot air stream (inlet temperature $\approx$340 °C). The core of the ring duct system is a manifold, in which classifier and blade systems are incorporated. The classifier uses the centrifugal forces of the ring structure and the air stream to separate heavy semidried material from the light dried material. The adjustable splitter blades are used to cut the particles. The dry product exits the system and gets conveyed to the product collection system. The moisture content of the dried gluten is usually below 10%. A small portion of dried gluten is drawn off and mixed with wet gluten to reduce moisture before starting drying. The semidried material returns back to the disintegrator, is mixed with fresh wet gluten and passes the ring system for another round. For energy saving, outlet air ($\approx$230 °C) can be recycled by returning to the dryer inlet. Altogether, all important conditions of drying (e.g., temperature of hot air, mixing ratio of wet and dried material, and number of cycles) are adjusted in such a way that the viscoelastic properties of the wet gluten are retained in

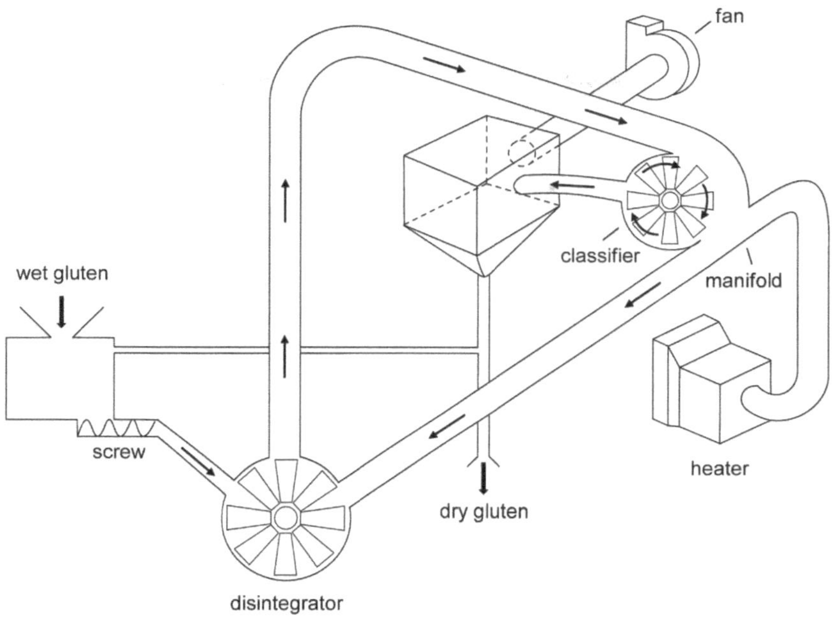

FIG. 5.2 Schematic illustration of a gluten ring dryer.

the final dried product ("vital gluten") as far as possible. The obtained material usually requires sieving followed by grinding of coarse particles to less than 100 μm size. Vital gluten is marketed as a cream-colored, free-flowing powder. When rehydrated, it should retain all intrinsic functionalities of native wet gluten.

5.2.2 Chemical composition

The essential composition, quality, and nutritional factors of commercial gluten are defined by the "Codex Standard for Wheat Protein Products Including Wheat Gluten" (CODEX STAN 163-1987, Rev. 1-2001). Accordingly, vital gluten should contain less than 10% moisture, more than 80% crude protein ($N \times 6.25$) on a dry weight basis, and less than 2.0% ash. The chemical analysis of 27 commercial gluten samples from international sources resulted in 72% crude protein ($N \times 5.7$), 10% starch, 7.5% water, 6.6% lipids, 0.8% fiber, and 0.8% ash on an average [6]. However, the range of variation in the compositions of the samples was extremely wide. A recent chemical analysis of a commercially available vital gluten sample revealed a similar base composition described above, except that the dietary fiber content was significantly increased (5%) (Table 5.2) [7].

The protein fraction, accounting for around 70% of vital gluten constituents, consists of numerous components that can be grouped into the salt-soluble albumins and globulins, the alcohol-soluble gliadins, the glutenins soluble in

TABLE 5.2 Chemical composition of a commercial vital gluten sample [7].[a]

Base composition (mg/g)		Protein composition (mg/g)			
Crude protein	719	ALGL	28	GLUT	214
Starch	108	GLIA	402	ωb	10
Water	62	ω5	16	HMW	55
Dietary fiber	50	ω1,2	28	LMW	149
Lipids	42	α	209	Residue	75
Ash	8	γ	149	GLUTEN	616
Residue	11				

[a] *ALGL, albumins/globulins; GLIA, gliadins; GLUT, glutenins; ω5, ω1,2, α,γ, gliadin types; ωb, glutenin-bound glutenins; HMW, high-molecular-weight glutenin subunits; LMW, low-molecular-weight glutenin subunits; GLUTEN, gliadins + glutenins.*

aqueous alcohols only after the reduction of disulfide bonds, and the insoluble residual proteins (see Section 2.2). The quantitative distribution of these groups can vary and depends on the quantitative composition of the flours used for gluten production, the intensity of washing steps, and the degree of heat treatment during drying. Typically, gliadins (≈56% of total gluten proteins) are predominant proteins followed by glutenins (≈30%) (Table 5.2) [7]. Albumins and globulins (≈4%) are strongly decreased in comparison to wheat flours due to the washing steps and lower than residual protein (≈10%). The proportions of single protein types correspond to those of the flours used for gluten production. Usually, α-gliadins (29%), γ-gliadins (21%), and low-molecular-weight glutenin subunits (LMW-GS) (21%) belong to the major protein types and ω5-gliadins (2.2%), ω1,2-gliadins (3.9%), glutenin-bound ω-gliadins (1.4%), and high-molecular-weight glutenin subunits (HMW-GS) (7.7%) to the minor protein types (Table 5.2). The amino acid composition of vital gluten is strongly determined by the high contents of gliadins and glutenins. Accordingly, glutamine (≈35%), proline (≈15%), and hydrophobic amino acids (leucine, phenylalanine, isoleucine, and tyrosine, together ≈15%) are dominant, whereas essential amino acids such as lysine, methionine, and tryptophan (together <3%) occur only in small proportions (Table 2.5). Thus gluten ranks comparatively low on the scale of nutritional quality among food proteins.

Gliadins are mainly in a monomeric state with molecular masses from around 28,000 to 55,000 (see Section 2.2). A small portion of gliadins are oligomers linked by interchain disulfide bonds with molecular masses between 100,000 and 600,000. The glutenin fraction consists of polymers linked by interchain disulfide bonds. Their molecular masses cover a wide range from around 600,000 to more than 10 million. Apart from covalent disulfide bonds, noncovalent bonds such as hydrogen bonds and to a lesser degree ionic and hydrophobic bonds contribute to the complex gluten structure and result in strong associations between single protein strands. The three-dimensional network of gluten is preformed in hydrated flour; after washing out starch, the proteinaceous network becomes very dense (Fig. 2.9).

Gluten proteins combine both hydrophilic and hydrophobic structural elements, which decisively determines solubility as well as water-binding and emulsifying capabilities. Hydrophilic elements are concentrated in the glutamine- and proline-rich repetitive sequence segments, whereas the nonrepetitive sequence segments contain the hydrophobic elements (see Section 2.2.1). Gluten is not soluble in water under normal conditions due to the hydrophobic segments of gluten proteins. Nevertheless, gluten absorbs water about twice its dry weight due to the hydrophilic segments. Thus it is possible to restore the physical properties of native wet gluten by rehydration of dry gluten. Hydrogen bonds between water and the side chains of glutamine, abundant in the hydrophilic segments, are mainly responsible for the high water-binding capacity. The presence of both hydrophilic and hydrophobic moieties mediates the emulsifying and foaming properties of gluten.

5.2.3 Physical properties

Rehydrated gluten is characterized by the combination of different physical features such as cohesivity, adhesivity, rheological properties (viscosity and elasticity), and film-forming capability that are ascribed to the unique structures of gluten proteins. However, the physical properties of rehydrated gluten can vary to a high extent and are mainly dependent on the protein composition of flours used for gluten production, the procedure of starch and gluten separation, and the drying process. Chemical modifications of functional groups of gluten proteins (see Section 5.2.4) allow a great variation of physical properties.

The unique physical properties of gluten are mainly based on the individual structures of proteins and the specific interactions between monomeric gliadins and polymeric glutenins. It is commonly accepted that both protein fractions account for the physical properties of gluten, but their functions are divergent. Hydrated gliadins have little elasticity and are less cohesive than glutenins. Hence, gliadins mainly contribute to the viscosity and extensibility of gluten. In contrast, hydrated glutenins are cohesive and elastic and are responsible for gluten strength (resistance to extension) and elasticity (ability to restore the original shape). Thus gluten appears to be a "two-component glue", in which gliadins act as plasticizer or solvent for glutenins. A proper ratio of gliadins and glutenins ($\approx$2:1) is necessary to achieve optimal physical properties. The careful selection of wheat varieties is therefore important for gluten preparation, because they can strongly vary in their gliadin/glutenin ratio [8]. Varieties with ratios much higher than 2:1 generate gluten with poor cohesivity, which hinders the agglomeration of gluten particles during the production process and the resulting soft gluten is not usable in many applications. On the other hand, ratios much lower than 2:1 yield in very strong gluten that may be useless, too. "Ancient" wheat species like emmer and einkorn have ratios much higher than those of common wheat (up to 14:1) [8] and are usually not suitable for gluten production.

When rehydrated, gluten proteins form a cohesive mass due to existing forces between particles that unite them. These forces result from strong associations of gluten proteins based on hydrogen bonds, ionic bonds, disulfide bonds and hydrophobic bonds. In addition to this pronounced cohesivity, gluten shows adhesive characteristics associated with important binding capacities that are useful for a number of special applications. Moreover, rehydrated gluten has viscous properties, i.e., it is a kind of tough fluid that allows gradual deformation by shear stress or tensile stress, but it simultaneously shows resistance to flow. The high content of bound water is the predominant cause of the fluid-like behavior of rehydrated gluten. The viscous properties are reflected by the extensibility and the resistance to extension of gluten. The interactions of both parameters can be documented by extension tests. The shape and area under the extension curve enable a clear differentiation of strong and soft wet gluten (Fig. 5.3) [9].

Elasticity is a particularly important physical property of wet or rehydrated gluten. Elasticity is the ability of a material to gradually resist a deforming force and to return to its original shape, when the force is removed. This can be impressively demonstrated by forming a ball from wet gluten and compressing it with a finger. After removing the finger, the ball form is largely restored. The elasticity of gluten is ascribed to the repetitive amino acid sequences of HMW-GS (see Section 2.2.1). These sequence segments are rich in glutamine (Q), proline (P), and glycine (G) included in the repetitive sequence QQPGQG. This sequence is frequently repeated (up to 73 times) and interspersed by hexapeptides such as YYPTSP and tripeptides such as QQP (for one-letter code of

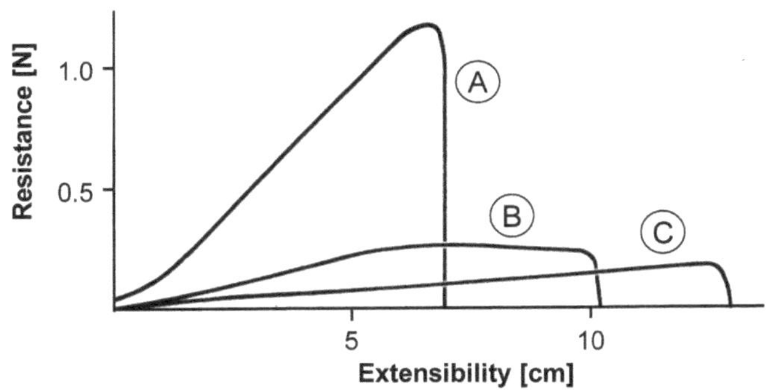

FIG. 5.3 Extensigrams of gluten from three common wheat varieties (A, Schirokko; B, Clement; and C, Maris Huntsman). *(Adapted from Belitz H-D, Kieffer R, Seilmeier W, Wieser H. Structure and function of gluten proteins. Cereal Chem 1986;63:336–41.)*

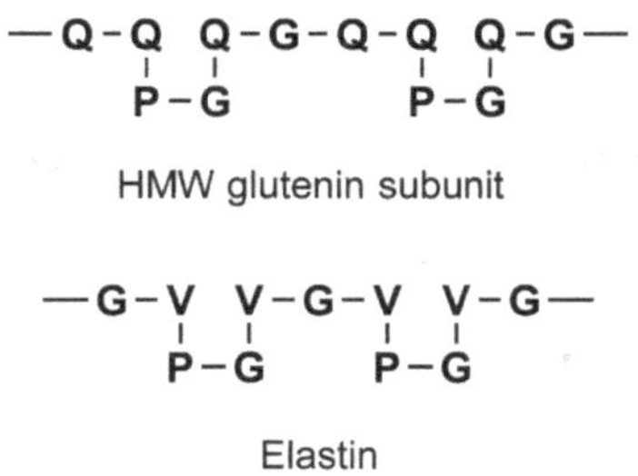

FIG. 5.4 Repetitive amino acid sequences of HMW glutenin subunits and elastin.

amino acids see Table 2.3). Sequence QPGQ generates so-called β-turns that are linked by GQ (Fig. 5.4). Regularly repeated sequences containing QPGQGQ units have been proposed to form a loose spiral structure (β-spiral) [10] similar to the repetitive sequences of elastin, an elastic protein of the connective tissue. In the β-turns of elastin, the hydrophilic glutamine (Q) of HMW-GS is replaced by hydrophobic valine (V) or other hydrophobic amino acids and the repetitive unit GVPGV comprises only five amino acids (Fig. 5.4) repeated up to 58 times and interspersed by nonrepetitive sequences. β-Spirals of HMW-GS containing hydrophilic glutamine are able to bind water, whereas those of elastin containing hydrophobic valine are water-repellent. These are important differences related to their different biological functions (storage protein versus fibrillar protein). Both spirals are thought to confer elastic properties to the proteins. Due to this special structural feature, HMW-GS have been proposed to contribute decisively to gluten elasticity [10].

Its film-forming property is another characteristic feature of gluten. Film-forming is especially important for the production of leavened bread (see Section 4.2.1.8.1), when carbon dioxide and water vapor expand the gluten network of the dough during fermentation and baking. The resulting gluten strands and embedded starch granules stabilize the generated gas bubbles, until the baking process is finished. The thermosetting properties of gluten under heat allow the irreversible change into an infusible, highly stable polymer network.

5.2.4 Modification of gluten

Chemical and physical treatments allow the conversion of gluten into products with strongly different properties. Gluten proteins contain a number of functional groups, present in the side chains of amino acid residues (e.g., amide, hydroxyl, amino, thiol, and disulfide groups) that can be modified by chemical reactions. A common procedure is the generation of a gluten product that is water-soluble or at least widely dispersible in water. This can be achieved by partial hydrolysis of peptide bonds either by enzymes (peptidases) or by acidic treatment. Different commercially available enzymes such as pepsin, trypsin, papain, and pronase are suitable to cleave in parts the peptide bonds of gluten proteins thereby reducing protein size and enhancing solubility. The resulting products have emulsifying and foaming properties that can be used for food and nonfood applications. Moreover, small amounts (1–2%) of the digests added to flour are suitable for improving the mixing properties of bread doughs. The conversion of the amide group of glutamine, abundant in gluten proteins, into the carboxyl group (glutamic acid) by diluted acids (e.g., weak hydrochloric acid) also increases solubility and yields products that have distinct emulsifying properties. They are frequently used to improve emulsification and foam stability in both food and nonfood systems. Treatment of gluten with strong acids like sulfuric and phosphoric acids results in material with greatly increased water-binding capacities.

Esterification by treatment with acids and alcohols (e.g., methanol or butanol) makes gluten more hydrophobic. So, foils produced from this modified gluten are more water-resistant than those from unmodified gluten. Solubility and other characteristics of gluten can also be altered by the reactions of hydroxyl, amino, and thiol groups of gluten proteins with α,β-unsaturated compounds such as acrylonitrile, methyl acrylate or acrylamide. Similarly, acylation reactions of the anhydrides of acetic, succinic, and maleic acids with amino, thiol, and phenolic hydroxyl groups of proteins provoke changes in solubility and functional properties of gluten. The reduction of disulfide bonds and subsequent reoxidation is another method to produce material with special properties.

Modifications of gluten by means of physical methods include extrusion, heat and high-pressure treatment, and UV irradiation. Examples for applications are given in the following. Fibrous structures in gluten can be induced by extrusion technology. Gluten proteins align during the extrusion process resulting in the formation of thin filaments or microfibrils. These assemble further to form macroscopic fiber strands that provoke the laminated, flesh-like appearance of textured gluten that is suitable to simulate meat fiber for instance. The tensile strength of foils, produced from gluten suspended in ethanol, can be enhanced by UV irradiation. The formation of cross-links between protein chains has been proposed as possible reaction.

Heat treatment strongly modifies the structure of gluten proteins. The effects, commonly referred to as heat denaturation of gluten, are characterized by the disruption of hydrogen and ionic bonds accompanied by the conversion of folded structures into randomly looped protein chains. Simultaneously, exchange reactions of thiol and disulfide groups give rise to highly aggregated protein complexes. Denaturation of gluten proteins starts at about 80 °C. The main effects on gluten proteins are decreased water-binding capability and the loss of all rheological properties typical of untreated gluten.

Structural and rheological properties of gluten can also be changed irreversibly by increased pressure, particularly in combination with increased temperature. Treatment of gluten with relatively low pressure (≈ 200 MPa) and temperature (≈ 30 °C) increases the proportion of ethanol-soluble proteins and decreases gluten strength [11]. Increased pressure and temperature significantly reduce the solubility of gliadins. For example, yields of gliadins extracted from

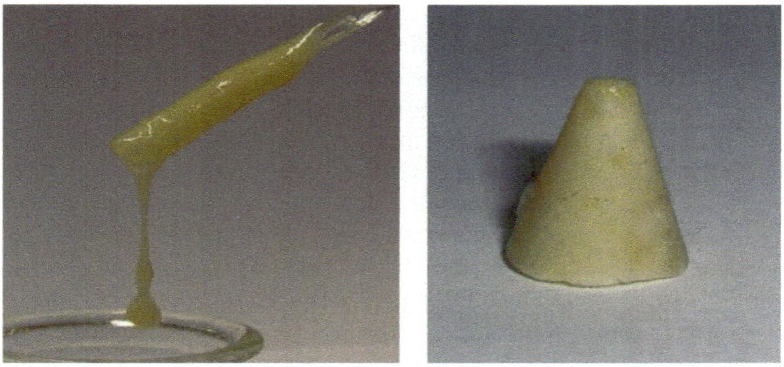

FIG. 5.5 Gliadins isolated from gluten before (left) and after treatment (right) with high pressure in the presence of cysteine (800 MPa, 70 °C, 10 min).

gluten are lowered from 65% under normal conditions to 19%, when gluten has been treated with high pressure (800 MPa) and temperature (60 °C) over 10 min. Cleavage of intrachain disulfide bonds and rearrangement of interchain disulfide bonds have been found to be involved in the pressure-induced reactions. This can impressively be shown for the gliadin fraction that has been extracted from gluten and treated with increased pressure and temperature in the presence of cysteine: The initial viscous gliadin material is readily transformed into a solid material (Fig. 5.5). Regarding rheological aspects, high pressure and temperature induce a drastic strengthening of gluten. Under extreme conditions (e.g., 800 MPa and 60 °C), gluten cohesivity is completely lost.

5.2.5 Food applications

Gluten offers many benefits to food production and thus has become an important commodity in international trade. Apart from its unique techno-functional properties, the low price and the absence of off-flavor make gluten especially attractive as food and feed additive. For most applications, the "vitality" of gluten is essential, which simply means the extent to which the physical and techno-functional properties of native wet gluten can be restored upon rehydration of ring-dried gluten. The major use of vital gluten has traditionally been in the baking sector but in the past decades, there has been a significant shift in gluten utilization to applications other than baking [12]. Water-binding, viscoelastic, and film-forming properties of gluten are now generally appreciated and useful for many other foods aside from baked goods such as meat and seafood, fortified breakfast cereals, and pet food. Substitutes of animal products based on gluten are of special interest for vegetarian and vegan people.

In a few cases, "devitalized" gluten is applied as an additive. Devitalization happens, when gluten proteins are denaturated, e.g., by heating at temperatures above 80 °C, which causes the loss of gluten-typical physical properties such as cohesivity and elasticity. For instance, devitalized gluten is used as a protein supplement in human and pet foods. An example is the high-protein supplementation of bread in the context of low carbohydrate diets ("protein bread"). In this case, the usage of vital gluten would make the dough too strong for mixing, whereas devitalized gluten increases the protein content with only small effects on dough consistency. In addition to vital and devitalized gluten, various forms of modified gluten are used to provide a wide range of further applications.

Fortification of wheat flour for bread dough preparation is still the main outlet of vital gluten [13]. It improves dough strength, mixing tolerance, and dough handling. Its strand-forming ability provides gas retention during fermentation and baking and enables the controlled expansion of dough for improved volume, uniformity, and texture of baked products. Its thermosetting properties contribute to structural rigidity and bite characteristics of products. Finally, its water absorption capacity improves product yield, softness, and shelf-life. Another important function of gluten is the adjustment of the flour protein level in order to meet the protein content specification for standardized baking. Gluten fortification of cheap, low-protein flour offers an attractive alternative to the blending with expensive high-protein flour. A common rule is that a flour with 10% protein fortified with 2% gluten has the same baking quality as a flour with around 11% protein. The trend to use gluten for flour fortification increased during the past decades. One possible explanation is that new wheat varieties that have been bred to give higher yields produce grains of lower protein content.

Aside from fortification of low-protein flours for bread production, vital gluten is commonly used for the improvement of baked or cooked specialty products. The supplementation with up to 10% gluten (based on flour) enables the production of hard rolls, high fiber breads, and multigrain breads of high quality. Gluten used at around 2% improves the strength of the hinge of presliced hamburger and hot dog buns and provides desirable crust characteristics, when the buns are stored in a steamer. Yeast-raised donuts supplemented with gluten have greater volume and increased strength during subsequent glazing. Gluten can also be used to strengthen pizza crust. It provides body and chewiness to the crust and reduces moisture transfer from the sauce to the crust. Additionally, different levels of added gluten enable the production of both thin and thick pizza crusts from the same flour. When common wheat farina instead of durum wheat semolina is used for pasta production, gluten addition increases resistance to breakage, reduces cooking loss and stickiness, and provides firmness of cooked pasta. Stability of pasta during canning and retorting is also improved. The rollability and pliability of corn tortilla dough is enhanced and cracking and breaking problems of tortillas are significantly reduced, when gluten is added.

In extruded and deep-fried snacks, the addition of gluten (1–2%) ameliorates crispiness, texture, and resistance to breakage during handling and shipping. Many snacks with high levels of added gluten are now being marketed worldwide including gluten balls in Europe, vegetarian items in the United States, Yachi-fu cakes in Japan, fried gluten items in China, and gluten wafers in Australia.

The water-binding and thickening properties of gluten are used to improve the quality of ice cream, instant pudding, soups, sauces, ketchup, marinades, and dressing. Due to the outbreak of bovine spongiform encephalopathy (BSE) and the subsequent efforts to replace gelatine, gluten found new applications in the production of some special foods such as chewing gum, chew candies, and fruit chews or as clarifying agent in must and wine production.

Gluten is frequently used as an additive for the production of breakfast cereals. Due to its binding properties, gluten helps to bind vitamin/mineral components to the fortified cereal or grain berry during processing and contributes to the strength and crispiness of the flaked product. Another example of application is the coating of dry roasted nuts with gluten, which provides adhesion of salt and other seasonings.

The binding, film-forming, and thermosetting characteristics of hydrated gluten are the basis for various applications in the manufacture of meat products. Gluten is effective in binding meat chunks or trimmings together to form special products such as extended ground meat, textured meat, restructured meat, canned hams, and poultry rolls. Adding gluten improves the slicing characteristics of loaf-type meat products and minimizes cooking losses during processing and/or preparations. Addition to breadings and batter mixes reduces cooking losses and improves yield and hot and cold adhesion. Gluten is also useful as a protein binder in sausages and other meat emulsion products. Another application is the production of strippable, edible coatings such as sausage casings.

The production of seafood analogs such as surimi is another field of gluten utilization. Surimi is a Japanese term for deboned, minced fish flesh that has been washed with chilled water, until it is odorless and colorless. Upon heating, surimi proteins form gels with an elastic, chewy texture that can be used for making analogs of crabs, scallop, caviar, shrimps, and other seafood. The addition of about 2% of gluten to the gel increases gel strength and reduces the development of an undesirable rubbery texture after frozen storage.

The increasing number of people, opting for a vegetarian or vegan diet, has enhanced the demand for substitutes of animal products that are frequently produced with the help of gluten. Since centuries, gluten has been documented in Asia as a substitute for meat ("mock ham" and "mock duck") and is still popular in the vegetarian cuisines of China, Japan, and other East and Southeast Asian countries. Corresponding gluten products are manufactured in various forms using different techniques. In China, for example, gluten balls are oil-fried and boiled in a savory soup or stew. Another gluten product is sausage-shaped, steamed, and sliced into medallions before eating. Spongy gluten is made by leavening raw gluten that is then baked or steamed. It is offered in fresh, frozen or canned forms. In Japanese cuisine, two types of gluten products are dominant. The raw type consists of gluten that is mixed with glutinous rice and steamed in large blocks. The products may be differently shaped and colored or made with sweetened gluten and filled with various fillings. The dry baked type contains leavened gluten that is formed into bread-like sticks and baked.

Cut into hard discs, it is typically added to soups. Seitan ("wheat meat") originating from Japan is increasingly popular for vegetarian and vegan diets in the Western countries. To prepare seitan, vital gluten is mixed with water to yield a viscoelastic mass and divided into pieces that are heat-treated by boiling or steaming. Typically, gluten pieces are boiled in a marinade consisting of soy sauce, algae, and different spices. Thereby, the product gets a meat-like texture and typical taste and flavor. Finally, seitan is roasted, deep-fried or baked and used like meat.

The viscoelastic properties of hydrated gluten are also exploited for the production of synthetic cheese with the characteristic texture and sensory properties of natural cheese. In products like American and Mozzarella imitation cheeses, gluten alone or in combination with soybean proteins can replace up to 30% of the more expensive sodium caseinate usually used.

Gluten is also applied as additive in the production of soy sauce. It mediates light color, slow browning rate, and pleasant flavor. Due to its high glutamine content, gluten has previously served as a valuable raw material for the production of monosodium glutamate, one of the most frequently used flavor enhancers worldwide. However, this process has now been widely replaced by biosynthetic methods.

5.2.6 Nonfood applications

Gluten, in native as well as in modified form, has many applications in the nonfood industry, for example as additive for feed as well as coatings, foils, adhesives, and detergents. Significant amounts of gluten are used for the production of animal feed, notably pet food (see Section 4.6). For example, the addition of gluten enables adjusting the protein content to the specified level and can contribute to the nutritional requirements of pet food at low costs. The high water absorption and excellent binding properties of gluten are used for meat-based canned products or pellets. In dog biscuits, gluten increases resistance to breaking during packaging and shipping. Gluten and modified gluten are suitable for calf-milk replacement and as feed and fat-filled powders for feeding piglets. Moreover, feed complemented with gluten offers desirable properties for aquatic organisms including fish and crustaceae. Its adhesive properties and water-insolubility provide pellets or granules with the necessary binding and stability in water and different treatments of gluten, for example, incorporation of air, determine whether surface- or bottom-feeding is desired.

The paper industry is the most important user for gluten. A number of paper types, for example, papers for printing, have to be coated and polished. In coating the raw paper, the applied pigments require cobinders that anchor them to each other and to the paper. Both synthetic polymers and natural binders are in use. Modified soy protein isolate has been the standard cobinder in industry among natural binders so far. However, gluten has been proven to be equal or even superior compared with soy protein regarding important functional properties such as binding power, water retention capacity, wet-picking resistance, whiteness, activation of optical brightness, and printability. To suit the specific requirements of a cobinder like residue-free solubility, ability to be incorporated into coating mixtures, binding power to the pigments, and microbiological stability, native gluten has to be modified. Modification with acrylonitrile in a strong alkaline medium has been shown to be especially effective [14]. For this purpose, gluten is dissolved in an aqueous sodium hydroxide solution and mixed with acrylonitrile. After the completion of the reaction, excess acrylonitrile is distilled off and the protein is precipitated in a weak acid (pH 4–5), gently air-dried, and ground. Similarly, succinylated gluten, admixed to a coating color formulation, is suitable as cobinder in coating color applied in paper and cardboard [15].

Nowadays, there is a great interest in plant polymers as substitutes for petroleum-based polymers, because they are readily biodegradable and renewable. Since gluten has unique inherent cohesive and viscoelastic properties and is insoluble in water, it exhibits a great potential for the substitution of conventional plastic. A fascinating feature of gluten is its film-forming ability and numerous studies on a laboratory scale have described the formation, functional properties, and applications of gluten films. In general, films can be simply made by suspending gluten and a plasticizer (e.g., glycerol dissolved in ethanol) in a volatile acid or base (e.g., formic acid or ammonium hydroxide). Acids have the advantage of causing more intense swelling of gluten proteins so that they can be dispersed more homogeneously. The suspension is then casted onto molds and gently dried (e.g., at 40 °C). Thickness and properties of films (e.g., cohesivity, extensibility, and elasticity) can be influenced in a wide range by selection of gluten concentration, process parameters (e.g., drying temperature), and additives. Physical and chemical modifications of gluten offer another possibility to adapt film properties to the desirable application. In particular, the treatment of rehydrated gluten with high pressure (e.g., 200–500 MPa) at different temperatures (e.g., 20–80 °C) allows the production of films with completely different physical characteristics, from soft and smooth to strong and hard rubber-like, from purely elastic to purely viscous and highly extensible [16]. Fig. 5.6 shows foils produced from gluten treated at 500 MPa and 50 °C for 10 min. Additives to gluten such as cysteine are additional keys for tailoring film properties.

To save energy for drying, modified films can be produced by mixing gluten with glycerol in the absence of water, molding the dispersion, and treating with high pressure. Carbon, glass, and plant fibers can be added to reinforce foils produced by casting or molding and the addition of colors (Fig. 5.6) may induce different appearances. Altogether, the film-forming properties of gluten allow the production of various biodegradable films obtained from a renewable source. However, the production is much more expensive than that of petroleum-based films and therefore industrial production is still not profitable. Exceptions are some niche products, for example, coated foods with the advantage that gluten films are edible. However, to protect individuals with wheat

FIG. 5.6 Extension test of a gluten film obtained after high-pressure treatment (500 MPa, 50 °C, 10 min) (left); gluten foils produced by high-pressure treatment and colored by food dyes (right).

hypersensitivities, Codex Alimentarius Standard 163-1987 prohibits the use of gluten films for foods that are gluten-free by nature such as fruits and cheese. In consequence, gluten films may be only used as a cover for gluten-containing foods such as bread, pizza, and biscuits.

The adhesive and binding characteristics of gluten are used to produce natural adhesives. Their properties can be modified by partial hydrolysis via enzymatic, acidic or alkaline treatment, by controlled cleavage of disulfide bonds, and by the addition of polyurethane and epoxy resins. Possible applications are medical bandages, adhesive tapes, envelopes, and postage stamps. The binding capabilities of gluten are additionally useful in environmental protection such as binding heavy metals in industrial processes, solidifying waste oil, and removing ink from waste paper.

Gluten is often used as additive for liquid medications, filler in pharmaceutic tableting, and water-binding agent in skin creams. Due to its emulsifying and foam-stabilizing properties, partially hydrolyzed gluten is frequently added to cosmetic articles such as soaps, shampoos, and toothpastes. Acid treatment with diluted hydrochloric acid for instance is the most commonly used method for hydrolysis. Thereby, glutamine, predominant in gluten proteins, is converted (deamidated) to glutamic acid accompanied by the limited cleavage of peptide bonds. Noncharged carboxamide groups of glutamine are replaced by negatively charged carboxyl groups, which bestow water-solubility and water-binding and emulsifying properties upon the material. The conditions that have to be chosen (degree of acidity, temperature, and treatment time) depend on the desired properties of the end product. Most gluten modifications show higher foaming properties than egg-white albumins and globulins and are nonirrigating like a number of synthetic emulsifiers.

5.3 Wheat germ oil

Wheat germ oil is produced from germs that are obtained from the grains either by special germ separators or by roller mills as separate grain fraction in the first two breaks (see Section 3.3) [17]. Germs contain around 26% of total grain lipids. Thus the germ is the compartment of the grain with the highest lipid concentration. Total lipids of the germ consist of 83% nonpolar lipids (triglycerides), 15% phospholipids, and 2% glycolipids (Table 2.7). In the first step of oil production, germs are conditioned by heating with steam in order to inactivate enzymes. Oil is then obtained by pressing and/or solvent extraction. Pressing is usually performed by means of an expeller or a screw press. The yield of the resulting crude oil is more than 90%. The predominant solvent, used for the extraction of oil from germ, is *n*-hexane; it can be removed from the extracted oil by distillation. The most economical technique for oil production is the so-called "prepress solvent extraction" process. First, germs are pressed at low pressure, leaving 15–20% of oil in the residue, followed by solvent extraction. The almost oil-free residue is steamed, dry-heated, and cooled; it serves as protein-rich animal feed. The crude oil obtained has to be refined by several steps: removal of phospholipids, deodoration with steam, removal of fatty acids with alkali, and decoloration with bleaching agents.

Triglycerides of wheat germ oil mainly consist of unsaturated fatty acids. Linoleic acid with 56% is predominant followed by palmitic acid (16%), oleic acid (13%), linolenic acid (8%), and stearic acid (0.8%) (Table 5.3) [18]. Apart from unsaturated fatty acids, wheat germ oil contains small amounts of phytosterols (β-sitosterol and campesterol) and α-tocopherol (vitamin E). With low amounts of saturated fatty acids, wheat germ oil belongs to the "healthy fats". The relatively high content of α-tocopherol additionally increases the nutritive value. Wheat germ oil is rich in spermidine and its consumption decisively contributes to the spermidine-mediated improvement of a number of age-related health problems (see Section 6.7). Wheat germ oil is preferably used for salad dressing, baking, and cooking. It is not suitable for deep frying because of its low smoke point (<230 °C). In addition, wheat germ oil is highly regarded in the manufacture of pharmaceuticals.

TABLE 5.3 Chemical composition of wheat germ oil [18].

Fatty acids (g/100 g)		Sterols/vitamin E (mg/100 g)	
Linoleic acid	55.7	β-Sitosterol	1606
Palmitic acid	16.4	Campesterol	611
Oleic acid	13.4	α-Tocopherol	174
Linolenic acid	7.8	Stigmasterol	19
Stearic acid	0.8		

5.4 Bran

Wheat bran comprises the outer layers of the grain including pericarp, testa, and aleurone layer (definition of millers). Bran is a major byproduct of wheat grain milling, when white flour is produced (see Section 3.3). The chemical composition (Table 5.4) is characterized by the high content (≈45%) of nonstarch polysaccharides (dietary fiber) that mainly consists of arabinoxylans (≈60%) and cellulose (≈30%). Additional characteristics are the relatively high contents of minerals (potassium and phosphorus), unsaturated fatty acids (linoleic and oleic acids), and vitamins (nicotinamide, pantothenic acid, and α-tocopherol). A disadvantage of bran is that the high lipid content can cause rancidity. Therefore bran is often heat-treated (stabilized) to prevent enzymatic fat oxidation.

Due to the valuable nutritional composition, wheat bran has been widely used as component for animal feed (see Section 4.6), in particular for livestock like horses, cattle, goats, pigs, and rabbits. In recent years, wheat bran has increasingly been in the limelight as dietary supplement for human nutrition. Many studies have shown that the consumption of wheat bran, containing a unique mixture of valuable bioactive components, improves bowel functions and reduces the risk of colon cancer, type 2 diabetes, and cardiovascular diseases

TABLE 5.4 Chemical composition of wheat bran [18].

Basic components (g/100 g)		Minerals (mg/100 g)	
Dietary fiber	45.1	Potassium	1340
Protein	14.9	Phosphorus	1142
Starch	13.4	Magnesium	480
Minerals	6.2	Calcium	67
Lipids	4.7	Iron	16
Sucrose	1.8	Manganese	13
Fatty acids (mg/100 g)		**Vitamins (μg/100 g)**	
Linoleic acid	2200	Nicotinamide (B_3)	18,000
Oleic acid	797	Pantothenic acid	2500
Palmitic acid	698	α-Tocopherol (E)	1600
Linolenic acid	50	Pyridoxine (B_6)	729
Stearic acid	42	Thiamine (B_1)	650
		Riboflavin (B_2)	510
		Folic acid (B_9)	195

(see Section 6.4). Due to valuable antioxidants, its intake can prevent the onset of various oxidative stress-related diseases. Wheat bran is therefore frequently used as food additive, for example for bread, other baked products, and breakfast cereals [19]. It has been introduced into various further market segments including functional foods, nutraceuticals, and pharmaceuticals. Increased consumer awareness of the health benefits of bran induced great demand and food-grade bran can now easily be purchased in drugstores, health food shops, and supermarkets. Taken together, wheat bran is no longer a useless waste product but can be used in many areas of applications.

5.5 Straw

Economical pressure requires better utilization of the whole wheat plant and not only of the grains. Straw (stems and leaves) makes up about half of the dry plant weight and is therefore an important target of usage. Today, it is normally separated from grains and chaff and collected in bales during the threshing procedure. The utilization of the obtained straw is surprisingly variable. Traditionally, straw is chopped into short pieces and used as part of animal feed, especially for cattle and horses, serving as roughage and as a source of warmth delivered by gut microorganisms that digest the material. Straw is further used as bedding for horses, cows, pigs, and rabbits. It is an excellent mulch material; therefore some of the straw should be left scattered on the field after harvest to improve the humus balance of the soil. Horticulture needs straw to cover the soil in order to protect strawberries and cucumbers for instance and cultivated mushrooms use straw for growing. In some parts of the world, straw is utilized for filling mattresses, manufacturing hats, and making waterproof, lightweight roofs. Straw is a good packing material and can be pulped for paper making. The importance of straw as a carbon-neutral renewable energy source is rapidly increasing. Processed into briquettes, it can be cofired with coal and fed to biogas plants. Since a couple of years, straw is increasingly used for the production of bioethanol (see Section 5.6).

5.6 Ethanol production

A small amount of wheat is utilized for commercial ethanol production specifically used as "bioethanol" for renewable motor fuel. Ground grains or flour are dispersed in tap water in a temperature-controlled stainless steel fermenter equipped with stirrers. α-Amylase is added to degrade starch into soluble dextrins. Glucoamylase is used to convert dextrins into glucose and yeast is added to transfer glucose into ethanol that is finally distilled from the fermenter. Optimal usage of the residue (stillage) is important for the commercial success of the ethanol fermentation process, because the production of ethanol alone is not economically competitive with processes used in the natural gas and petroleum industry. Fermentation predominately consumes starch; thus other constituents such as proteins and dietary fiber are enriched in the stillage that can be separated into solids and liquid matter by centrifugation. The chemical composition of the dried solids shows protein contents of around 40% and dietary fiber contents of around 35% [20]. These data recommend the application of the solid fraction as both high protein and high dietary fiber source. The liquid matter contains only about 5% of solids and the evaporation of large amounts of water is expensive. Therefore most water is usually removed by using ultrafiltration or reverse osmosis and thus large savings of energy and costs are possible. The material obtained after drying contains around 45% protein and can be used as a high protein source.

Production of ethanol from grain-based materials is often viewed as competing with food production. Therefore straw being rich in cellulose (40–50%) has been shown to be a promising alternative raw material for biofuel production ("cellulosic ethanol") [21]. The process technology includes the enzymatic hydrolysis of cellulose into glucose by cellulases, for example, produced by genetically modified yeast strains followed by fermentation of glucose into ethanol and distillation. A number of pilot plants have been built worldwide during the last decade and commercial-scale plants are now in operation with capacities of 100 and more million liters of ethanol per year.

References

[1] Shevkani K, Singh N, Bajaj R, Kaur A. Wheat starch production, structure, functionality and applications—a review. J Food Sci Technol 2017;52:39–58.

[2] Van der Borght A, Goesaert H, Veraverbeke WS, Delcour JA. Fractionation of wheat and wheat flour into starch and gluten: an overview of the main processes and the factors involved. J Cereal Sci 2005;41:221–37.

[3] Maningat CC, Seib PA. Understanding the physicochemical and functional properties of wheat starch in various foods. Cereal Chem 2010;87:305–14.

[4] Cornell HJ, Hoveling AW. Wheat—chemistry and utilization. London, Basel: Technomic Publishing CO., INC.; 1998. p. 221–53.

[5] Belitz H-D, Grosch W, Schieberle P. Food chemistry. 4th ed. Berlin: Springer; 2009. p. 876.

[6] Wadhawan CK. Fundamental studies on vitality of gluten for breadmaking. Ph.D. Thesis, Winnipeg: University of Manitoba; 1988.

[7] Wieser H, Scherf KA. Preparation of a defined gluten hydrolysate for diagnosis and clinical investigations of wheat hypersensitivities. Nutrients 2018;10:1411.

[8] Wieser H, Koehler P. Is the calculation of the gluten content by multiplying the prolamin content by a factor of 2 valid? Eur Food Res Technol 2009;229:9–13.

[9] Belitz H-D, Kieffer R, Seilmeier W, Wieser H. Structure and function of gluten proteins. Cereal Chem 1986;63:336–41.

[10] Shewry PR, Halford NG, Tatham AS. High molecular weight subunits of wheat glutenin. J Cereal Sci 1992;15:105–20.

[11] Kieffer R, Schurer F, Köhler P, Wieser H. Effect of hydrostatic pressure and temperature on the chemical and functional properties of wheat gluten. I. Studies on gluten, gliadin and glutenin. J Cereal Sci 2007;45:285–92.

[12] Day L, Augustin MA, Batey IL, Wrigley CW. Wheat gluten uses and industrial need. Trends Food Sci Technol 2006;17:82–90.

[13] Ortolan F, Steel CJ. Protein characteristics that affect the quality of vital wheat gluten to be used in baking: a review. Compr Rev Food Sci Food Safety 2017;16:369–81.

[14] Kempf W, Bergthaller W. New industrial applications of chemically modified wheat gluten. In: Pomeranz Y, editor. Wheat is unique. St. Paul: AACC; 1989. p. 541–61.

[15] Kersting H-J, Lindhauer MG, Bergthaller W. Application of wheat gluten in non-food industry—wheat gluten as a natural cobinder in paper coating. Ind Crops Prod 1994;3:121–8.

[16] Koehler P, Kieffer R, Wieser H. Effect of hydrostatic pressure and temperature on the chemical and functional properties of wheat gluten. III. Studies on gluten films. J Cereal Sci 2010;51:140–5.

[17] Ghafoor K, Özcan MM, AL-Juhaımı F, Babıker EE, Sarker ZI, Ahmed IA, et al. Nutritional composition, extraction, and utilization of wheat germ oil: a review. Eur J Lipid Sci Technol 2017;119:1600160.

[18] Souci SW, Fachmann W, Kraut H. Lebensmitteltabelle für die Praxis. 5th ed Stuttgart: Wissenschaftliche Verlagsgesellschaft; 2011.

[19] Hemdane S, Jacobs PJ, Dornez E, Verspreet J, Delcour JA, Courtin CM. Wheat (*Triticum aestivum* L.) bran in bread making: a critical review. Compr Rev Food Sci Food Safety 2016;15:28–42.

[20] Wu YV. Utilization of by-products of wheat-based alcohol fermentation. In: Pomeranz Y, editor. Wheat is unique. St. Paul: AACC; 1989. p. 657–74.

[21] Tishler Y, Samach A, Rogachev I, Elbaum R, Levy AA. Analysis of wheat straw biodiversity for use as a feedstock for biofuel production. Bioenergy Res 2015;8:1831–9.

Chapter 6

Nutritional value of wheat

Wheat-based foods have been staple foods since thousands of years and are an essential part of the nutrition of the global population. Apart from energy supply mainly by starch, wheat grains are the source of important "healthy" nutrients and contain low amounts of "antinutritive" and "unhealthy" compounds. Particularly whole grain products have a high nutritive value and countless clinical studies have shown that their regular consumption is associated with numerous health benefits. Altogether, the consumption of wheat-based foods is highly recommended owing to their excellent nutritional profiles. In the last 10 years however, wheat has received much negative attention in Western countries. Several pseudoscientific books and numerous media reports fueled the overall impression that wheat consumption generally has adverse health effects. Hence, the popularity of a wheat-free diet has increased considerably in the USA and other Western countries. In consequence, many people without any health problems voluntarily abstain from wheat-containing foods, although it is obvious that subjects under a strict wheat (gluten)-free diet show signs of poor supply with vitamins, minerals, and dietary fiber. The present chapter will illustrate the benefits of wheat consumption regarding the diversified contributions of grain constituents to a healthy nutrition.

6.1 Nutrition and health

Since mankind exits, the first aim of nutrition has been to fill the stomach with food, i.e., knowingly to satisfy hunger and thirst and unknowingly to provide the body with energy and nutrients essential for life. When a surplus of foodstuffs was available, these were selected for consumption according to their appearance, flavor, and taste and the enjoyment they provided. The nutritional value of foodstuffs did not play any decisive role in science and research until the 19th century. The situation changed within the course of the 20th century. Diseases, caused by nutritional deficiencies, were recognized and appropriately treated and nutritional science focused on identifying essential food factors, defining their chemical structures and reactions, and determining minimum requirements according to sex, age, and physical activity. One after the other, the importance of proteins, carbohydrates, lipids, minerals, vitamins, and other minor components for an adequate nutrition was recognized. Thereby, a close scientific link between diet and health was established. After the Second World

Wheat—An Exceptional Crop. https://doi.org/10.1016/B978-0-12-821715-3.00006-X

War, nutritional deficiencies, previously a serious problem, were gradually replaced by diseases due to dietary excess and imbalances and overconsumption of energy-dense, nutrient-poor foods, which especially happened in developed countries. Currently, there are more overweight than underweight people not only in the developed countries but also in those developing countries, where a rapid economic growth takes place. In connection with this unfavorable development, certain dietary components such as fat and sugars have become a major concern.

In the past, research on nutrition was an underdeveloped area of science that often devoted more efforts to animal feed than to human food. More recently, research on the links between health and nutrition has increased enormously. Healthy nutrition has become one of the fad topics among the population of Western countries and has been intensely fueled by the media. Countless "healthy" diets, partly sound, partly silly or even dangerous, have been propagated and many controversial opinions currently exist among both common people and experts. The following simplified guidelines, proposed by national and international authorities, agencies, and organizations, may serve as a basis for a healthy diet:

(1) Reduce total calories and, in particular, fat and sugar calories when overweight.
(2) Reduce saturated fatty acids, cholesterol, simple sugars, salt, and alcohol.
(3) Correlate carbohydrate consumption with physical activity.
(4) Increase dietary fiber and unsaturated fatty acids.
(5) Increase consumption of vegetables, fruits, whole grain products, and sea fish.

6.2 Wheat as a staple food

Wheat-based products have been staple foods since thousands of years and a major source of nutrients for people in many regions of the world. Global wheat consumption amounts to around 66 kg per capita and year, with the European continent having the highest (108 kg) and Africa the lowest (45 kg) consumption [1]. Wheat is often seen mainly as a source of carbohydrates that provide energy and this is certainly important. But apart from energy supply, wheat-based foods are substantive sources of important nutrients, when consumed as a major component of the diet. They include proteins, dietary fiber, vitamins, minerals, and phytochemicals, all important contributors to a healthy diet. On the other hand, wheat has low contents of free sugars, sodium, fat, and saturated fatty acids and is free of cholesterol, ingredients with an unhealthy image. Therefore the consumption of wheat, especially whole grain products, is highly recommended because of its excellent nutritional profile. "Ancient" wheat species like einkorn, emmer, and spelt have been suggested to have more health benefits when compared with "modern" durum and common wheats. Recent studies however demonstrated that they differ little in the contents of valuable components and thus ancient wheats do not appear to be "healthier" than modern wheats [2].

Establishing the contribution of wheat (as well as of other food sources) to health is, in part, difficult and the reliability of nutritional studies depends on the quality of test systems applied (e.g., large-scale dietary intervention studies, short-term studies, and nonhuman systems). The benefits of vitamins and minerals are well-documented and will not be discussed in detail in this section. Some harmful deficiencies are described in Section 8.7.2. Studies on other wheat components are partly less convincing and sometimes doubtful. Health claims, suggested by official bodies [e.g., the Food and Drug Administration (FDA) in the USA and the European Food Safety Authority (EFSA) in the EU] and approved by legal authorities, have the highest standard of evidence for the relationship between a food or a food component and health. Eleven out of 250 health claims, approved in the EU, concern wheat (see below) and related cereals (rye, barley, and oats). These claims particularly establish a clear relationship between the consumption of whole grain wheat products and health.

The distribution of single nutrient classes in wheat grains is highly unbalanced. All components with established or proposed health benefits (dietary fiber, vitamins, minerals, and phytochemicals) are found in their highest concentrations in the germ and outer layers (pericarp, testa, and aleurone layer) of the grain. Thus significant quantities of nutrients are lost, when grains are milled to produce white endosperm flour. Whole grain flour, containing all parts of the grain, or flour of high extraction rates (80–95%) should therefore be used for food production and consumption whenever possible to retain high levels of valuable nutrients. Unfortunately, white flour products currently account for a high proportion (at least 90%) of wheat products consumed in the Western world and it is difficult to change consumer preferences. Further efforts are therefore necessary to convince people of the advantage of eating whole grain products.

Table 6.1 shows the differences between whole grain flour and white flour in the concentrations of the main nutrient classes [3]. Highly significant differences

TABLE 6.1 Average contents (%) of nutrients in whole grain and white flour [3].

Compound	Whole grain	White flour
Starch	58.2	71.0
Dietary fiber	13.3	4.0
Protein	10.6	9.8
Lipids	1.8	1.0
Minerals	1.7	0.4
Sugars[a]	0.6	0.4

[a] *Sucrose, fructose.*

are particularly found in the contents of dietary fiber and minerals, components that are important for a healthy diet. Similarly, the contents of vitamins are much higher in whole grain flour than in white flour (Table 2.8). Not only the grain milling conditions but also further food processing have a great influence on the nutritional value of the end product. For example, baking, cooking, steaming, and frying can distinctly decrease the biological activities of vitamins and phytochemicals, whereas germination and fermentation can introduce or increase valuable ingredients such as vitamins and phenolic compounds. In the following sections, the contributions of different grain constituents (starch, dietary fiber, proteins, lipids, minerals, vitamins, and phytochemicals) to the nutritional value of wheat are described.

6.3 Starch

Starch is the main component of wheat grains; the content in whole grain flour is around 58% and in white flour around 71% (Table 6.1). The primary physiological function of starch for both wheat plants and humans is the same: supplying the organism with energy. According to the Food and Agriculture Organization (FAO) of the United Nations, cereals (mainly wheat and rice) provide almost 50% of the food energy for the human population on a global basis. On average, whole grain wheat flour supplies 297 kcal/100 g and white flour 335 kcal/100 g [3]. In those countries, where wheat plays a prominent role in nutrition, wheat starch provides up to 20% of the energy intake supplied by the diet. The slow and steady delivery of glucose into the bloodstream during digestive starch degradation is considered advantageous, because it avoids large fluctuations in blood sugar. Thus a constant supply of glucose for the brain and throughout the body is ensured. Before modern energy drinks were introduced, top athletes such as tennis players, cyclists, and marathon runners were preferably accustomed to eating starch-rich food, notably spaghetti, before starting exercise. Hence, a regular and long-lasting supply of energy was guaranteed.

A precondition for energy production is the enzymatic degradation of starch into glucose during the digestion process. Native starch, having an intact granule structure, shows low digestibility. Amylopectin is less susceptible to digestion than amylose due to its semicrystalline and branched structure. However, wheat grains and flour, used for human nutrition, are almost exclusively heat-treated, mostly by baking or cooking. In consequence, native starch granules are destroyed by swelling and gelatinization and the availability of amylose and amylopectin for digestion greatly improves. The ability of humans to digest starch developed long before cereals were introduced into the diet, because starch-containing fruits and tubers had been important components of the human diet since hundred thousands of years. In consequence, the digestion of starch in the human digestive tract is well established. Starch degradation starts during chewing by salivary amylases, combined with the reduction of the particle size of foods, and is intensely continued in the small intestine by amylases and glucoamylase secreted by the pancreas. Thus starch is quite well digested and the main end product, glucose, is readily absorbed by the intestine. The extent of starch digestion however depends on many inherent factors (e.g., food processing, food matrix, viscosity of the digest, and interactions with proteins and lipids) and plays a crucial role in regulating the post-prandial blood glucose and insulin concentrations (see Section 6.4.4).

A small portion of starch (so-called resistant starch) escapes complete digestion in the small intestine. Resistant starch is defined as the portion of starch and degradation products of starch that reach the large intestine of healthy adult humans. Resistant starch mainly originates from foods that contain whole grains, only partially milled grains (e.g., in muesli) or modified starches. Physiologically, resistant starch is comparable with dietary fiber. Its health benefits are well established [4, 5]: increase of fecal bulk, reduction of both glycemic load and risk of colorectal cancer, and formation of beneficial short chain fatty acids in the colon (see below). Retrograded starch (e.g., present in cooked foods after cooling) is also partially resistant to digestion. In this case, amylose is less digestible than amylopectin because of its faster retrogradation rate.

6.4 Dietary fiber

According to the Codex Alimentarius Commission [6], dietary fiber (DF) is defined as carbohydrate polymers with ten or more monomeric units that are not hydrolyzed by endogenous enzymes in the small intestine of humans. Additionally, nondigestible carbohydrates with 3–9 monomeric units (e.g., fructo-oligosaccharides) can also be included into the DF content. The definitions of DF, presented by the EU in 2011 and FDA in 2016, additionally emphasize the physiological effects that are beneficial to human health and have been demonstrated by generally accepted scientific significance. The term "dietary fiber" should be distinguished from the term "crude fiber" that has been internationally defined as all constituents of plants resistant to hydrolysis by acid and subsequently by alkali. DF of wheat is enriched in the outer part of grains (pericarp, testa and aleurone layer) termed "bran" by millers. Pericarp/testa and aleurone layer consist of around 56% and 20% DF, respectively, whereas the concentration in the starchy endosperm is much lower (≈3.5%).

Wheat grains contain about 10–13% DF that predominantly consists of arabinoxylans, cellulose, β-glucans, fructans, and arabinogalactan peptides (see Section 2.1.2). Whole grain flour contains more DF (≈13%) than white flour (≈4%), where only arabinoxylans are present in significant amounts. Common structural features of DF compounds are β-glycosidic linkages of monosaccharide units that are resistant against degradation by most human glycohydrolases. Therefore they majorly remain undigested in the small intestine and are partially fermented by the colon microbiota. It has been estimated that at least 50% of cellulose and around 80% of the other nonstarch polysaccharides are metabolized in the colon. Residual DF is excreted.

Interest in the relationship between DF and health was fueled in the 1970s by reports of people in rural regions of Africa, who consumed primarily foods rich in DF and very rarely showed signs of constipation, diverticular disease, colon cancer, diabetes mellitus, and arteriosclerosis. In contrast, these diseases were found to be frequent in the populations of North America and Western Europe, who consumed foods low in DF. Since that time, many extensive reviews have been written regarding the effects of DF on health. In 2014, the Scientific Advisory Committee on Nutrition (SACN) of the United Kingdom published one of the most comprehensive reviews on health benefits of cereal DF [7]. For example, foods with high DF content have been proposed to improve gastrointestinal transit, to increase the beneficial intestinal microbiota, to reduce the level of cholesterol, and to stabilize the blood sugar level. Low consumption of DF has been associated with certain complaints and diseases such as constipation, obesity, colon cancer, type 2 diabetes, arteriosclerosis, coronary disease, and stroke. However, not all of these effects have been conclusively proven in controlled experiments and further well-designed studies on human nutrition are required. Nevertheless, the intake of DF is commonly considered to have multiple health benefits (see below). EFSA recommends a daily intake of 25 g and FDA 38 g for men and 25 g for women. However, the real daily intake in both Europe and the USA is only about 15–18 g and should therefore be increased. The supplementation of foods with DF appears to be helpful and should be promoted.

In conclusion, wheat DF is one of the most commonly consumed fibers worldwide. Sound statements on the relation between health and wheat DF can be focused on the following specific effects: (1) regulating colonic functions, (2) protecting against cancer, (3) normalizing serum lipid level, (4) attenuating the post-prandial glucose response, and (5) suppressing appetite. Additionally, several publications report positive effects of DF, especially of β-glucan and arabinoxylans, on the immune system [8, 9], but further *in vivo* studies are necessary to confirm these findings.

6.4.1 Colonic functions

The improvement of colonic functions is the best- and longest-known effect of DF and is clearly approved as health claim by regulatory authorities. Many human studies with controlled diets showed that wheat bran fiber contributes to a valuable increase in fecal bulk [ESFA claim 2010; 8(10):1817]. The effect has been related to the water-binding capacity of DF. Fecal bulk is additionally increased by the growth of bacteria induced by the fermentation of arabinoxylans and resistant starch that serve as a source of energy for the colonic microbiota. Cellulose, linked with the phenolic polymer lignin, also contributes to feces weight, because it is quite resistant to degradation by colon bacteria.

Detailed scientific studies demonstrated that DF improves bowel functions by mechanical effects, reduces intestinal transit time [ESFA claim 2010;

8 (10):1817], and relieves constipation. Cereal DF appears to be more effective than that of fruits and vegetables. Constipation is the most frequent irregularity of the colon. It involves hard feces, excreted with difficulty, and increased transit time. In most cases, constipation can be eliminated by an increased intake of DF. Similarly, a number of studies have shown that long-term increased consumption of DF can reduce or eliminate the symptoms of diverticular disease that is characterized by a saccular herniation of the bowel wall often combined with high intraluminal pressure. DF has been further shown to stimulate the growth of beneficial intestinal microbiota like bifidogenic and lactic acid bacteria. Thus DF is an excellent source of prebiotics and its prebiotic effect is now reasonably well accepted. There is increasing evidence that the products of colonic fermentation of DF positively influence the body as a whole through effects of immunological, neuronal, and hormonal feedbacks from the colon to the upper digestive tract.

6.4.2 Cancer

Evidence for the protective effect of wheat DF against cancer is not yet clear. A number of studies have reported decreased risks of colon, small intestinal, pancreatic, prostate, and breast cancer. The effects on colorectal cancer are most evident [10]. The relation between DF intake and prevention of colorectal cancer has been suggested since more than 40 years. Most clinical investigations, based on epidemiological, observational, and intervention studies, in fact established the association between the beneficial effects of DF and the reduced risk of colorectal cancer. Some other studies however failed to support this association. Taken together, scientific evidence has a distinct bias in favor of the beneficial role of DF regarding colonic cancer.

The mechanisms, by which DF protects against colon tumor development has not been fully explored. Most of the proposed mechanisms are based on the physiological functions of DF [11]:

(1) The transit time of fecal material is reduced by DF so that the contact with potentially carcinogenic compounds is shorter.
(2) The concentrations of harmful agents and tumor promotors are diluted by the increase of fecal bulk.
(3) DF reduces the concentration of harmful bile acids. Bile acids and especially their metabolites are known as potent carcinogens. Water-soluble fibers behave like hydrocolloids; they hold a lot of water and form highly viscous solutions that allow binding bile acids and other harmful organic compounds and removing them from the intestine.
(4) Fermentation of DF (e.g., arabinoxylans) by colonic bacteria produces short-chain fatty acids, e.g., acetic, propionic, and butyric acid. Short-chain fatty acids have a range of protective effects relevant for a reduction of cancer risk. For example, butyric acid has the ability to promote a normal phenotype in colonic epithelial cells by repairing DNA. Moreover, short-chain

fatty acids reduce the conversion of bile acids into secondary carcinogens (e.g., deoxycholic acid and lithocholic acid) that would promote cell proliferation and growth of adenomas.

(5) Several antioxidants, disengaged from wheat bran into the colon, prevent the development of cancer.

Ferulic acid and, to a lesser degree cinnamic acid, released from arabinoxylans by bacterial enzymes, have been proposed to have antioxidant, antimutagenic, and antiinflammatory effects. Nonfibrous components of wheat bran such as alkylresorcinols, sphingolipids, steroids, lignans, and phenolic compounds were additionally suggested to be responsible for the observed anticancer effects [11]. These compounds exhibit significant antioxidative capabilities such as scavenging free radicals, chelating metal ions, and activating antioxidative enzymes. Using human colon cancer cells, alkylresorcinols were identified as the major anticancer components of wheat bran [12].

6.4.3 Cardiovascular diseases

Cardiovascular diseases (CVD) include all diseases of the heart and blood vessels. Arteriosclerotic diseases are known best; they are considered one of the most frequent causes of death in the Western population. The main risk factors are high blood pressure, high intake of saturated fats and cholesterol, type 2 diabetes, overweight, tobacco consumption, and physical inactivity. High blood lipid and cholesterol levels have been shown to be important risk factors in the development of CVD. In this context, the influence of diets on the incidence of CVD has been the subject of many studies and the potential effects of DF play an important role in these investigations. In a recent metaanalysis (45 cohort studies and 64 publications), the relation between whole grain intake and CVD were summed up [13]. Thereby, a clear inverse association between the intake of whole grain products and prevalence of CVD was established. When comparing consumers with a low whole grain intake (<90 g/day) to those with a high whole grain intake (>90 g/day), CVD risks were reduced by ≈30% in cases of high whole grain intake. Other studies demonstrated that the consumption of whole wheat foods significantly reduced the serum levels of triglycerides, total cholesterol, and low density lipoprotein (LDL) cholesterol. Therefore a number of professional organizations and governmental agencies have developed recommendations on the positive role of DF in lipid metabolism particularly associated with a reduction of serum lipid levels including cholesterol.

A number of mechanisms have been proposed for the effects of DF on CVD. The high viscosity of DF solutions reduces or delays the absorption of lipids in the small intestine. The increase in fecal excretion of bile acids is another explanation. For example, the gel-forming viscous properties of DF have been shown to be associated with bile acid excretion and lower serum cholesterol levels. Among fiber components, β-glucans especially contribute to the maintenance of normal blood cholesterol levels and these effects were approved as health claim [ESFA claim 2011; 9(6):2207].

6.4.4 Post-prandial glycemic response

Numerous epidemiological and intervention studies revealed significant correlations between increased whole grain consumption and the reduced occurrence of type 2 diabetes (T2D) [14]. In contrast, elevated consumption of refined flour significantly increases the risk of T2D. Quantitative metaregression demonstrated a significant linear inverse relationship between whole grain intake and T2D [15]. For instance, consuming three servings of whole grain foods (45 g of whole grain ingredients daily) would induce a 20% relative reduction in the T2D risk as compared to consuming a half serving. Many studies demonstrated that DF as part of whole grain products, has the ability to reduce insulin and glucose response significantly. Investigations with whole grain bread showed attenuated glucose (−7% to −28%) and insulin (−15% to −31%) responses relative to white bread controls. Health claims report that the consumption of arabinoxylans as part of a meal significantly contributes to the reduction of the blood glucose increase after a meal (ESFA claim 2011; 96:2205).

The mechanisms that underlie the effects of whole grain consumption on T2D are not well defined at the moment. Some but not all of the protective effects assigned to DF may be explained by slower gastric emptying, binding of carbohydrates in the small intestine, reduced concentrations of inflammatory markers that increase T2D risk (e.g., plasminogen activator inhibitor type 1 and C-reactive protein), and increased serum concentration of adiponectin, a cytokine that increases insulin sensitivity [16]. Further research is needed to explore details of the mechanisms behind the effects of a whole grain diet on T2D risk.

6.4.5 Obesity and weight loss

Obesity is a major nutrition-related health problem in the Western population. Consumption of whole grain flour has been associated with decreased obesity and weight loss when assessed by epidemiological studies. However, clinical studies on obesity and weight loss that compared the effects of whole grain flour and white flour resulted in contradictory conclusions. Only a limited number of prospective studies on whole grains and DF reported small but significant reductions in weight gain. The general conclusion is that the effect of DF may be modest at best [17]. Weight loss, observed with the intake of whole grain flour, was related to the ability of viscous DF to bind high-caloric compounds like fat and sugars in the small intestine and thus to reduce their digestibility, so that they escape absorption. Moreover, DF has been suggested to induce faster satiety and a feeling of fullness combined with avoiding excessive consumption of calories. The tendency for enhanced satiety might also be affected by the increased particle size in whole grain flour (>250 μm), since particle size is associated with satiety. Taken together, there is currently limited evidence that DF significantly reduces the risk of developing obesity or effectively aid weight loss.

6.5 Proteins

The protein content of wheat grains mostly ranges between 9% and 14%. Whole grain flour contains about 1% more protein than white flour (Table 6.1). Globally, wheat products are one of the leading sources of plant proteins. On average, people in developed countries obtain about 25% of dietary proteins from wheat on average. The total protein harvested in wheat is higher than that harvested in legume crops, another important source of plant protein. The intake of proteins is obligatory for human life, because some amino acids (so-called essential amino acids) cannot be synthesized by the body and their lack would negatively affect important physiological functions. Table 6.2 includes the average contents of essential amino acids in whole grain flour and white flour [3]. The values indicate that the starchy endosperm (white flour) contains less essential amino acids because of the high proportions of the nonessential amino acids glutamine and proline in gluten proteins, whereas the other parts of the grain contain higher amounts of essential amino acids due to the more balanced amino acid compositions of albumins and globulins (Table 2.5). Nevertheless, each type of wheat flour contributes adequate amounts of essential amino acids to the daily requirements [18] (Table 6.2). The consumption of 200 g whole grains plus white flour covers much more than 50% of the daily requirements for most essential amino acids. The only limiting amino acid is lysine that may be not sufficient for infants and children, who require more lysine. This is not a

TABLE 6.2 Content of essential amino acids in whole grain and white flour and their recommended daily intake [3, 18].

Amino acid	Whole grain flour (mg/100 g)	White flour (mg/100 g)	Daily requirement (mg/kg body weight)
Histidine	280	220	10
Isoleucine	540	460	20
Leucine	920	820	39
Lysine	380	240	30
Methionine[a]	220	170	15
Phenylalanine + Tyrosin	1050	870	25
Threonine	430	320	15
Tryptophan	150	120	4
Valine	620	490	26

[a] *+ Cysteine.*

real problem, because wheat proteins are usually consumed in combination with other proteins, for example from dairy products, eggs, meat, and vegetables, with complementary amino acid profiles. Altogether, wheat is an important source of proteins and due to the higher contents of total proteins and essential amino acids, whole grain flour should be preferred to white flour.

6.6 Lipids, minerals, and vitamins

Wheat grains contain 1.5–2.5% lipids; most of them are concentrated in the germ. The major lipid classes are nonpolar lipids, glycolipids, and phospholipids (Table 2.7). The minor lipid components are fat-soluble vitamins and phytochemicals (see below and Section 2.5). The majority of wheat lipids are present as triglycerides, whereas free fatty acids occur only in traces. The spectrum of esterified fatty acids consists nearly completely of C_{16}- and C_{18}-compounds (Fig. 2.10). The average composition in wheat grain consists of 62% linoleic acid, 21% palmitic acid, 11% oleic acid, 5% α-linolenic acid, and 1% stearic acid. Linoleic and α-linolenic acids are essential fatty acids, i.e., they must be ingested, because the body cannot synthesize them. In summary, 22% of fatty acids are saturated and 78% are unsaturated resulting in a ratio of 0.28. Even if the absolute amount of lipids in wheat grains is low (≈2%), the ratio of saturated to unsaturated fatty acids is very valuable regarding health effects. White flour contains around 1% lipids; the ratio of saturated to unsaturated fatty acids is 0.34.

The total mineral (ash) content of whole grains amounts to around 2 g/100 g. The minerals are enriched in the aleurone layer of the grains (about 60% of the total content), whereas the starchy endosperm contains only about 20% of total minerals. The average contents of major minerals, presented in Table 2.8, amount to 380 mg potassium, 342 mg phosphorus, 97 mg magnesium, and 33 mg calcium per 100 g. Other minerals such as iron, manganese, zinc, and copper are trace elements (<4 mg/100 g). The daily intake of wheat products, containing 100 g whole grain and 100 g white flours, considerably contributes to the adult requirement for phosphorus (≈60%), iron, and zinc (≈40% each). Calcium and phosphate are important in building and maintaining bones. Iron is an essential part of hemoglobin (blood) and myoglobin (muscle) and iron and zinc are components of numerous metabolic enzymes such as peroxidases and catalases (iron) and dehydrogenases and carboxypeptidases (zinc). Selenium is an essential micronutrient for humans but is deficient in at least a billion people worldwide. Wheat may help to reduce deficiency, because it is a major source of dietary selenium (2.1 μg per 100 g in whole grains).

Information on the bioavailability of minerals in diets that contain wheat-based foods, is contradictory. The bioavailability may be restricted by phytic acid (inositol hexaphosphoric acid) [19]. It occurs in wheat grains in amounts of 0.4–1.4% of dry matter and is mainly found in parts of the bran (aleurone layer) and the germ, whereas the starchy endosperm is free of phytic acid. Phytic acid binds bivalent cations such as calcium, magnesium, and iron and impedes their absorption in the intestinal tract. Wheat processing can have remarkable reducing effects on the content of phytic acid. For example, fermentation, extrusion cooking, and baking cause significant losses of phytic acid. But altogether, there is considerable uncertainty on the negative effect of wheat phytic acid on the bioavailability of minerals, its level necessary for effects, and the clinical relevance. Therefore future studies are urgently needed. In contrast to the possible negative implications, phytic acid can act as antioxidant through the formation of complexes with various metals that reduce metal-catalyzed redox reactions and associated oxidative damage.

Wheat is an important source of vitamins, especially vitamins of the B-group and vitamin E. These vitamins are not distributed uniformly throughout the grain but are enriched in the outer part and the germ. Thus white flour contains much less vitamins than whole grains (Table 2.8). The only fat-soluble vitamin, present in wheat grains in a significant amount, is α-tocopherol (vitamin E). It is enriched in the germ, therefore only whole grain flour has significant amounts ($\approx 1.4\,mg/100\,g$). α-Tocopherol is a radical scavenger and protects important bioactive compounds such as lipids, enzymes, hormones, vitamin A, and ubiquinone from harmful oxidation.

The vitamin B complex is predominant among vitamins. B vitamins belong to the group of water-soluble vitamins and often occur together. The highest content in wheat grains (100 g) has been found for nicotinamide (B_3) (5.1 mg). Nicotinamide is part of nicotinamide adenine dinucleotide (NAD^+) or its phosphorylated form ($NADP^+$). NAD^+ and $NADP^+$ are coenzymes in a wide variety of enzymatic redox reactions, the most notable being glycolysis, the citric acid cycle, and the electron transport chain. The classical picture of nicotinamide deficiency is pellagra that mainly affects skin, digestion, and nervous system. Pantothenic acid (B_5) (1.2 mg) is integrated in the synthesis of coenzyme A, the main carrier of acetyl and other acyl groups in cell metabolism. Thiamin (B_1) (0.46 mg) in the form of its pyrophosphate is the coenzyme of several important enzymes such as pyruvate dehydrogenase, α-ketoglutarate dehydrogenase, and transketolase. Thiamine deficiency results in beri-beri, a disease with neurological and cardiac symptoms. Pyridoxal phosphate, the active form of pyridoxine (B_6) (0.27 mg), functions as a component of many enzymes and is required to synthesize amino acids, carbohydrates, and lipids. Riboflavin (B_2) (0.1 mg) is the cofactor for a variety of flavine enzyme reactions especially important in protein metabolism. Tetrahydrofolate, the active form of folic acid (B_9) (0.09 mg), is a component of enzymes that transfers single carbon units like formyl or hydroxymethyl residues. Examples for important functions are contributions to the production and maintenance of new cells and to DNA and RNA synthesis and their protection against structural changes. Biotin (B_7) is part of carboxylating enzymes such as acetyl-CoA-carboxylase and pyruvate carboxylase and plays an important role in fatty acid biosynthesis.

The daily consumption of food that in total may contain 100 g whole grain flour plus 100 g white flour covers a considerable percentage of the daily adult requirement of thiamin (≈47%), nicotinamide (≈43%), pyridoxine (≈33%), and folic acid (≈32%). As far as known, the constituents of the bran do not interfere with the bioavailability of vitamins. An exception is nicotinamide that in parts is bound to the bran, so that only 10–20% may be bioavailable. Vitamins vary in their sensitivity to light, temperature, and moisture. Therefore processing of wheat products can have negative effects on the vitamin content.

In summary, the consumption of wheat-based products, especially those containing whole grain flour, significantly contributes to an important supply with essential fatty acids, minerals, and vitamins.

6.7 Phytochemicals

Wheat grains are rich sources of various phytochemicals such as phenolic acids, terpenoids, tocopherols, and sterols (see Section 2.5) that decisively contribute to health-promoting effects [20]. Phenolic acids are the major group of phytochemicals and mainly include ferulic acid (around 50%), caffeic acid, and *p*-coumaric acid (around 20% each); they occur either in a free form or as conjugates with fiber components. The outer layers of the grain contain most of phenolic acids. Free phenolic acids and a small part of conjugated phenolic acids are absorbed in the small intestine. The major part of ferulic acid and to a lesser degree other phenolic acids are bound to arabinoxylans and are released by fermentation in the colon. The importance of phenolic acids for human health is not completely clear. Phenolic acids are known to exhibit strong antioxidant activity [21], but the evidence for human health is still under discussion. Clinical studies on ferulic acid proposed that it improves vascular functions [22] and *in vitro* studies with human colon cancer cells indicated a high antitumor activity of phenolic acids [23]. However, the *in vivo* efficacy and the underlying mechanisms are still unknown. Health effects of phenolic acids are therefore not yet accepted to support health claims.

Among terpenoids, phytosterols (e.g., β-sitosterol and campesterol) have well-studied cholesterol-lowering effects and reduce the risk of coronary heart disease; corresponding health claims have been evaluated by EFSA and are approved in Europe. The tocol fraction of wheat consists of α-, β-, and γ-tocopherol and tocotrienols. They all are often summarized as vitamin E, but only α-tocopherol has significant biological activity. This activity is mainly based on its antioxidative properties that stabilize or regulate many physiologically important reactions. Tocopherols and tocotrienols may have further biological activities (e.g., anticancer effects and prevention of neurodegeneration) [24], but the effects are of less scientific significance. Current evidence for an important beneficial bioactivity (e.g., antioxidant activity and prevention of colon cancer) of alkylresorcinols is weak and mostly based on *in vitro* tests.

Betaine (*N,N,N*-trimethylglycin) and choline are methylating compounds that reconstitute harmful demethylated products in the body. For example, a high plasma homocysteine level (called hyperhomocysteinemia) is a serious risk factor for cardiovascular diseases. Betaine and choline convert homocysteine into the essential amino acid methionine. In this way, both compounds are involved in numerous other methylation reactions that improve health. Humans are not able to synthesize betaine and obtain it almost solely from the diet. Wheat grains contain one of the highest reported betaine levels among plants with levels of around 1–3 mg/g enriched in the bran and almost absent in the starchy endosperm [25]. The intake of betaine has been associated with a number of health biomarkers, e.g., decreased plasma homocysteine and LDL cholesterol [26]. EFSA has accepted a health claim related to betaine but only for products containing at least 500 mg per serving.

Spermidine (*N*-(3-aminopropyl)butan-1,4-diamine) (Fig. 2.13F) is a regulator of cell growth, differentiation, and apoptosis. Its physiological functionality plays an important role in counteracting aging and promoting longevity. Autophagy, a process used to eradicate damaged cell organelles and unused proteins, is considered the main mechanism of action of spermidine at the molecular level. Wheat germs contain the highest spermidine content (243 mg/kg) among foodstuffs. For this reason, wheat germ extracts have been used to study the impact of spermidine on health and the safety and tolerability of dietary supplementation [27]. Different benefits of spermidine on health have been described [28]: High levels of dietary spermidine are linked to reduced blood pressure, resistance to stress, lower incidence of cardiovascular disease, and reduced cancer-related mortality. Spermidine has also been associated with antiaging properties and decreases the occurrence of age-related pathology and loss of locomotor ability. In conclusion, the consumption of whole grain wheat products decisively contributes to the spermidine-mediated improvement of a number of age-related health problems.

6.8 Wheat processing

Countless studies have demonstrated that the type of wheat grain milling has an enormous impact on the content and composition of bioactive compounds in the products obtained and thus on their health benefits. Flours with high extraction rates, notably whole grain flour, contain much more valuable compounds than white flour. However, also other technological processes apart from milling influence the content of bioactive compounds and can be used in food production to improve nutritional values of wheat-based foods. Sourdough fermentation of bread doughs has been reported to reduce the digestibility of starch combined with a decrease of glycemic load, to increase the bioavailability of minerals by degradation of mineral-binding phytic acid, and to release bound ferulic acid suggested to have health benefits [29, 30]. Sprouting of wheat grains has been described to increase the amounts of bioactive components such as folates, phenolic compounds, tocopherols, and nicotinamide [31, 32]. Although germinated grains are not suitable for making conventional products like bread or pasta, they can be used for soups, porridges, and breakfast cereals. When bran is applied for fortification of white flour, the reduction of particle size by ultrafine bran grinding before incorporation increases the bioavailability of endogenous vitamins, minerals, and phytochemicals [33].

References

[1] Food and Agriculture Organization (FAO) Statistics.

[2] Shewry PR, Hey S. Do "ancient" wheat species differ from modern bread wheat in their contents of bioactive components? J Cereal Sci 2015;65:236–43.

[3] Souci SW, Fachmann W, Kraut H. Lebensmitteltabelle für die Praxis. 5th ed. Stuttgart: Wissenschaftliche Verlagsgesellschaft; 2011. p. 241–3.

[4] Hasjim J, Lee SO, Hendrich S, Setiawan S, Ai Y, Jane JL. Characterization of novel resistant starch and its effects on postprandial plasma-glucose and insulin responses. Cereal Chem 2010;87:257–62.

[5] Birt DF, Boylston T, Hendrich S, Jane JL, Hollis J, Li L, et al. Resistant starch: promise for improving human health. Adv Nutr 2013;4:587–601.

[6] Codex Alimentarius Commission: Codex CAC/GL 2-1985, revised 2010.

[7] Scientific Advisory Committee on Nutrition Draft Carbohydrate and Health Report 2014.

[8] Rice PJ, Adams EL, Ozment-Skelton T, Gonzles AJ, Goldman MP, Lockhart BE, et al. Oral delivery and gastrointestinal absorption of soluble glucans stimulate increased resistance to infectious challenge. J Pharmacol Exp Ther 2005;314:1079–86.

[9] Li W, Zhang S, Smith C. The molecular structure features-immune stimulatory activity of arabinoxylans derived from the pentosan fraction of wheat flour. J Cereal Sci 2015;62:81–6.

[10] Wcislo G, Szarlej-Wcislo K. Colorectal cancer prevention by wheat consumption: a three-valued logic—true, false, or otherwise? In: Watson RR, Preedy VR, Zibadi S, editors. Wheat and rice in disease prevention and health. London, Waltham, San Diego: Academic Press, Elsevier; 2014. p. 91–111.

[11] Sang S, Zhu Y. Bioactive phytochemicals in wheat bran for colon cancer prevention. In: Watson RR, Preedy VR, Zibadi S, editors. Wheat and rice in disease prevention and health. London, Waltham, San Diego: Academic Press, Elsevier; 2014. p. 121–9.

[12] Zhu Y, Conklin DR, Chen H, Wang L, Sang S. 5-Alk(en)ylresorcinols as the major active components in wheat bran inhibit human colon cancer cell growth. Bioorg Med Chem 2011;19:3973–82.

[13] Aune D, Keum N, Giovannucci E, Fadnes LT, Boffetta P, Greenwood DC, et al. Whole grain consumption and risk of cardiovascular disease, cancer, and all cause and cause specific mortality: systematic review and dose-response meta-analysis of prospective studies. BMJ 2016;353. i2716.

[14] Nieman KM, Rains TM, Dicklin MR, Maki KC. Wheat fiber in postprandial metabolic profile and health. In: Watson RR, Preedy VR, Zibadi S, editors. Wheat and rice in disease prevention and health. London, Waltham, San Diego: Academic Press, Elsevier; 2014. p. 59–66.

[15] Chanson-Rolle A, Meynier A, Aubin F, Lappi J, Poutanen K, Braesco V. Systematic review and meta-analysis of human studies to support a quantitative recommendation for whole grain intake in relation to type 2 diabetes. PloS One 2015;10:e0131377.

[16] Aune D, Norat T, Romundstad P, Vatten LJ. Whole grain and refined grain consumption and the risk of type 2 diabetes: a systematic review and dose-response meta-analysis of cohort studies. Eur J Epidemiol 2013;28:845–58.

[17] Svihus B. Nutritive and digestive effects of starch and fiber in whole wheat. In: Watson RR, Preedy VR, Zibadi S, editors. Wheat and rice in disease prevention and health. London, Waltham, San Diego: Academic Press, Elsevier; 2014. p. 81–7.

[18] FAO/WHO/UNO. Protein and amino acid requirements in human nutrition. WHO Press; 2007. p. 150.

[19] Ahmed A, Randhawa MA, Sajid MW. Bioavailability of calcium, iron, and zinc in whole wheat flour. In: Watson RR, Preedy VR, Zibadi S, editors. Wheat and rice in disease prevention and health. London, Waltham, San Diego: Academic Press, Elsevier; 2014. p. 67–80.

[20] Zhu Y, Sang S. Phytochemicals in whole grain wheat and their health-promoting effects. Mol Nutr Food Res 2017;61(7).

[21] Beta T, Nam S, Dexter JE, Sapirstein HD. Phenolic content and antioxidant activity of pearled wheat and roller-milled fractions. Cereal Chem 2005;82:390–3.

[22] Badawy D, El-Bassossy HM, Fahmy A, Azhar A. Aldolase reductase inhibitors zopolrestat and ferulic acid alleviate hypertension associated with diabetes: effect on vascular reactivity. Can J Physiol Pharmacol 2013;91:101–7.

[23] Drankhan K, Carter J, Madl R, Klopfenstein C, Padula F, Lu Y. Antitumor activity of wheat with high orthophenolic content. Nutr Cancer 2003;47:188–94.

[24] Piironen V, Lampi AM, Ekholm P, Salmenkallio-Martilla M, Liukkonen KH. Micronutrients and phytochemicals in wheat grain. In: Khan K, Shewry PR, editors. Wheat: chemistry and technology. 4th ed. St. Paul: AACC; 2009. p. 179–222.

[25] Corol DI, Ravel C, Raksegi M, Bedo Z, Charmet G, Beale MH, et al. Effects of genotype and environment on the contents of betaine, choline, and trigonelline in cereal grains. J Agric Food Chem 2012;60:5471–81.

[26] Price RK, Keaveney EM, Hamill LL, Wallace JM, Ward M, Ueland PM, et al. Consumption of wheat aleurone-rich foods increases fasting plasma betaine and modestly decreases fasting homocysteine and LDL-cholesterol in adults. J Nutr 2010;140:2153–7.

[27] Schwarz C, Stekovic S, Wirth M, Benson G, Royer P, Sigrist S, et al. Safety and tolerability of spermidine supplementation in mice and older adults with subjective cognitive decline. Agin s0 g 2018;8:19–33.

[28] Madeo F, Eisenberg T, Pietrocola F, Kroemer G. Spermidine in health and disease. Science 2018;359:eaan2788.

[29] De Angelis M, Rizzello CG, Alfonsi G, Arnault P, Cappelle S, Gobbetti M. Use of sourdough lactobacilli and oat fibre to decrease the glycaemic index of white wheat bread. Br J Nutr 2007;96:1196–205.

[30] Lopez HW, Krespine V, Guy C, Messager A, Demigne C, Ramesy C. Prolonged fermentation of whole wheat sourdough reduces phytate level and increases soluble magnesium. J Agric Food Chem 2001;49:2657–62.

[31] Koehler P, Hartmann G, Wieser H, Rychlik M. Changes in folates, dietary fiber, and proteins in wheat as affected by germination. J Agric Food Chem 2007;55:4678–83.

[32] Zilic S, Basic Z, Sukalovic V, Filipovic M. Can the sprouting process applied to wheat improve the contents of vitamins and phenolic compounds and antioxidant capacity of the flour? Int J Food Sci Technol 2014;49:1040–7.

[33] Hemery Y, Rouau S, Lullien-Pellerin V, Barron C, Abecassis J. Dry processes to develop wheat fractions and products with enhanced nutritional quality. J Cereal Sci 2007;46:327–47.

Chapter 7

Principles of wheat hypersensitivities

Wheat products are staple foods all over the world. Wheat provides a number of valuable nutrients such as proteins, dietary fiber, vitamins, and minerals and supplies up to 20% of the energy intake of the global population. In the last ten years however, wheat has received much negative attention. Several pseudoscientific books such as "Wheat Belly" (W. Davis) [1], "Grain Brain" (D. Perlmutter) [2], and "Gluten, comment le blé moderne nous intoxique" (J. Venesson) [3] and numerous media reports fueled the overall estimation that wheat consumption makes people sick, stupid, fat, and addicted. Most reports against wheat intermingle scientific evidence with hypothetical, controversial, and partly wrong statements [4–6]. Altogether, there is no scientific indication for eliminating wheat from the diet for the vast majority (90–95%) of the population. Only the remaining 5–10% have to avoid wheat products, because they suffer from wheat hypersensitivities. The spectrum and classification of wheat-related disorders are shown in Fig. 7.1. Most of these hypersensitivities are mediated by the gastrointestinal tract after oral wheat intake. Examples are celiac disease (CD), dietary wheat allergy (WA), wheat-dependent exercise-induced anaphylaxis (WDEIA), and non-celiac gluten sensitivity (NCGS) [7,8]. Further hypersensitivities to wheat can be caused by allergen contact with the respiratory tract (respiratory allergy) or the skin (contact urticaria). The following sections present an overview on the principles of wheat hypersensitivities including human immunity, precipitating proteins, digestive processes, intestinal permeability, and blood-brain barrier.

7.1 Human immunity

Right from the beginning of their existence humans have been confronted with numerous enemies, not only wild animals, insects, worms, and other vermin but also microorganisms such as pathogenic fungi, bacteria, and viruses. To protect their bodies humans have developed complex systems to defend both the outer surface (skin) and the body openings (nose, mouth, lungs, stomach, intestine, and vagina) against harmful invaders. On the first line, skin in the form of multiple layers of ectodermal tissue prevents pathogens from entering the body and mucous membranes are covered with a mucous layer that traps particles and microorganisms.

Wheat—An Exceptional Crop. https://doi.org/10.1016/B978-0-12-821715-3.00007-1

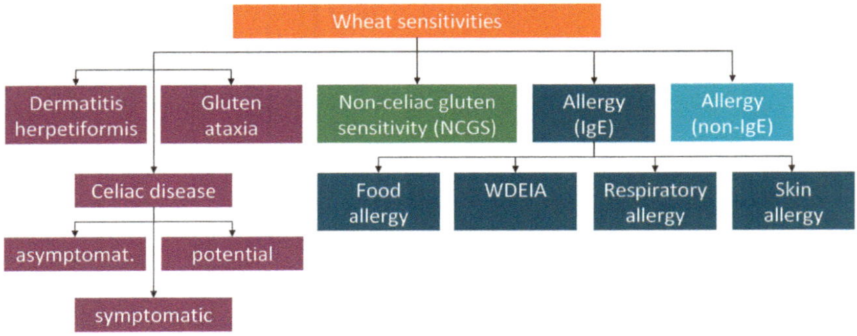

FIG. 7.1 Spectrum and classification of wheat-related disorders. *WDEIA*, wheat-dependent exercise-induced anaphylaxis.

To protect body openings like the respiratory tract coughing and sneezing as well as secreted mucus remove pathogens. Chemical barriers also protect against pathogens. For example, antimicrobial compounds are secreted by the skin and the respiratory tract and gastric acid and peptidases serve as chemical defenses in the stomach. Last but not least commensal microbiota in the intestinal tract act as biological barrier by competing with pathogenic microorganisms and eliminating competitor strains. Even if these outer barriers are breached, skin epidermis and the mucosa of the respiratory and gastrointestinal tracts have been equipped with an immune-reactive tissue containing numerous special components that identify and destroy dangerous agents. Foreign proteins and glycoproteins have been selected by the immune system as targets for its defensive activity.

Human immunity has developed extremely complex systems that can be roughly classified into the innate and the adaptive immunities and the allergies. They include numerous humoral and cellular components. The fast innate response has low specificity and no memory, i.e., it does not confer long-lasting immunity. White blood cells (leukocytes) are important cellular contributors to the innate system. Examples for these cell types are T-lymphocytes, phagocytes (macrophages, dendritic cells, and neutrophils), natural killer cells, mast cells, basophils, and eosinophils. These cells identify and kill pathogens either through direct contact or by engulfing. Moreover, some of these cells are also important for the activation of the adaptive immune system. The major humoral component of the innate response is the so-called complement, a complex glycoprotein system of the blood. It contains over twenty different proteins that bind to the surface of pathogens directly or to antibodies that have attached to the pathogens and initiate a rapid killing process. In addition, cytokines (e.g., interleukins) are released by activated immune cells.

The slow adaptive immune system permits a delayed but highly specific response to a pathogen and is endowed with long-term memory. The immunological memory is maintained by memory cells that help eliminate those pathogens that infect the body more than once. This process is well known as basis of vaccination. The major cellular components of the adaptive system are T- and B-lymphocytes. Both lymphocyte types carry receptor proteins on their surfaces that bind specific antigens resulting in the activation of the cells. T-lymphocytes release cytokines and chemokines (e.g., interferons, interleukins, tumor necrosis factor, and matrix metalloproteinase) and B-lymphocytes produce antigen-specific immunoglobulins (Ig) like IgA, IgG, and IgE antibodies. Both T- and B-cell-mediated pathways destroy pathogens. When T- and B-lymphocytes are activated and begin to replicate, some of them become long-life memory cells.

Allergic responses are roughly divided into IgE-mediated and non-IgE-mediated reactions; mixed allergies are of less importance. The mechanism of IgE-mediated immune responses includes two phases [9]: (1) sensitization upon first exposure to an allergen and (2) immune reaction upon reexposure to an allergen. B cells that circulate in the blood and lymphatic system bind the allergen on their surface via B-cell receptors. Once naive B cells encounter their cognate allergen and receive help from T-helper cells, they become activated and differentiate further into many fully functional "effector B lymphocytes" (sensitization). Again exposed to the allergen they produce and secrete large amounts of IgE antibodies (immune reaction). The constant (Fc) region of IgE antibodies is bound to FcεR receptors on the surface of mast cells and basophils that are filled with granules containing a variety of reactive substances (Fig. 7.2). When an allergen is bound to the allergen-sensitive antibodies, the resulting complex activates mast cells and basophils. A characteristic feature of the immune reaction is that allergens require two IgE-binding epitopes to facilitate the cross-linking of two neighboring IgE antibodies necessary for activation (Fig. 7.2). When activated, mast cells and basophils start degranulation and release histamine, heparin, and other inflammatory mediators (e.g., cytokines, leukotrienes, and prostaglandin-D2) followed by inflammatory reactions manifested in allergic symptoms.

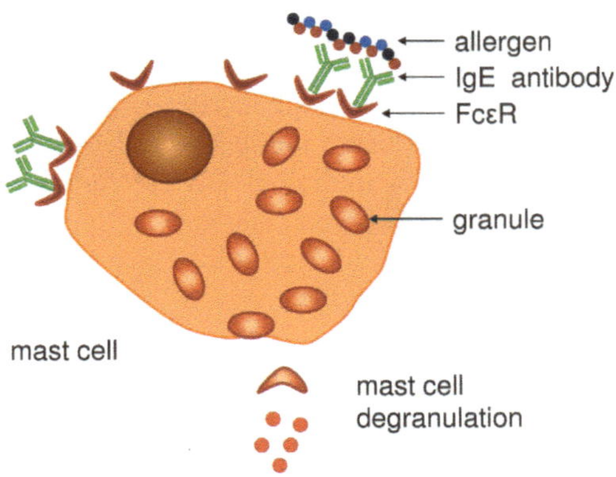

FIG. 7.2 Schematic picture of IgE-mediated basophil activation.

Much less is known about non-IgE-mediated (cellular) allergies that are poorly defined both clinically and scientifically [10]. They occur less frequently than IgE-mediated allergies and primarily affect the gastrointestinal mucosa. The immune reaction does not require prior exposure to the allergen. Unlike IgE-mediated allergies non-IgE-mediated allergies are rarely life-threatening. The delayed onset of symptoms (4–28 h after food ingestion) and the variability in presentation at different sites are typical, which makes the clinical association between the offending food and symptoms difficult. The pathomechanism of non-IgE-mediated allergy is not fully understood and may include contributions from T cells, B cells, macrophages, mast cells, and eosinophils. Specifically, the activation of both mast cells and eosinophils and the release of their mediators (e.g., tumor necrosis factor-γ and eosinophil cationic protein) are important components. Some of these mediators have a synergistic effect on IgE-mediated mast cell activation. Altogether, further studies are needed for a better understanding of non-IgE-mediated allergies, which will lead to the development of diagnostic tests and treatments [11].

In the course of evolution, the immune system has learnt to differentiate between the body's own proteins and food proteins, both necessary for life, on the one hand and foreign proteins from harmful organisms on the other hand. In simple terms, the immune system is tolerant to welcome proteins and intolerant to unwelcome proteins. The acquisition of tolerance is a complex process and far from being understood. In principle, tolerance consists of two phases within the immune response: the maintenance of homeostasis (maintaining internal conditions) and the suppression of immune responses by antigen-specific regulatory T-lymphocytes. However, the immune system of some people cannot differentiate harmful agents from certain harmless agents probably resulting from two different phenomena. The first could be that tolerance never developed and the second that prior tolerance has been lost at some point in life. Lost tolerance to the body's own proteins results in autoimmune diseases and lost tolerance to food proteins results in food hypersensitivities.

Until the Neolithic age, humans were not exposed to wheat for several hundred thousand years. Only about 10,000 years ago, farming and consumption of wheat together with barley started in the Middle East, gradually crossed Europe and Asia, reached the American continent and is now the basis of staple foods worldwide. However, it appears that a number of individuals did not adapt to this "new" food item and are immunologically intolerant to specific wheat proteins at the skin, the respiratory tract, and especially in the small intestine (body`s largest immune organ and the largest interface with external surroundings) resulting in wheat hypersensitivities.

7.2 Wheat proteins

Grain proteins are the unique precipitating factors of wheat hypersensitivities. They represent the second predominant class of wheat constituents after carbohydrates (see Section 2.2). The protein content of dry wheat grains varies in a wide range between 9% and 14% in the vast majority of cultivars. Around 70 million metric tons of proteins are produced annually in the grains of wheat plants. The protein fraction of grains contains several hundred different components. According to divergent solubilities, they have been classically grouped into albumins soluble in water, globulins soluble in diluted salt solutions, gliadins soluble in aqueous alcohols, glutenins only soluble in aqueous alcohols after the cleavage of disulfide bonds, and residual proteins insoluble in conventional extraction solvents [12] (see Section 2.2). Wheat gluten, mainly consisting of gliadins and glutenins, is the term for the rubber-like proteinaceous mass that remains, when wheat dough is washed with water to remove soluble constituents and starch granules.

An alternative classification of wheat proteins is based on their different functions:

1. Storage (gluten) proteins, comprising gliadins and glutenins, provide the embryo with nitrogen and amino acids during the first stage of germination (70–80% of total grain proteins).
2. Metabolic (e.g., enzymes) and protective proteins (e.g., enzyme inhibitors) are predominantly present in the albumin and globulin fractions (15–20%).
3. Structural proteins such as membrane proteins and lipoproteins correspond to the residual proteins (≈5%).

Storage proteins and a considerable portion of albumins and globulins are involved in different wheat hypersensitivities [13]. A common feature of most of these immunogenic proteins is the resistance to gastrointestinal digestion, so that protein fragments (peptides) survive in an appropriate size necessary for immune responses. CD and related disorders, WDEIA, and NCGS are triggered by gluten proteins, whereas IgE-mediated disorders (WA and contact urticaria) are predominantly caused by metabolic and protective proteins [14].

7.3 Digestion of proteins

Wheat-mediated hypersensitivities, caused by exposure through inhalation or skin contact (respiratory allergy and contact urticaria), are triggered by native wheat proteins. In contrast, wheat proteins that induce hypersensitivities after oral intake (dietary WA, WDEIA, CD, and NCGS) are degraded by gastrointestinal enzymes and in fact, peptides resulting from digestion are the precipitating factors. Amino acid sequences of proteins and specificities of digestive enzymes are key factors for the formation of immunogenic peptides.

After ingestion, chewing, insalivation, and swallowing, food constituents reach the digestive tract. Its oroanal length amounts to around five meters, two-third of which refers to the small intestine. The inner diameter of the small intestine is ≈2.5 cm and that of the large intestine ≈4.8 cm. The mucosa of the small intestine is enlarged around 1.6 times by plicae circulares (intestinal wrinkles). Morphometric data, obtained by light and electron microscopy of biopsies, demonstrated that villi and microvilli together amplify the small intestinal surface area by 60–120 times [15]. The mean total mucosal surface of the digestive tract interior amounts to around 32 m^2, of which about only 2 m^2 refers to the large intestine. Thus the total area of the human intestinal mucosa is not in the order of a tennis lawn as frequently mentioned by textbooks but is in the order of half a badminton court.

After intake, food proteins are usually broken down into small fragments (oligopeptides and amino acids) by specific digestive enzymes beginning in the oral cavity by means of saliva and microorganisms and then in the gastrointestinal tract. Pepsin is active in the stomach, whereas trypsin and chymotrypsin are released by the pancreas and activated in the small intestine. Enzymes, bound to the brush border of the intestinal mucosa such as amino peptidases and dipeptidyl peptidases care for the rest of digestion. In consequence, most food proteins are almost completely cleaved into amino acids, di- and tripeptides, before they pass the intestinal epithelium. The size of these degraded compounds is too small for any induction of innate or adaptive immune responses.

However, a number of food proteins are resistant to extensive digestive fragmentation. Thus relatively long protein fragments (polypeptides) reach the intestinal mucosa and either pass the epithelial barrier (see Section 7.4) or are metabolized by intestinal bacteria. Among these resistant proteins are numerous wheat proteins, in particular gluten proteins and protective proteins (e.g., enzyme inhibitors). Due to the limited digestibility of gluten proteins, the earliest theory of CD pathogenesis was based on the "missing enzyme hypothesis", i.e., an unknown digestive enzyme is missing in the intestine of patients with CD. Today, it is commonly accepted that there is no difference between healthy people and patients, suffering from wheat hypersensitivities, in their non-ability to digest gluten proteins. Numerous studies demonstrated that the high frequency of the amino acid proline in gluten proteins makes these proteins resistant to digestion and moreover, heat processes like baking even reduce the digestibility of gluten proteins. In consequence, protein fragments with a length of up to 100 amino acid might reach the intestinal mucosa [16]. Partly, they permeate through the small intestinal enterocytes, partly they arrive in the large intestine, where they are available for bacterial metabolism.

In summary, many of the wheat proteins triggering hypersensitivities are poorly digested by gastrointestinal enzymes, but this fact is not sufficient for hypersensitivities, because there is no fundamental difference between healthy and affected individuals. The first pathogenic step is the passage of the culprit protein fragments through the intestinal epithelium.

7.4 Intestinal permeability

The intestinal tract is not only responsible for digestion and absorption of nutrients and electrolytes and for water homeostasis, but it has also an important barrier function [17,18]. Its functioning is dependent on epithelial layer integrity, intestinal microbiota homeostasis, and the gut-associated immune-active tissue. The single layer of epithelial cells normally permits the absorption of nutrients and water and restricts the passage of harmful pathogens (e.g., luminal noxious molecules). The understanding of the control mechanisms that regulate barrier functions has increased significantly over the last decades. Today, a disturbed intestinal barrier function ("leaky gut") is considered a key factor in the development of intestinal inflammation and plays an important role in both the pathogenesis and perpetuation of various intestinal diseases such as inflammatory bowel diseases and food hypersensitivities. It is not fully understood, how harmful agents permeate the enterocyte layer and reach the immune-active lamina propria, the thin layer of connective tissue located below the enterocytes. There is however evidence that this occurs by two routes: the paracellular and the transcellular pathways. The epithelial passage of the paracellular pathway occurs between two enterocytes and that of the transcellular pathway through the enterocytes. The most striking difference between both routes is that agents, crossing through the paracellular pathway, reach the submucosa unmodified, whereas agents may be metabolized through the transcellular pathway. Moreover, some ulcer forms can break the epithelial cell lining and permit direct access of luminal contents to lamina propria. For example, enteric virus infection or medications like non-steroidal antiinflammatory drugs frequently cause small-intestine ulceration.

Among the most important structures of the paracellular route are the epithelial pore-forming "tight junctions" [19]. They are composed of about 30 proteins that provide continuous belt-like cell-to-cell junctions and regulate paracellular permeability (Fig. 7.3). Major protein types are claudins 1–12, occludin, junctional adhesion molecule A, tricellulin, and zonulin 1. The function

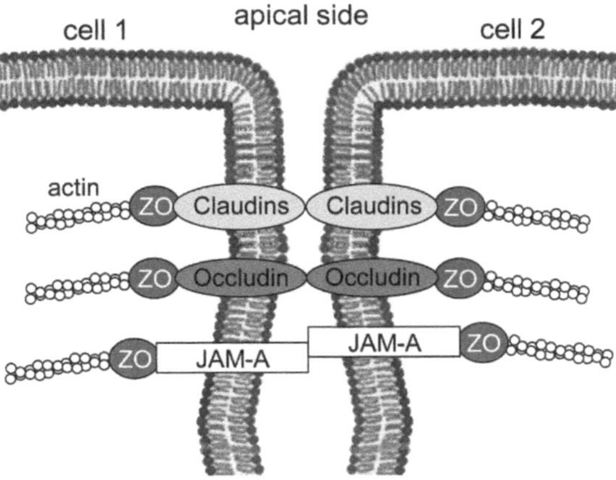

FIG. 7.3 Schematic representation of the basis structural components of tight junctions (ZO, zonulin; JAMA-A, junctional adhesion molecule A).

of these proteins can be subdivided into "tightening" that strengthens barrier properties and "leaking" that selectively mediates permeability. Zonulin-1, an analog of zonula occludens toxin of *Vibrio cholera*, has been described as "the only physiological modulator of intercellular tight junctions that is involved in trafficking of macromolecules and therefore in tolerance/immune response balance" [20]. The increased formation of leaking zonulin-1 has been proposed to be one of the initial factors for the pathogenesis of CD.

The transcellular pathway is characterized by three components: endocytosis, transcytosis, and exocytosis. Endocytosis means an active transport of molecules (e.g., peptides) through the apical membrane of enterocytes. The entrance into the cells is enabled by several mechanisms such as clathrin-dependent or caveolae-mediated endocytosis or by direct interaction with the membrane [21]. Transcytosis includes the transport of molecules captured in vesicles from the apical to the basolateral membrane passing different endocytic compartments (early endosomes, lysosomes, and late endosomes) of the cells. Lysosomes contain numerous hydrolytic enzymes that degrade peptides, carbohydrates, lipids, and nucleic acids. The final step of the transcellular pathway is the active transportation of molecules out of the cells (exocytosis). The transport is mediated by vesicles that carry the molecules through the basolateral membrane and secret them into the extracellular environment. During transcellular epithelial transport, the agents are almost totally degraded by the lysosomal system of the enterocytes. Thus an excessive activation of the submucosal immune system is likely avoided. The products of protein digestion enter into the bloodstream primarily as free amino acids.

In healthy individuals, the transcellular pathway is apparently predominant. Increased paracellular permeability might be the consequence of mucosa damaging factors (e.g., inflammation) and the activation of leaking tight junction proteins (e.g., zonulin-1) [19,22]. External factors that provoke an increased permeability are known to be non-steroidal antiinflammatory drugs (e.g., acetylsalicylic acid), excessive alcohol consumption, psychological stress, and high-intensity exercise [23,24]. Recent studies have begun to address the role of the intestinal microbiota in influencing intestinal permeability. For example, beneficial gut microbiota such as *Lactobacillus* and *Bifidobacterium* species may assist in maintaining intestinal barrier function [25].

The lactulose/mannitol (LAMA) test has been used as gold standard for determining the permeability of the intestinal epithelium. The test consists of the oral administration of the sugar alcohol mannitol (molecular mass: 182) and the disaccharide lactulose (molecular mass: 342) followed by subsequent measurement of these substances in the urine. Mannitol is believed to cross the epithelium of an intact intestinal mucosa, whereas lactulose crosses only, when the mucosa is injured. The ratio of lactulose to mannitol, measured in the urine, is taken as a parameter for intestinal permeability. Though useful in clinical studies on the overall condition of the intestinal tract, the LAMA test does not necessarily correlate with the permeability for macromolecules like bacterial toxins or food allergens [26].

7.5 Blood-brain barrier

The blood-brain barrier (BBB) separates the circulating blood from the brain extracellular fluid in the central nervous system in a highly selective way. The BBB is present along all capillaries at the interface between blood and brain, where the endothelial cells are stitched together by tight junctions. These are composed of a number of transmembrane proteins that are anchored into the endothelial cells and control the passage of substances. For example, the passage of water, oxygen, carbon dioxide, glucose, amino acids, and hormones is admitted, whereas the passage of potential neurotoxins, for instance, is restricted.

As described in Sections 8.7.6, 8.7.7, and 10.2, both CD and NCGS cause neurological and psychiatric complaints. It has been hypothesized that gluten peptides, having permeated the gut barrier and circulating in the blood stream, are able to cross the BBB and initiate neuroinflammation or affect neurotransmission (communication between two neurons) within the nervous system and stimulate the endogenous opiate system of the brain [27]. In the 1970s, peptides with opioid activity, similar to that of morphine and other opioids, have been identified in the brain. The source of opioid peptides was found in peptic digests of wheat gluten and milk casein [28]. These peptides are called exorphins because of their exogenous origin and morphine-like activity. The so-called gliadorphins and casomorphins have been proposed to be produced in the stomach, survive in the intestinal tract, be absorbed by the intestinal mucosa, reach the bloodstream, cross the BBB, and interact as opiates with opiate receptors of the brain [29]. Gliadorphins are small peptides with amino acid sequences like YPQPQPF, GYYPT, YGGW, YGGWL, and YPISL (one-letter code for amino acids see Table 2.3). Exorphins have been proposed to have negative effects on attention, learning, social interactions, and brain maturation and promote the etiology of schizophrenia and autism [30,31] (see Section 8.7.7). Selected strains of *Bifidobacterium longum* and *Bifidobacterium bifidum* showed high degradative activity in food-derived opioid peptides and can be used as probiotic microorganisms to eliminate these peptides [32].

References

[1] Davis W. Wheat belly. New York: Rodale Inc; 2011.

[2] Perlmutter D. Grain brain. New York, Boston, London: Little, Brown and Company; 2013.

[3] Venesson J. Gluten, comment le ble moderne nous intoxique. Vergeze: Thierry Souccar Editions; 2013.

[4] Jones J. Wheat belly—an analysis of selected statements and basic theses from the book. Cereal Foods World 2012;57:177–89.

[5] Brouns FJ, van Buul VJ, Shewry PR. Does wheat make us fat and thick? J Cereal Sci 2013;58:209–15.

[6] Gaesser GA, Angadi SS. Navigating the gluten-free boom. JAAPA 2015;28:1–7.

[7] Sapone A, Bai JC, Ciacci C, Dolinsek J, Green PH, Hadjivassiliou M, et al. Spectrum of gluten-related disorders: consensus on new nomenclature and classification. BMC Med 2012;10:13.

[8] Ludvigsson JF, Leffler DA, Bai JA, Biagi F, Fasano A, Green PH, et al. The Oslo definitions for coeliac disease and related terms. Gut 2013;62:43–52.

[9] Hawrylowicz CM, O'Garra A. Potential role of interleukin-10-secreting regulatory T cells in allergy and asthma. Nature Rev Immunol 2005;5:271–83.

[10] Yu Y, Blokhuis BR, Garssen J, Redegeld FA. Non-IgE mediated mast cell activation. Eur J Pharmacol 2016;778:33–43.

[11] Caubet JC, Szajewska H, Shamir R, Nowak-Wegrzyn A. Non-IgE-mediated gastrointestinal food allergies in children. Pediatr Allergy Immunol 2017;28:6–17.

[12] Wieser H, Koehler P, Konitzer K. Celiac disease and gluten—multidisciplinary challenges and opportunities. London, Waltham, San Diego: Academic Press, Elsevier; 2014; 99–119.

[13] Scherf KA. Immunoreactive cereal proteins in wheat allergy, non-celiac gluten/wheat sensitivity (NCGS) and celiac disease. Curr Opin Food Sci 2019;25:35–41.

[14] Tatham AS, Shewry PR. Allergens in wheat and related cereals. Clin Exp Allergy 2008;38:1712–26.

[15] Helander HF, Fändriks L. Surface area of the digestive tract—revisited. Scand J Gastroenterol 2014;49:681–9.

[16] Wieser H, Koehler P, Konitzer K. Celiac disease and gluten—multidisciplinary challenges and opportunities. London, Waltham, San Diego: Academic Press, Elsevier; 2014, p. 48.

[17] Schumann M, Siegmund B, Schulzke JD, Fromm M. Celiac disease: role of the epithelial barrier. Cell Mol Gastroenterol Hepatol 2017;3:150–62.

[18] Odenwald MA, Turner JR. The intestinal epithelial barrier: a therapeutic target? Nat Rev Gastroenterol Hepatol 2017;14:9–21.

[19] Suzuki T. Regulation of intestinal epithelial permeability by tight junctions. Cell Mol Life Sci 2013;70:631–59.

[20] Fasano A. Zonulin and its regulation of intestinal barrier function: the biological door to inflammation, autoimmunity, and cancer. Physiol Rev 2011;91:151–75.

[21] Barone MV, Zimmer KP. Endocytose and transcytose of gliadin peptides. Mol Cell Pediatrics 2016;3:8.

[22] Fasano A. Intestinal permeability and its regulation by zonulin: diagnostic and therapeutic implications. Clin Gastroenterol Hepatol 2012;10:1096–100.

[23] Bjarnason I, Scarpignato C, Holmgren E, Olszewski M, Rainsford KD, Lanas A. Mechanisms of damage to the gastrointestinal tract from nonsteroidal anti-inflammatory drugs. Gastroenterology 2018;154:500–14.

[24] Karhu E, Forsgard RA, Alanko L, Alfthan H, Pussinen P, Hämäläinen E, et al. Exercise and gastrointestinal symptoms: running-induced changes in intestinal permeability and markers of gastrointestinal function in asymptomatic and symptomatic runners. Eur J Appl Physiol 2017;117:2519–26.

[25] Sultana R, McBain AJ, O'Neill CA. Strain-dependent augmentation of tight junction barrier function in human primary epidermal keratinocytes by *Lactobacillus* and *Bifidobacterium* lysates. Appl Environ Microbiol 2013;79:4887–94.

[26] Vojdani A. For the assessment of intestinal permeability, size matters. Altern Ther Health Med 2013;19:12–24.

[27] Catassi C, Elli L, Bonaz B, Bouma G, Carroccio A, Castillejo G, et al. Diagnosis of non-celiac gluten sensitivity (NCGS): the Salerno experts' criteria. Nutrients 2015;7:4966–77.

[28] Zioudrou C, Streaty RA, Klee WA. Opioid peptides derived from food proteins. The exorphins. J Biol Chem 1979;254:2446–9.

[29] Pruimboom L, de Punder K. The opioid effects of gluten exorphins: asymptomatic celiac disease. J Health Popul Nutr 2015;33:24.

[30] Knivsberg AM, Reichelt KL, Høien T, Nodland M. A randomised, controlled study of dietary intervention in autistic syndromes. Nutr Neurosci 2002;5:551–61.
[31] Cade R, Privette M, Fregly M, Rowland M, Sun Z, Zele V, et al. Autism and schizophrenia: intestinal disorders. Nutr Neurosci 2000;3:57–72.
[32] Sakurai T, Yamada A, Hashikura N, Odamaki T, Xiao JZ. Degradation of food-derived opioid peptides by bifidobacteria. Benef Microbes 2018;9:675–82.

Chapter 8

Celiac disease

Celiac disease (CD), also known as celiac sprue, nontropical sprue, and gluten-sensitive enteropathy, is one of the most frequent food hypersensitivities worldwide. It can be defined as an inflammatory disease of the upper small intestine in genetically predisposed individuals triggered by the storage (gluten) proteins of wheat, rye, barley, and possibly oats. Previously, CD was considered to be a rare childhood disorder, but it is actually a frequent condition that affects around 1% of the Western population and may develop at any age. It is now increasingly found in areas of the developing world such as North Africa, Middle East, China, and India. The clinical features are manifold and comprise both intestinal and extraintestinal symptoms affecting nearly any part of the body. The immunological mechanisms of CD are complex and involve adaptive and innate responses. The diagnosis is complicated and therefore not always straightforward and frequently delayed. A strict gluten-free diet (GFD) is currently the only effective treatment.

The interest in CD among physicians and the general population has strongly increased during the last two decades. Intense multidisciplinary studies have contributed to a substantial progress in understanding CD and now provide detailed insights into the different features of the disorder. The present chapter covers the most important aspects of CD including epidemiology, genetics, clinical manifestations, triggering factors and cofactors, pathomechanisms, and diagnosis. The treatment and follow-up of CD patients is described in Chapter 12.

8.1 History

The earliest description of an intestinal disorder similar to CD was documented by Aretaeus of Cappadocia, a physician practicing in Rome and Alexandria during the second century BC. He wrote: "If diarrhea does not proceed from a slight cause of only one or two days duration and if in addition the general system be debilitated by atrophy of the body, the celiac sprue of chronic nature is formed." He called these patients "koiliakos" according to the Greek term "koilia" that simply means abdomen. The remains of a young lady, found in the archaeological site of Cosa (Italy) and dating from the first century BC, impressively confirmed the existence of CD in antiquity. She was characterized by clinical signs of malabsorption typical of CD such as short height, osteoporosis, dental enamel hypoplasia, and anemia [1]. Moreover, she carried the gene HLA-DQ2.5 that is associated with the highest risk of CD.

Wheat—An Exceptional Crop. https://doi.org/10.1016/B978-0-12-821715-3.00008-3

The first accurate clinical description of CD was presented at the end of the 19th century by Samuel Gee, an English physician and pediatrician. He used the term "celiac affection" and defined the disease as "a kind of chronic indigestion that is met within persons of all ages, yet especially occurs in children between one and five years old". Even without knowing the precipitating factor (gluten), he already recommended a dietary treatment: "If the patient can be cured at all, it must be by means of the diet". After that, different dietary treatments were recommended during the subsequent decades. For example, the exclusion of all sources of carbohydrate-containing foods (bread, cereals, and potatoes) except bananas resulted in an improvement of CD symptoms. Nevertheless, many CD patients were affected by serious clinical features and 15–20% of children with CD died of the so-called celiac crisis due to chronic diarrhea, electrolyte and metabolic abnormalities, and severe weight loss.

It was not until the year 1950 that Willem K. Dicke (Fig. 8.1), a Dutch pediatrician, identified wheat as the triggering factor of CD, a milestone in the history of CD. Dicke observed a decline of CD in The Netherlands, when there was a

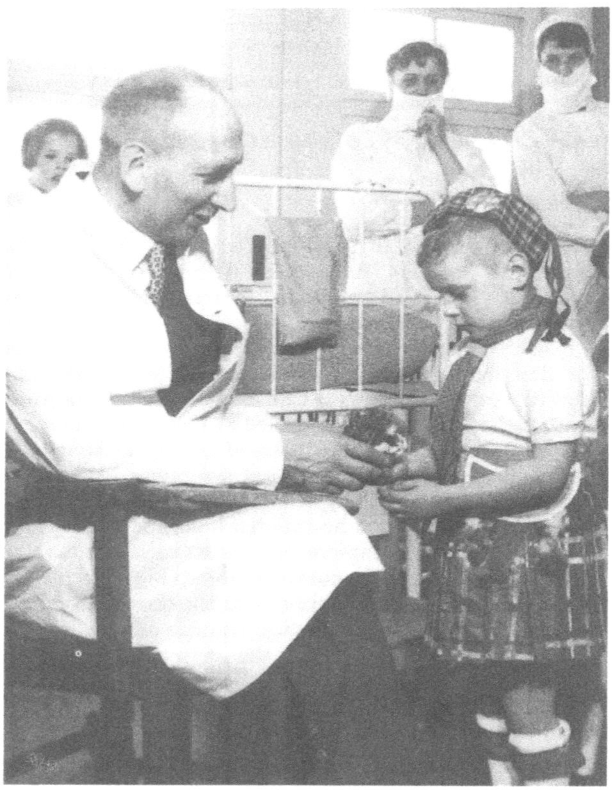

FIG. 8.1 Willem Karel Dicke (1905–1962).

scarcity of wheat and bread towards the end of World War II and an increase of CD afterwards, when wheat products were available again. He concluded that the ingestion of wheat was responsible for the symptoms of CD in affected children. Subsequent detailed studies by Dicke's group revealed that wheat gluten bore the toxic components and that rye and barley were also harmful to CD patients. Since that time, all cereal proteins that trigger CD were called "gluten" and a GFD was successfully introduced as the conventional treatment of CD. Shortly afterwards, John W. Paulley was the first to demonstrate the abnormalities of the mucosa tissue obtained from the small intestine of the CD patients. This finding was confirmed by the implementation of the peroral biopsy of the intestinal mucosa by Margot Shiner in 1956 and William H. Crosby in 1957, a procedure that has been the gold standard of CD diagnosis to this day.

8.2 Epidemiology

Formerly, CD was thought to be a rare disease with a prevalence from around 1:500 to 1:1000 affecting almost exclusively people of European origin. Since the 1980s however, modern serological screening, followed by small intestinal biopsy, has revealed a prevalence between 1:60 and 1:200 in most Western populations, which corresponds to an average prevalence of around 1% [2] (Table 8.1).

Recent epidemiological studies have indicated that CD is common not only in Western countries but also in North Africa, the Middle East, and South and East Asia. A systematic review and meta-analysis, including 96 international articles, showed a pooled global prevalence of 1.7% based on positive serology and 0.7% based on biopsy-confirmed CD [4]. Only native African and Japanese people appear to be hardly affected by CD. The highest rate of CD (≈6%), described in the world so far, was found among the Sarahawi refugees of Berber-Arabic origin in North Africa [5]. Due to the international food relief program during the Civil War in the late 1970s, the nomadic population was exposed to high amounts of wheat for the first time and CD became a major concern.

The prevalence in developing countries may be underestimated because of lack of awareness and low suspicion of the disease. For example, it is estimated that as many as 10 million people in India and perhaps a similar number of individuals in China currently have undiagnosed CD. In summary, CD is one of the most frequent food hypersensitivities worldwide. It can occur at any age and is more common in women than in men (ratio between 2:1 and 3:1). The prevalence among first-degree relatives is clearly elevated (10–20%) and the rate in monozygotic twins is approximately 80%.

A significant increase of CD prevalence has been observed during the last decades and this appears to be a true rise, not only due to increased awareness and modern diagnostic techniques [6]. This fact was impressively demonstrated with the help of frozen blood serum samples collected from more than 9000 male recruits in the US Air Force during the period 1948–1952. The samples

TABLE 8.1 Prevalence of CD in selected Western countries [2].

Country	Prevalence (%)
Argentina	0.6
Australia	0.4
Brazil	0.5
Finland	1.7
Germany[a]	0.2
Ireland	0.8
Italy	0.9
Netherlands	0.5
New Zealand	1.2
Portugal	0.7
Spain	0.8
Sweden	1.6
United Kingdom	1.1
USA	0.6

[a] *Later studies on children and adolescents found a prevalence of 0.9% [3].*

were analyzed later by using modern CD marker tests (transglutaminase and endomysium serum antibodies) showing positive results for 0.2% of the probands [7]. Starting in 2006, a second group, consisting of 5500 men with birth years similar to the recruits (mean age 70 years) and a third group, consisting of 7200 young men (mean age 37 years), were selected for serum sampling. Corresponding serum tests showed 0.8% and 0.9% positive results, respectively. Thus the study indicated a fourfold increase of CD markers over fifty years. Similar Finnish investigations with around 7000 male and female persons revealed an increase from 1.05% in period 1978–1980 to 1.99% in period 2000–2001, a nearly twofold increase over 20 years [8]. Population-based studies in North America and Denmark confirmed a continued increase in incidence and prevalence of CD during the past three decades [9,10].

What are the reasons for the notable increase of CD prevalence? Major causes that are being discussed are protein compositions of wheat altered by breeding and modern processing, increased gluten consumption, and changed environmental conditions. A popular theory proposes that modern wheat breeding, specifically the development of semidwarf phenotypes (see Section 1.3), may have led to a higher content of gluten in the grains and higher wheat sensitivity.

Several profound scientific studies did not support this claim [11]. For example, investigations on wheat landraces that were not subjected to breeding practices and a survey of data from the 20th and 21st centuries for the USA demonstrated that breeding did not lead to an overall increase in protein and gluten contents [12]. These findings were confirmed by studies on 60 German winter wheat varieties, admitted from 1891 to 2010 and cultivated at identical conditions in the period 2014/2015 [13]. The quantitative protein analysis revealed that neither crude grain protein nor gluten proteins increased in this long period (Table 8.2). In contrast, there is even a tendency of decreasing gluten content in the period from 1891 to 2010. These data clearly reflect the breeding aim of enhancing grain yield, which usually results in higher starch and lower protein contents [12]. Moreover, the study showed that the ratio of gliadins to glutenins and thus the amount of gliadins decreased. Because gliadins are considered to be the major precipitating factor of CD, the lower contents of gliadins refute the supposed enhanced CD toxicity of modern wheats. The quantitation of immunogenic and toxic gluten peptides additionally indicated no significantly difference between old and modern wheat cultivars [14,15]. A number of landraces of common wheat that were not subjected to breeding practices presented even

TABLE 8.2 Average protein composition of German wheat varieties approved from 1891 to 2010 and harvested in 2015 [13].[a,b]

Year of admission	CP (mg/g)	ALGL (mg/g)	GLUTEN (mg/g)	GLIA/GLUT
1891–1900	101.0	17.2	64.3	2.46
1901–1910	106.7	17.9	71.1	3.38
1911–1920	98.8	17.5	64.4	2.53
1921–1930	107.8	18.4	70.8	2.93
1931–1940	115.6	18.6	75.6	2.74
1941–1950	120.7	18.6	89.5	2.62
1951–1960	100.3	16.2	66.8	2.18
1961–1970	110.2	16.8	74.8	2.54
1971–1980	116.1	16.8	71.5	2.97
1981–1990	106.1	15.8	69.4	1.89
1991–2000	81.5	15.8	45.7	1.67
2001–2010	85.5	16.0	49.5	2.07

[a] *Mean value of five varieties.*
[b] *ALGL, albumins + globulins; CP, crude protein; GLIA, gliadins; GLUT, glutenins; GLUTEN, gliadins + glutenins.*

higher amounts of CD-toxic epitopes when compared to modern varieties [16]. The characterization of gluten proteins and CD-related immunogenic epitopes in different *Triticeae* cereals (wheat, rye, and barley) confirmed that domestication and breeding contributed to decrease the content of gluten [17].

Apart from increasing grain yield, the enhancement of plant resistance against diseases and parasites has been one of the most important breeding aims. Therefore protective proteins like amylase/trypsin inhibitors that are believed to activate the innate immunity [18] may have increased by breeding during the past decades. However, a scientific proof of an increase of immunogenic wheat components is still missing. Likewise, there is no scientific evidence that modern food processing (e.g., short and nonacidic fermentation of doughs or addition of vital gluten) resulted in a higher gluten toxicity. Epidemiological data demonstrate that countries, in which long-fermented sourdough breads are common (e.g., Finland and Sweden), have even higher CD prevalences (Table 8.1) than countries (e.g., Italy and Spain), where almost exclusively short-fermented yeast-leavened breads are consumed.

It is true that the worldwide production of wheat and thus the daily intake of gluten may have increased in parts of the world. Therefore a "gluten overload" or an "overdose effect" was suggested as possible culprit for the increased CD prevalence. However, the consumption of bread, the major supplier of gluten, even decreased in developed countries over the last decades. Thus a "gluten overload" is probably not the cause for the increased CD prevalence in developed countries. A number of developing countries particularly in Asia however intensified wheat cultivation and import resulting in a wide-spread consumption of wheat products. The greatest changes can be observed in traditionally rice-consuming countries such as India and China with the consequence that the prevalence of CD might increase significantly.

Changes in the environment and modern lifestyle (see Section 8.5) are also considered a major cause. According to the so-called "hygiene hypothesis", the living environment in the Western world is too clean regarding the exposure to pathogens so that the immune system is not permanently challenged (e.g., by infections) and thus accidentally reacts against harmless food and pollen proteins. The altered intestinal microbiota theory assumes that "good" (commensal) bacteria decreased and "bad" (pathogenic) bacteria increased over the past 50 years leading to more infections and altered intestinal permeability and thus to more cases of CD. The increased exposure to antibiotics may also have contributed to these unfavorable conditions. Interestingly, children born by cesarean section may have a modestly increased risk of later CD compared to children born vaginally. Different initial microbiota (skin bacteria versus vaginal bacteria) were proposed to be the reason [19]. In general, the results on the mode of delivery are conflicting so far and more studies are needed to judge the influence of changed environmental conditions on CD prevalence.

One of the most striking differences to former epidemiological data is that a considerable portion of newly diagnosed CD cases has been found in individuals aged over fifty. Approximately a quarter of all diagnoses are now made at the age of 60 years or more and a fifth at 65 years or over [20]. The factors, leading to the increased development of CD in elderly persons, are not yet known. The widespread use of medications that increase intestinal permeability ("leaky gut") are suspected to induce CD more frequently. For example, acetylsalicylic acid (aspirin) and other nonsteroidal antiinflammatory drugs are known to affect the gastrointestinal mucosa, which may lead to an increased permeability for antigenic agents, e.g., gluten peptides (see Section 8.5). Thus the raised consumption of such drugs by elderly persons could contribute to increased CD prevalence. Similarly, medications that inhibit stomach acid secretion have increased in recent decades. A Swedish study indicated that the exposure to these antisecretory medications (proton pump inhibitors) was strongly associated with the subsequent diagnosis of CD [21]. Detailed research is urgently needed to verify the link between CD prevalence and medications. Apart from special drugs, high-intensity exercise, excessive alcohol consumption, and psychological stress may contribute to increased intestinal permeability and thus to higher sensitivity against gluten. All these factors are attributes of modern life.

In summary, there is no doubt that CD prevalence has increased in developed countries during the last five decades. There is no scientific proof yet that the qualitative and quantitative composition of gluten decisively changed by breeding, cultivation or processing of wheat. It appears more plausible that several environmental factors that modify intestinal permeability [22] and microbiota composition [23] activate the CD-specific immune system in a higher percentage of genetically predisposed individuals. Much more research is necessary to investigate the reasons for increased CD prevalence with the aim to reduce the risk of developing CD.

8.3 Genetics

As many other diseases, CD has a unique genetic architecture. New insights have enabled major advances in the fields of genetics and genomics [24]. CD mainly is associated with the human leukocyte antigen (HLA) class II alleles HLA-DQ2 and HLA-DQ8 at the major histocompatibility complex (MHC) at locus p21 on the short arm of chromosome 6. Almost all CD patients (90–95%) possess one of these HLA genes; the large majority is DQ2-positive and the remainder (≈5%) is DQ8-positive. The absence of these genes is a rather reliable negative predictor, even if a few CD patients are DQ2- and DQ8-negative. Two main isoforms, DQ2.5 (predominant) and DQ2.2 (rare) were identified among DQ2 alleles.

Around 30% of the general population however carry HLA-DQ2 or -DQ8 genes. The reasons, why most individuals with a genetic predisposition (HLA-DQ2/8) are not affected by CD, are unclear; it is not known why some develop CD and most do not. Additional genes were considered to contribute to the activation of the disease and further genome research, mainly focused on non-HLA-genes, were performed. A number of new loci, including many immune-related candidates, were identified, unfortunately with little consensus. The CD-related contribution of all known non-HLA genes together was estimated to account for < 10%. Therefore a number of international genome-wide association studies (GWAS) have started to uncover genetic components contributing to CD. Results demonstrated that approximately 50 genome loci were associated with CD apart from the well-known HLA-genes. Both HLA- and non-HLA-genes can now explain about 50% of the genetic variation of the disease. However, it is still too early to use genetics in clinical practice for predicting individual risks. An important result from the GWAS was the strong genetic overlap with autoimmune diseases (see Section 8.8). Besides identifying further CD-related genes, gene-gene and gene-environment interactions need to be investigated in the future.

HLA-DQ2 and -DQ8 genes code for heterodimeric proteins (α- and β-chains) that are expressed on the surface of antigen-presenting cells (e.g., dendritic cells and macrophages). The combined α- and β-chains bind immunogenic gluten peptides in a "peptide groove" and present them to T-lymphocytes in the lamina propria followed by the promotion of the adaptive immune response (see Section 8.9.1). The level of gluten peptides, presented by antigen-presenting cells to T-lymphocytes, critically influences CD development. Moreover, the different ability of DQ2.5-, DQ2.2-, and DQ8-coded heterodimers to form stable complexes with gluten proteins is closely related to the risk for CD: The HLA-DQ2.5 genotype is associated with a very high risk followed by -DQ8 (high) and -DQ2.2 (low) [25].

Because CD has increasing prevalence in the elderly people, older adults (age ≥50 years) with seropositive CD were evaluated whether the disease is associated with shorter leukocyte telomere length [26]. The comparison with seronegative controls demonstrated that CD in older subjects was significantly associated with enhanced telomere attrition. The role of epigenetics (e.g., methylation or histone modification) and the occurrence of genetic defects have not been explored until now, even though these changes may play an important part in CD susceptibility.

8.4 Causative proteins

As described briefly in Section 8.1, the English pediatrician Samuel Gee was the first, who recognized CD as a syndrome of malabsorption triggered by some unknown foodstuffs. For a long time afterwards, carbohydrates (starch) were suspected of bearing the toxic factor. Only in 1953, the group of Willem K. Dicke found that wheat gluten, the proteinaceous constituent of wheat dough, and not starch was the factor triggering CD [27]. This milestone in CD research led to countless investigations on the toxicity of cereal proteins and over the course of time, methods for isolation, characterization, and toxicity testing of proteins were improved step by step.

8.4.1 Chemistry of gluten proteins

The proteins of CD-toxic cereal grains (wheat, rye, barley, and oats) can be roughly grouped according to their different solubilities into albumins and globulins soluble in diluted salt solutions, prolamins soluble in aqueous alcohols, and glutelins insoluble in salt solutions and aqueous alcohols. After the reduction of disulfide bonds, glutelins are transformed into monomeric subunits that are soluble in aqueous alcohols like prolamins. Prolamins and glutelins are the major protein fractions of the grains and represent the storage proteins. Their only function is to provide the embryo with nitrogen and amino acids during the first stage of germination. The storage proteins of the CD-toxic cereals, collectively called "gluten" or "gluten proteins" in the field of CD, were given common names: wheat gliadins (prolamin fraction) and glutenins (glutelin fraction), rye secalins, barley hordeins, and oat avenins (only prolamin fraction).

The storage proteins consist of numerous components that can be divided into three groups based on homologous amino acid sequences and similar molecular weights (MW) (Table 8.3). Each group can be subdivided into single protein types. The high-molecular-weight (HMW) group (MW: 70,000–90,000) contains three closely related protein types: HMW glutenin subunits (HMW-GS) of wheat, HMW-secalins of rye, and D-hordeins of barley. The medium-molecular-weight (MMW) group (MW: 40,000–50,000) consists of the related ω1,2-gliadins of wheat, ω-secalins of rye, and C-hordeins of barley; ω5-gliadins are unique to wheat. The low-molecular-weight (LMW) group (MW: 22,000–35,000) contains the related γ-gliadins of wheat, γ-40k-secalins of rye, γ-hordeins of barley, and avenins of oats. α-Gliadins and LMW glutenin subunits (LMW-GS) of wheat, γ-75k-secalins (MW: around 50,000) of rye, and B-hordeins of barley have partly unique structures. Within each type, the number of components is quite different; for example, HMW-GS consists of 3–5 components and α-gliadins consists of more than 20 components. Proteins of the LMW group as well as ω-secalins and C-hordeins belong to the major protein types (> 17% of total storage proteins), while those of the HMW group and ω-gliadins belong to the minor components (< 12%) (Table 8.3) [28].

The molecular weights of gluten proteins range from around 22,000 (avenins) to 88,000 (HMW-GS), which corresponds to around 200–800 amino acid residues, respectively. The amino acid compositions are characterized by high contents of glutamine (26–53 mol%) and proline (11–29 mol%), which together amount to a maximum of around 70 mol% (proteins of the MMW group) [28]. Hydrophobic amino acids such as leucine, tyrosine, and phenylalanine are of further importance. High glycine contents (17–20 mol%) are typical of the HMW group. The contents of essential amino acids like lysine, methionine, and tryptophan are extremely low (< 1 mol%). The amino acid sequences of each type can be divided into several segments with completely different structural features [28]. The schematic architecture of wheat gluten proteins is presented in Section 2.2.1. Segments with repetitive sequences, rich in the amino acids

TABLE 8.3 Classification and approximate proportions (in % of total gluten proteins) of gluten proteins from CD-toxic cereals [28].[a]

Group	Wheat	Rye	Barley	Oats
HMW	HMW-GS	HMW-Secalins	D-Hordeins	–
	11%	9%	5%	–
MMW	ω1,2-Gliadins	ω-Secalins	C-Hordeins	–
	4%	18%	36%	–
	ω5-Gliadins	–	–	–
	3%	–	–	–
LMW	γ-Gliadins	γ-40k-Secalins	γ-Hordeins	Avenins
	27%	25%	32%	100%
	α-Gliadins	–	–	–
	33%			
	LMW-GS	γ-75k-Secalins	B-Hordeins	–
	22%	48%	27%	

[a] *GS, glutenin subunit; HMW, high-molecular-weight; LMW, low-molecular-weight; MMW, medium-molecular-weight.*

glutamine and proline (up to 73 mol%), are most characteristic for all gluten proteins. Table 8.4 includes selected repetitive sequences of wheat proteins showing the predominance of the amino acids glutamine (Q) and proline (P).

8.4.2 Testing toxicity

Numerous *in vivo* and *in vitro* methods were developed to identify CD toxicity and immunogenicity of cereals, cereal proteins, and isolated or synthesized peptides. Initially, CD patients were challenged *in vivo* by feeding tests followed by measurement of malabsorption (xylose or fat absorption) or examination of intestinal biopsies. However, these tests were quite unsatisfactory, because high amounts of challenge material (more than 1 g of protein) were needed, the optimal duration of challenge was uncertain, and only a few patients and controls could be tested. The introduction of oral challenge by direct instillation onto the small intestine and the judgment of biopsies afterwards allowed the reduction of challenge material and the improvement of test results. However, the strain on test persons was still immensely high.

The subsequent development of *in vitro* systems was a key step for testing toxicity. Organ cultures of intestinal tissue of CD patients are now considered as

TABLE 8.4 Selected repetitive sequences of wheat gluten protein types.

Protein type[a] (sequence position)	Sequence[b]
ω5 (192–221)	QQFPQQQFPQQQFPQQQFPQQEFPQQQQFP
ω1,2 (207–236)	QQPQQPFPLQPQQPFPQQPQQPFPQQPQQS
α (45–74)	QPQPFPSQQPYLQLQPFPQPQLPYPQPQPF
γ (50–79)	QPQQPQQPFLQPQQPFPQQPQQPFPQTQQP
LMW (27–56)	QQQPFPQQQPPFSQQQPSFSQQQPPFSQQQ
HMW (295–324)	QQPGQGQQPGQLQQPAQGQQPGQGQQGQQP

[a] *HMW, high-molecular-weight glutenin subunits; LMW, low-molecular-weight glutenin subunits: ω5, ω1,2, α, γ, gliadin types.*
[b] *One-letter-code for amino acids, see Table 2.3.*

the most reliable *in vitro* model and may mirror the *in vivo* situation. Intestinal tissues are taken from the small intestine as part of the diagnostic procedure and incubated in a culture medium containing the material to be tested. The measurement of CD-specific effects (e.g., morphology of biopsies, enzyme activity, and levels of cytokines) after 24 or 48 h of incubation can be used as markers for toxicity. Compared to *in vivo* challenge, a disadvantage is that the tissues lack blood circulation, nervous system, and connection to lymphatic organs.

Although the organ culture system is the best *in vitro* model for detecting toxic effects, it was widely replaced by studies on T-lymphocytes (T cells) that identify only immunogenic effects. Gluten-sensitive T-cell lines and clones, derived from the small intestinal mucosa or the peripheral blood of CD patients, can be used to perform a large number of experiments with very low amounts of challenge material (< 1 mg). A frequently applied method is the T-cell proliferation test: Gluten-sensitive T cells are treated with transglutaminase 2 (TG2) in the presence of antigen-presenting cells (e.g., B cells) and incubated with the putative antigen and radioactive tritiated thymidine. The proliferation of T cells measured by scintillation and/or the increase of CD-specific cytokines (e.g., interferon-γ and interleukins) are taken as parameters for immunogenic effects. The disadvantages of T-cell tests are that they do not necessarily reflect the small intestine *in vivo* due to the lack of any connection to other factors participating in the immune response and that variability in their reactivity against the target antigen is high. Moreover, the measured immune response does not always correspond to the toxic effect demonstrated by *in vivo* or organ culture tests. For example, the prolamin fraction of oats (avenins) was shown to be immunogenic in T-cell tests but nontoxic in organ culture tests [29,30].

A number of further *in vitro* assays were applied as screening tests for CD activity: leukocyte migration inhibition factor, macrophage procoagulant activity, agglutination of leukemia K 562 cells, caco-2 cells originating from human colon adenocarcinoma, and T84 cells derived from a lung metastasis of colon cancer. Aside from human intestinal material, cultures of the immature intestine of rat or chicken fetus were used to test CD toxicity. However, all these tests have not gained common acceptance.

The level of CD-specific toxicity and immunogenicity would be an important parameter for the characterization of CD-active cereals, protein fractions, and peptides. However, only a few attempts, including organ culture tests and T-cell tests, were made to quantitate the effects of harmful agents and a standardized method to determine the degree of toxicity/immunogenicity is still missing.

8.4.3 Toxicity of cereals

Early investigations in the 1950s led to the conclusion that wheat, rye, and barley are harmful to CD patients. Later, wheat was extensively studied confirming its undoubted toxicity and due to the high consumption worldwide, wheat has been considered as the main food component triggering CD. Testing of rye and barley has been rather limited, but their strong botanical and chemical similarities with wheat support their CD toxicity. A number of reports on oat toxicity during 1953–1976 revealed different effects, either harmful or harmless. In these studies however, oat samples were not tested for contamination with wheat, rye or barley. Studies afterward included a larger number of patients that were challenged with oats proven to be pure. In summary, the results demonstrated that most CD patients tolerate oats and do not show any serological immune response and damage of small intestinal mucosa [31]. However, individuals with newly diagnosed CD should observe a stabilization phase on a GFD, during which all symptoms of CD have resolved, before introducing uncontaminated oats to the GFD. Owing to the limited number of cases sensitive to oats, no exact numeral estimate of the size of this subgroup is available. Most likely, this subgroup may constitute $<1\%$ of CD patients. In accordance with this fact, Codex Alimentarius stated that the allowance of oats that are not contaminated with wheat, rye or barley may be determined at the national level. The incorporation of oats into a GFD could be beneficial for CD patients, because it provides high fiber, vitamin B, magnesium, zinc, and iron contents, increases palatability of a GFD, and has positive effects on cardiovascular health.

Whereas wheat, rye, barley, and possibly oats are considered harmful to CD patients, other common cereals (corn, rice, sorghum, and millet), rarely used cereals (e.g., teff, ragi, and Job's tears) and all plants outside of the cereal family have come to be regarded as safe, because no contrary evidence from diets including these crops has been encountered over decades. The taxonomy of plants might provide useful guidance in judging CD-specific toxicity. All toxic cereals (wheat, rye, and barley) are found in a single tribe (*Triticeae*) within the grass family (*Poaceae*) (Fig. 1.1). Oats, controversially judged (see above), belong to the same subfamily (*Pooideae*) but another tribe (*Aveneae*). All nontoxic cereals such as corn, rice, sorghum, and millet are more distant from the tribe *Triticeae* and show separate evolutionary lines within the grass family. All plants outside the grass family (e.g., potatoes, quinoa, buckwheat, and amaranth) are assumed to be safe for CD patients and can be used for the production of gluten-free products.

The "ancient" wheat species emmer (tetraploid) and einkorn (diploid) were proposed to contain fewer protein epitopes toxic for CD patients than hexaploid common wheat because of the absence of the D- and D/B-subgenomes, respectively [32]. Indeed, they have fewer gluten protein components and CD-active epitopes, but all gluten protein types are present and in even higher amounts (Table 2.4) [33]. While some older studies on ancient wheat species proclaimed reduced CD toxicity (whatever that means), new findings based on *in vitro* and *in vivo* tests provided evidence for their CD toxicity [14,34–36]. Therefore ancient wheats are not safe for individuals with CD. In 2016, the first "ultralow gluten" barley (gluten content of grains <5 mg/kg) created by breeding was reported [37]. It was proposed to be suitable to produce food or malt-based beverages for consumption by CD patients and is now used for the production of gluten-free beer. Regarding oats, their storage proteins (avenins) contain only a few CD-active epitopes. Differences in these epitopes among oat varieties indicate that selection and breeding of completely safe varieties for all CD patients may be a realistic possibility [38].

Many questions will arise regarding the approach towards breeding and selection of gluten-free lines from wheat, rye, barley, and oats. If CD-safe cultivars can be developed at all, they may have poorer yields and product characteristics. It is unlikely that these specially modified plants would ever replace conventional plants. Therefore they must be grown in alternative small-scale production systems only. Thus the economic efficiency of such an approach is limited and the higher price might be prohibitive for CD consumers.

8.4.4 Toxicity of cereal proteins and peptides

Early investigations on wheat proteins revealed that albumins/globulins were harmless and wheat gluten, consisting of gliadins and glutenins, was harmful for CD patients. The question, how much gluten can be tolerated by CD patients, is difficult to answer. The reasons are that patients differ individually regarding the tolerable level, the time for developing symptoms is variable and may be very long, and the clinical procedure for testing (two biopsies at least) is a serious challenge for test persons. Nevertheless, a couple of *in vivo* studies provided an estimation of safe amounts of gluten. The most extensive study, performed by Catassi and coworkers, included a double-blind placebo-controlled multicenter trial with 39 CD patients in remission [39]. The results demonstrated that most patients tolerated 10 mg gluten per day, whereas 50 mg were sufficient to cause significant damages of the intestinal mucosa. Based on these findings, the revised Codex Alimentarius Standard for Gluten-Free Foods of the year 2008 laid down that the gluten level in foods should not exceed 20 mg/kg. The total daily intake of gluten should not be beyond 20 mg/day [39], which approximately corresponds to one hundredth of a slice of bread.

Numerous studies demonstrated that the gliadin fraction of wheat bears the precipitating factor of CD, whereas the effect of the glutenin fraction was described controversially as either nontoxic, weakly toxic or as toxic as gliadins. The initial findings on the toxicity of the glutenin fraction were questionable, because the fraction was more or less contaminated with modified gliadins covalently linked to glutenins. The conclusion that glutenins are indeed toxic, are now based on studies on purified gliadin-free glutenin subunits [40]. Findings on rye and barley proteins (secalins and hordeins) were derived from similarities to wheat without intensive testing. Structurally homologous to wheat proteins, they also were associated with CD toxicity. Avenins, the prolamins of oats, were differently judged in organ culture and T-cell tests [29,30]. The glutelin fraction of oats does not belong to the class of storage proteins and has not been evaluated toxicologically up to now.

Even harsh physical and chemical treatments of gluten proteins do not abolish CD toxicity. For instance, strong heating, (e.g., baking bread) does not remove toxicity showing that the steric structures of the proteins do not play any role in CD development. Similarly, enzymatic digestion in the human gastrointestinal tract has an insufficient detoxifying effect showing that the structural integrity of proteins is not necessary. This additionally indicates that not only intact proteins but also protein fragments (peptides) are CD-active. Detailed studies on peptides demonstrated that a minimum length of nine amino acid residues is required for harmful effects of gluten peptides (see Section 8.9.1).

A number of *in vivo* and *in vitro* studies were performed to determine the CD toxicity of single gluten protein types; but the results were incomplete and mainly restricted to wheat proteins. A-gliadin, a group of aggregable α-type gliadins, was the first protein proven to be toxic by means of instillation into the small intestine followed by biopsy [41]. Afterwards, ω- and γ-gliadins were also proposed to contain the toxic factors. The protein preparations however were not necessarily pure and the ω-gliadin preparation was not differentiated into ω5- and ω1,2-gliadins. Among glutenins, HMW-GS were either isolated from wheat flour or produced by genetic engineering via transgenic *Escherichia coli*, transgenic yeast or transgenic corn. Both *in vivo* and *in vitro* studies showed clearly that HMW-GS can exacerbate CD just like gliadins [40]. LMW-GS from wheat as well as all secalin types from rye were not intensely tested for CD-toxicity until now. Regarding hordeins, *in vitro* assays demonstrated that all four types of barley storage proteins (D-, C-, B-, and γ-hordeins) activated gluten-sensitive T cells, but D- and C-hordeins were most active [42]. Though information on the CD toxicity of single gluten proteins was and still is full of gaps and unsatisfactory, studies were strongly reduced because of the availability of synthetic peptides and *in vitro* tests. These studies complemented the data available and confirmed that all gluten protein types have the potential to cause CD.

Since the 1950s, it has been known that gluten peptides, produced by gastrointestinal enzymes (e.g., pepsin, trypsin, and chymotrypsin), and not intact gluten proteins are the actual triggering factors of CD. However, it was not until the year 1984 that the first CD-toxic peptide was isolated from a gliadin fraction and identified as a fragment derived from α-gliadins including the amino acid sequence positions 3–55 [43]. The peptide was characterized by high contents of glutamine and proline indicating that these amino acid residues decisively contributed to CD toxicity. Since that time, numerous research groups focused their studies on testing peptides either isolated from gluten proteins or synthesized. A database compiled from more than 60 publications (http://www.allergenonline.org) includes now more than 1000 CD-active gluten peptides derived from wheat, rye, barley, and oat proteins. These peptides are subdivided into "toxic" peptides characterized by *in vivo* or organ culture tests and "immunogenic" peptides characterized by T-cell tests. The discrepancy between the numbers of toxic and immunogenic peptides is extraordinary. So far, 1014 peptides are classified as immunogenic, seven peptides as toxic, and five peptides as toxic and immunogenic. Most of these peptides occur in the glutamine- and proline-rich sequence segments of the proteins. Table 8.5 shows a small selection of the origin, nomenclature, and amino acid sequences of peptide epitopes active in CD [44].

The following structural properties of peptides are important for the initiation of CD:

(1) Glutamine and proline are predominant in the amino acid sequences.
(2) A peptide length of at least nine amino acid residues forming these peptides are necessary.
(3) These peptides are exceptionally resistant to the digestion by gastric, pancreatic, and intestinal brush border enzymes.
(4) Specific glutamine residues are preferably deamidated by TG2, which plays a major role in CD pathogenesis.
(5) Due to the high proline content, these peptides adapt a steric structure necessary for the initiation of the CD-specific immune response.

Provided that these five preconditions are fulfilled, the simultaneous binding of gluten peptides to the HLA-DQ heterodimers of antigen-presenting cells and presenting to the receptors of T cells can occur (see Section 8.9.1). Altogether, the unique amino acid sequences of gluten proteins reflect very well their specific role in triggering CD.

The so-called 33-mer peptide from α2-gliadin (position: 56–88; sequence: LQLQPFPQPQLPYPQPQLPYPQPQLPYPQPQPF) is the most prominent example among CD-active epitopes. The 33-mer peptide is resistant to cleavage by human gastrointestinal peptidases and is widely called the most immunodominant gluten peptide, because it contains three overlapping CD-active epitopes, namely PFPQPQLPY (DQ2.5-glia-α1a), PYPQPQLPY (DQ2.5-glia-α1b), and PQPQLPYPQ (DQ2.5-glia-a2) that initiate a strong immune response.

TABLE 8.5 Origin, nomenclature, amino acid sequences and anchor positions (p1–p9) of epitopes present in selected CD-active gluten peptides [44].[a]

		Sequence[b]/anchor positions[c]								
Origin	**Epitope**	**p1**	**p2**	**p3**	**p4**	**p5**	**p6**	**p7**	**p8**	**p9**
DQ2.5-restricted										
α-Gliadin	DQ2.5-glia-α1a	P	F	P	Q	P	**Q/E**	L	P	Y
α-Gliadin	DQ2.5-glia-α2	P	Q	P	**Q/E**	L	P	Y	P	Q
γ-Gliadin	DQ2.5-glia-γ2	I	Q	P	**Q/E**	Q	P	A	Q	L
ω1,2-Gliadin	DQ2.5-glia-ω1	P	F	P	Q	P	**Q/E**	Q	P	F
γ-Secalin	DQ2.5-sec-1	P	F	P	Q	P	**Q/E**	Q	P	F
γ-Hordein	DQ2.5-hor-2	P	Q	P	**Q/E**	Q	P	F	P	Q
Avenin	DQ2.5-ave-1a	P	Y	P	E	Q	**Q/E**	Q	P	F
DQ2.2-restricted										
α-Gliadin	DQ2.2-glia α2	Q	Y	S	Q	P	**Q/E**	Q	P	I
LMW-GS	DQ2.2-glut-L1	P	F	S	**Q/E**	Q	**Q/E**	Q	P	V
DQ8-restricted										
α-Gliadin	DQ8-glia-α1	**Q/E**	G	S	F	Q	P	S	Q	**Q/E**
γ-Gliadin	DQ8-glia-γ1b	**Q/E**	Q	P	Q	Q	P	Y	P	**Q/E**
HMW-GS	DQ8-glut-H1	**Q/E**	G	Y	Y	P	T	S	P	Q/E

[a] *HMW-GS, high-molecular-weight glutenin subunit; LMW-GS, low-molecular-weight glutenin subunit.*
[b] *One-letter-code for amino acids, see Table 2.3; glutamine residues (Q), preferably deamidated by TG2 to glutamic acid (E), are shown in bold.*
[c] *See Section 8.9.1.*

The quantitative analysis of 49 common wheat and 2 spelt flours by liquid chromatography tandem mass spectrometry indicated that all flours contained the 33-mer peptide ranging from 91 to 603 μg/g flour [45]. In contrast, the peptide was absent from tetra- and diploid wheat species (durum wheat, emmer, and einkorn), most likely because of the absence of the D-subgenome that encodes α2-gliadin.

8.5 Environmental factors

In addition to the genetic predisposition (see Section 8.3), environmental factors besides gluten proteins have been associated with the development of CD [46]. Infections by a variety of pathogens, including adenovirus 12, hepatitis C virus, and rotavirus, were proposed to contribute to the onset of CD. Viral infections were suggested to break oral tolerance to dietary proteins and incite the sensitivity to food such as found in CD [47]. Additionally, microbial infections may play an indirect role in initiating CD by creating a proinflammatory environment and a disruption of the intestinal barrier that facilitates gluten peptide penetration. Repeated gastrointestinal infections in early life were found to be associated with increased risk of CD in later life [48,49]. Children born in the summer were proposed to have a higher risk of developing CD than children born in winter, possibly because the introduction of gluten at around 6 months of age occurs in winter accompanied by the seasonal peak of infections. A nationwide study of 29,000 CD patients and 140,000 controls however indicated only a small nonsignificant risk of summer birth for CD and general infection disease exposure early in life is unlikely to be a major cause of CD [50].

An imbalanced small intestinal microbiota may also be associated with the development of CD [51]. The intestinal microbiota comprises bacteria that promote regulatory immune responses and bacteria that promote inflammatory immune responses. Numerous studies have shown differences in the composition of the intestinal microbiota between CD patients and healthy people. For example, decreases in regulatory bacteria (e.g., *Faecalibacterium prausnitzi* and bifidobacteria) as well as expansions of potentially pathogenic bacterial populations (e.g., *Escherichia coli* and *Staphylococcus* spp.) were reported in adult CD patients [52]. Moreover, the intake of antibiotics is suspected to influence the microbiota negatively. Altogether, knowledge about the role of intestinal microbiota in CD pathogenesis is scarce and more studies are needed to comprehensively investigate the role of intestinal microbiota and its metabolites in CD.

Children born by cesarean section may have a modestly increased risk of later CD possibly due to a different microbiota. Their intestinal flora resembles skin bacterial communities, whereas children born vaginally have microbiota that resemble their mother's vaginal flora. However, cohort studies on children with CD indicated that the prevalence of CD at 5 years of age was not significantly different between children born by cesarean or vaginal delivery [53,54]. The effects of acute caesarean section versus elective cesarean section were discriminated by Kristensen and Henriksen [55]. The results demonstrated that only children, delivered by acute cesarean section, had an increased risk of CD compared to vaginal delivery. Another study, involving 6596 children who developed CD before 15 years of age, showed that elective cesarean delivery in combination with repeated maternal urinary tract infections during pregnancy were associated with an increased risk of CD onset during childhood [56].

The hygiene hypothesis, postulating that a high level of hygiene in the industrialized world contributes to increased allergic and autoimmune disorders [57], has become very popular during the last decades. This hypothesis was used to explain the differences in the prevalence of CD in Finland (1–2%) and the neighboring Russian Karelia ($\approx$0.2%) [58]. Lack of exposure to infectious agents, symbiotic microorganisms, and parasites does not allow the immune system to use its full capacity towards pathogenic agents and in consequence, immune reactions are directed to harmless agents like food proteins. Parasitic helminths, common inhabitants in the human intestine previously, are thought to play an important role in modulating the host`s immune system. A 52-week study in fact demonstrated that the festination of CD patients with the hookworm *Necator americanus* resulted in a promoted tolerance to the intake of gluten [59] (see Section 12.5). Much more investigation and statistical evaluation are necessary to clarify, whether cleaner hygienic conditions promote the development of CD.

Previous studies suggested a protective role of a moderate introduction of gluten in children between 4 and 7 months of age as well as breastfeeding at the time of gluten introduction. In consequence, several position papers of international committees recommended avoidance of both early (<4 months) and late (>7 months) introduction of gluten and gradual introduction of small amounts of gluten, while the child is still breast-fed. Subsequent multicenter studies however demonstrated that all these associations were questionable [60,61]. Neither early nor delayed gluten introduction appeared to influence CD risk and breastfeeding did not protect against the development of CD. Anyway, consumption of large quantities of gluten should be avoided during the first weeks after gluten introduction.

Agents that increase intestinal permeability are assumed to be important cofactors of CD, because they enable immunogenic gluten proteins/peptides to reach the lamina propria unmodified and in high amounts. Apart from infections and inflammation, certain drugs, notably nonsteroidal antiinflammatory drugs (NSAIDs), are known to provoke severe injuries of the gastrointestinal mucosa combined with damage to tight junctions and increase in intestinal permeability [62]. The best investigated example among NSAIDs is acetylsalicylic acid (ASA) (aspirin). ASA has been taken for the relief of pain, fever, and inflammation for more than one hundred years. Today however, an additional frequent indication of ASA is for the prevention of cardiovascular events (e.g., coronary heart disease, cerebrovascular disease, and peripheral artery disease). ASA is now daily used for these indications by millions of people worldwide, mainly above the age of 50. However, the regular use of ASA, even in low doses below 500 mg, is associated with serious side effects not only in the stomach but also in the upper intestine [63]. Apart from less severe symptoms, bleeding, ulcer perforation, and obstruction are induced by ASA. Due to these mucosal injuries, increased intestinal permeability and the release of tissue transglutaminase from damaged cells, an important cofactor in the pathogenesis of CD, provide optimal conditions for the development of CD. It is therefore not surprising that a notable part of newly diagnosed CD patients are elderly individuals. Unfortunately, statistic evidence of the association of CD with the intake of NSAIDs is not yet available. Additionally, excessive alcohol consumption, physical activity, and even psychic stress may increase intestinal permeability, but a relationship with CD is not yet proven. Besides NSAIDs, medications that inhibit stomach acid secretion and have been increasingly prescribed in recent years, were suspected of contributing to the development of CD. For example, patients, who took both proton pump inhibitors and histamine-2 receptor antagonists, had a higher rate of CD than those, who took proton pump alone or histamine-2 receptor antagonist alone [21].

In summary, many studies on the influence of environmental factors on CD development were performed. However, they are in parts disappointing and many questions remain unanswered. Primary prevention of CD by consideration of environmental factors is currently not possible [64]. The only identified risk factors are the HLA genotype and gluten intake. Future prospective multicenter studies with long-term follow-up are necessary to investigate the effect of different intervention strategies such as intrauterine and perinatal exposures as well as environmental influences later in life including drug exposure, microbial infections, and the microbiome [60].

8.6 Serology and pathology

CD is characterized by a disease-specific antibody response to gluten proteins and TG2. These antibodies belong to the Ig classes A and G; they bind to their specific antigen in the diseased intestinal mucosa and also appear in the blood of CD patients. Their production starts early after gluten intake and is stopped after the establishment of a GFD. Antibodies against gliadins (AGAs) were described first; their specificity according to CD however was fairly poor, because they could also be detected in nonceliac enteropathies as well as in 5–10% of healthy people. The selective deamidation of gliadins by TG2 in the lamina propria induces the production of antibodies against deamidated gliadin peptides (DGPAs) that are highly sensitive and specific in CD [65]. Apart from gliadins, HMW-GS were reported to cause the production of antibodies specific for CD patients [66]. Additionally, antibodies against reticulin, a fiber in the connective tissue, and endomysium (EMAs), a protein in the collagenous matrix of human and monkey esophagus, were detected in patients with CD. In 1997, TG2 was identified by Dieterich and coworkers as CD-specific autoantigen and it was demonstrated that TG2 was also the corresponding autoantigen in reticulin and endomysium [67]. The formation of antibodies against TG2 (TGAs) was absolutely unexpected, but since that time it has been clear that the immune response in CD has an autoimmune component. DGPAs and TGAs are most CD-specific and therefore they are used for the diagnosis of CD (see Section 8.10.1).

Up to 10% of diagnosed CD patients are seronegative. A critical summary of the most recent work on seronegative CD was presented by Schiepatti and coworkers [68]. Seronegative CD is defined by the negativity of all CD-specific antibodies in the presence of a positive histology of duodenal biopsy samples and HLA-DQ2/8 alleles. Patients with seronegative CD do not display a different clinical profile and benefit from GFD as patients with seropositive CD [69]. CD patients with selective IgA deficiency (see Section 8.8) are only positive for IgG antibodies. IgA deficiency is a common primary immunodeficiency and defined as a distinct decrease in serum IgA levels in the presence of normal levels of other immunoglobulin isotypes.

Regarding pathological features, CD is the most important cause of intestinal villous atrophy, crypt hyperplasia, and chronic inflammatory cell infiltrate in the lamina propia. It is classically characterized by damage of the mucosa of the upper small intestine (duodenum and proximal jejunum). In healthy individuals, the surface of the mucosa is covered with millions of finger-like projections (villi). The outer part of the villi consists of the so-called brush border, where countless tiny microvilli are located. The brush border is equipped with a complex enzyme/carrier system for the digestion and absorption of nutrients. In CD, the intake of gluten proteins induces the flattening of the mucosa characterized by a loss of normal villous architecture and brush border (Fig. 8.2).

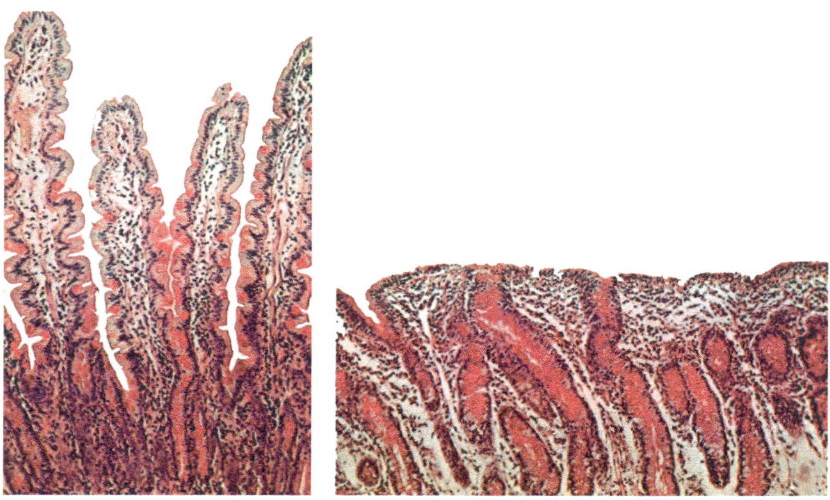

FIG. 8.2 Cross-section of normal (left) and flattened mucosa (right). *(Reprinted with permission from the German Celiac Society.)*

In consequence, the uptake of nutrients is strongly reduced in the affected areas by the loss of mucosal enzymes and carriers and by the decrease of the luminal surface. In full-blown atrophic mucosa, the surface may be reduced from the area of half a badminton court to that of a sheet of paper. Besides the flattening of the mucosa, the increase in intraepithelial lymphocytes (IELs) is a typical feature of CD. Numbers of more than 30 IELs per 100 enterocytes are considered pathologic.

Depending on the severity of the damage, the histological appearance of the affected mucosa can vary from mild partial villous atrophy to a total absence of villi. In 1992, Michael Marsh introduced a grading scheme to classify the degree of intestinal damage that was modified by Oberhuber and coworkers [70] and is now commonly used for the diagnosis of CD (see Section 8.10.2). The different types of conditions were grouped as follows:

Type 0: normal preinfiltrative mucosa with a villous height (vh) to crypt depth (cd) ratio of > 3:1 and with < 30 IELs per 100 enterocytes.
Type 1: infiltrative mucosa with normal vh/cd ratio (> 3:1) and an increased number of IELs (> 30).
Type 2: infiltrative-hyperplastic mucosa with normal vh/cd ratio and an increased number of IELs (> 30).
Type 3: destructive mucosa with an increased number of IELs (> 30) and a vh/cd ratio of < 3:1 or 2:1 (type 3a), < 1:1 (type 3b) or completely flat mucosa (type 3c).
Type 4: atrophic mucosa with only a few crypts and nearly normal IEL counts (usually observed in patients with refractory CD).

8.7 Clinical features

Originally, CD was regarded as a rare pediatric disorder of the small intestine characterized by chronic diarrhea and malabsorption of nutrients. Today, it is recognized as one of the most frequent food hypersensitivities in Western populations and can affect children and adults at any age. The clinical manifestations of CD are highly heterogeneous and can range from asymptomatic appearance to full-blown symptoms due to the multisystem nature of CD. It is therefore important for all physicians to be aware of the many potential clinical presentations of CD. Patients can be classified into three groups: The first group comprises "symptomatic" patients. They have typical intestinal and/or extraintestinal symptoms, a significantly increased level of CD-specific serum antibodies, and a flat intestinal mucosa. A small subgroup of symptomatic patients encompasses subjects with refractory CD characterized by persistence of symptoms and villous atrophy despite a strict GFD (see Section 8.7.8). The second group consists of "asymptomatic" (previously named "silent") patients that have atypical, minimal or even missing complaints commonly associated with CD, but they have increased CD-specific antibody levels and a flat duodenal mucosa. In such cases, other parts of the small intestine (jejunum and ileum) might compensate for the lost ability to absorb nutrients. The third group comprises patients with "potential" CD. They have no symptoms and a normal mucosa but are characterized by the presence of CD-specific antibodies. Potential CD is considered as a prestage of active CD that entails the risk of developing the full-blown disease later. Many individuals with asymptomatic or potential CD have not yet been diagnosed for CD.

The clinical picture of symptomatic CD patients is characterized by intestinal and extraintestinal symptoms. The classical presentation of CD, consisting primarily of intestinal symptoms like chronic diarrhea and abdominal pain, is more common in young children, whereas the majority of adolescent and adult patients also present extraintestinal symptoms. In principle, extraintestinal manifestations can affect any body part, from skin, bones, and liver to the central nervous system. The rates of typical extraintestinal symptoms in 157 pediatric (< 18 years) and 171 adult patients with biopsy-proven CD were determined by Jericho and coworkers [71]. 60% of pediatrics and 62% of adults presented extraintestinal symptoms. In pediatrics, short statue (33%), fatigue (28%), and headache (20%) were most common among symptoms, while iron deficiency anemia (48%), fatigue (37%), and headache/psychiatric disorders (24%) were predominant in adults.

The severity of symptoms is dependent on the location, length, and degree of mucosal damage that can be patchy in nature. Symptoms develop gradually or they may appear suddenly and dramatically; they can range from very mild to potentially life-threatening. Some individuals only have one or two symptoms and others have many symptoms simultaneously. Most CD symptoms can be alleviated by a GFD. However, some CD-mediated complications may be irreversible, unless CD is treated in time. For example, growth retardation, osteoporosis or abnormal dentition will remain permanent, if not treated early. The so-called celiac crisis, frequently described previously and characterized by explosive watery diarrhea, marked abdominal distension, metabolic and electrolyte disturbances, and shock, is now observed rarely, but it should be considered in patients with apparently chronic diarrhea [72].

The temporal course of symptomatic, serological, and histological responses to gluten challenge was shown by the group of Leffler et al. [73]. Twenty adult CD patients on a GFD were challenged with 3.0 or 7.5 g gluten per day for 14 days and changes in symptoms, levels of antibodies, intestinal permeability, and histology of the intestinal mucosa were judged at different points in time, also after gluten intake was stopped. Gastrointestinal symptoms were enhanced significantly by day 3 and returned to baseline by day 28. Serum titers of TGAs and DGPAs increased slightly to day 14 and markedly by day 28. Permeability, determined by the ratio of lactulose to mannitol in urine samples, did not change significantly. The investigation of intestinal biopsies showed a significant reduction in the ratio of villous height to crypt depth and an increase in IELs from day 1 to day 14.

8.7.1 Gastrointestinal complaints

Gastritis (inflamed stomach) is frequently associated with untreated CD. Lymphocytic gastritis, chronic active gastritis, and chronic inactive gastritis are common findings and show an improvement upon GFD. Indigestion, burping, belching, and feelings of fullness can also result from impaired stomach functions caused by CD. Gastrointestinal dysmotility, gastroesophageal reflux, nausea, and vomiting are further complaints. A relationship between *Helicobacter pylori* and CD could not be established.

The inflammation and damage of the intestinal mucosa can lead to a broad spectrum of intestinal complaints. Chronic diarrhea is the classical symptom of CD in both children and adults. Loose watery stools, occurring either several times per day or episodic with periods of normal stools between, are typical. Food constituents that cannot be absorbed by the flattened mucosal surface (e.g., sugars and fat), stimulate the colon to secrete rather than absorb water resulting in diarrhea. Bacterial overgrowth in the small intestine, typical of active CD, is another cause of diarrhea. Some CD patients have periods of constipation either in isolation or changing with periods of diarrhea. Malabsorption of fat induces steatorrhea characterized by fatty stools with a greyish, greasy appearance and foul odor. Steatorrhea is considered to be more typical of active CD than diarrhea. Therefore the examination of stool fat after *in vivo* challenge with gluten was one of the first diagnostic tools in the history of CD.

Undigested nutrients are fermented in the small intestine and the colon by bacteria leading to excess gas production combined with abdominal pain, bloating, flatulence, and esophageal reflux. Abdominal pain is one of the most frequently reported intestinal symptoms of CD in children and adults and abdominal distension is a typical sign of CD in children (Fig. 8.3). Loss of the enzyme lactase, usually produced by the brush border of the villi, results in lactose intolerance and is also combined with abdominal pain, flatulence, and diarrhea.

8.7.2 Manifestations of malabsorption

Various nutritional deficiencies induced by malabsorption can lead to a number of extraintestinal disorders in CD patients. The degree of malabsorption is dependent on the size of the affected area of the small intestine. Many CD patients show distinct effects of vitamin and mineral deficiencies. Malabsorption of fat-soluble vitamins (A, D, E, and K) as well as folic acid and vitamin B_{12} are common, whereas deficiencies of other water-soluble vitamins are less often observed. Vitamin A plays an essential role in vision and retinal functions and is necessary for normal growth and sexual reproduction. Severe deficiencies of vitamin A can result in inhibited growth, loss of taste, and night blindness. Examined by the Schirmer test and break-up time tests, the eyes of a number of children with active CD compared to controls did show decreased anterior chamber depth, decreased anterior chamber volume, lower values of

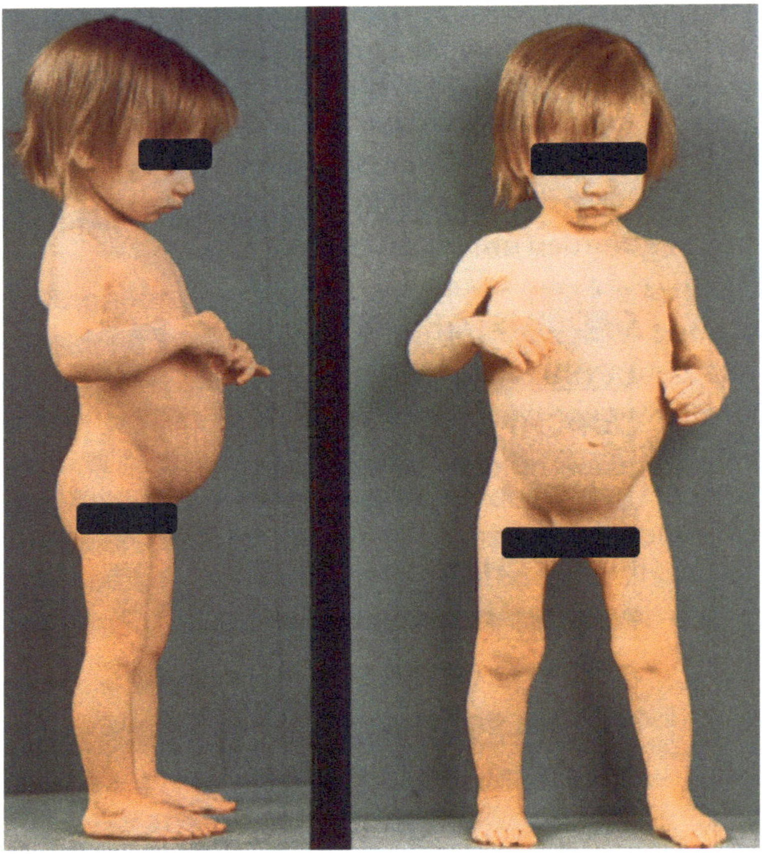

FIG. 8.3 Child with abdominal distension typical of CD. *(Reprinted with permission from the German Celiac Society.)*

Schirmer and break-up time test as well as lower retinal nerve fiber layer [74]. Vitamin D is an important factor in the intestinal absorption of calcium and its deficiency observed in CD causes soft, deformed bones in children and bone problems in adults. Vitamin E is an antioxidant that protects cells from damage and its lack is associated with neurological problems and anemia frequently found in CD patients. Vitamin K is responsible for blood clotting and deficiency results in impaired synthesis of clotting factors and abnormal bleeding. Low levels of folic acid can be found in almost half of untreated CD patients. Folic acid is necessary for the formation and growth of blood cells and cell division. Its deficiency plays a role in neurological defects, anemia, blood disorders, and gastrointestinal disturbances and is proven to cause fatigue, apathy, and depression. The lack of vitamin B_{12} has the potential to cause irreversible neurological distortions.

Similarly, the levels of minerals, including iron, calcium, magnesium, copper, zinc, and selenium, can be low in CD patients. Iron deficiency anemia is one of the most common extraintestinal presentations in teenagers and adults with active CD. Anemia symptoms can appear as pallor, shortness of breath, chest pain, and heart attack. Lack of calcium, combined with vitamin D and magnesium deficiencies, is frequently associated with reduced bone density, bone pain and fracture, osteopenia, osteoporosis, and dental enamel defects. In older people, one third of untreated CD patients has osteopenia and one third has osteoporosis. Thus testing for CD is recommended for all individuals with osteoporosis, even in the absence of intestinal symptoms. Early diagnosis and treatment of CD during childhood may protect CD patients from osteoporosis. For malnourished patients, supplementation with vitamin D and calcium results in significant improvement. Oral manifestations are common findings in pediatric patients, thus making the dentist an integral part of the diagnostic team. For example, dental enamel defects, recurrent aphthous stomatitis, delayed tooth eruption, multiple caries, angular cheilitis, atrophic glossitis, dry mouth, and burning tongue may be related to untreated CD [75]. Copper deficiency can cause different forms of anemia and lack of zinc results in poor tissue healing, dermatitis or loss of taste. Deficiencies of vitamins and minerals caused by malabsorption can contribute to a number of further complaints, e.g., hair and nail problems, reproductive disorders as well as neurological and psychiatric manifestations (see below).

Generalized malabsorption is the most common cause of unexplained failure to thrive normally, slow growth rate, and weight loss in children. Remarkably, CD is the most frequent cause of slow growth rate in Western countries and much more common than growth hormone deficiency. Delayed sexual maturation of adolescents is also a known complication of CD. Young adults with CD are at increased risk of iron deficiency anemia and early atherosclerosis. In older adults, malabsorption may contribute to various types of arthritis such as rheumatic arthritis and osteoarthritis and muscular diseases. Fatigue, frequently reported by CD patients, can be a symptom of both malabsorption and intestinal inflammation. Historically, unintentional weight loss was considered a classical symptom of CD. Later findings however suggest that untreated CD patients are more likely to be overweight in Western populations highlighting the importance of considering CD regardless of the body mass index (BMI).

8.7.3 Skin, hair, and nail symptoms

Dermatitis herpetiformis (DH), also known as Duhring disease, is the cutaneous counterpart of CD ("skin CD") [76]. DH was the first and best-described extraintestinal manifestation of gluten sensitivity. The prevalence is much lower than that of CD ranging from 1:2000 to 1:10,000. Like CD, there is a strong association with HLA-DQ2 and -DQ8 genes. In contrast to CD, males have a higher prevalence of DH than females. Characteristic symptoms include intense itching and burning of the skin caused by herpetiform clusters of pruritic urticated papules and vesicles especially on the elbows, upper forearms, buttocks, and knees but also on other parts of the body (Fig. 8.4). In children, the rash may only be present as purpura on the palms of their hands. Typically, granular IgA deposits are present in the skin and inflammatory cells, cytokines are found in the lesions and antibodies to epidermal transglutaminase (eTG) occur in the serum. The majority of DH patients has subclinical villous atrophy in the small intestine identical to CD and the remainder shows signs of mucosal inflammation. Studies, including 393 patients with DH, demonstrated that 29% had severe villous atrophy, 41% partial villous atrophy, and 30% normal villous architecture [77]. The degree of damage in the small intestine correlated with the serum level of IgA TGAs. Though the majority of DH patients has intestinal lesions, only a minority of patients has intestinal symptoms that are usually mild. The diagnosis of DH is based on the determination of IgA antibodies in the skin and by serological tests using IgA antibodies against eTG. After establishing

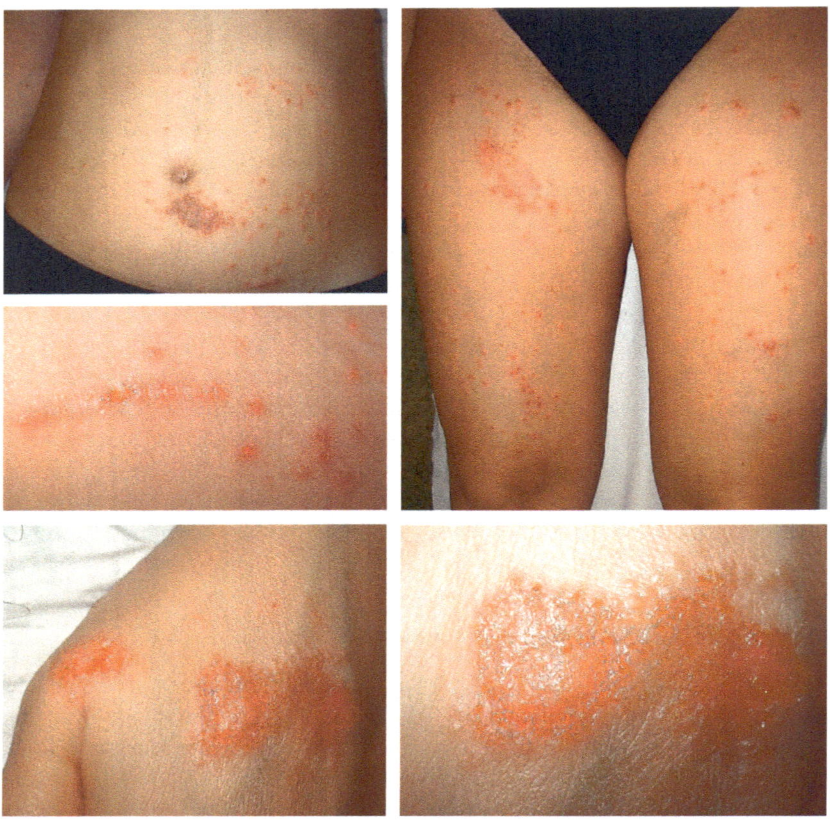

FIG. 8.4 Typical lesions in a patient with dermatitis herpetiformis.

the diagnosis, a GFD is recommended for treatment. Because the improvement takes several months, dapsone and sulfapyridine are used as primary medications to treat acute DH.

Over the last few years, there have been multiple reports of the association between CD and skin affectations apart from DH that may improve with a GFD [78]. Examples are chronic urticaria, atopic dermatitis, aphthous stomatitis, and rosacea. Early studies showed contradictory results with regard to the cooccurrence of CD and psoriasis. Four retrospective cohort studies with 12,912 cases of psoriasis and 24,739 controls however demonstrated an approximately threefold increased risk of CD among patients with psoriasis [79]. Psoriasis is a chronic relapsing dermatosis characterized by cutaneous lesions and less commonly pustulation involving the entire body. A genetic predisposition has been clearly demonstrated, but the mechanism of possible links between CD and psoriasis is still unclear. Both diseases are characterized by T cell-mediated immune reactions. An association with high levels of IgA AGAs and an improvement of skin lesions on a GFD are further clinical features of psoriasis. Therefore patients with psoriasis should be investigated for CD.

Individuals with untreated CD may have thin and slow growing hair as well as patches of hair loss. The color of hair might change due to intestinal malabsorption of pantothenic acid or manganese. The relationship between alopecia areata and CD and the potential benefit of a GFD on hair loss or regrowth is well known [80]. Other CD patients have nails that are brittle, malformed, and break easily. Nails grow slowly and can have color changes and horizontal ridges. Nail symptoms are partially or fully caused by nutrient deficiencies including vitamins, minerals, essential fatty acids, and amino acids.

8.7.4 Liver, kidney, and cardiovascular diseases

A significant minority ($\approx 10\%$) of CD patients has been diagnosed with several bile and liver disorders such as sclerosing cholangitis, biliary cholangitis, hepatitis B and C, autoimmune hepatitis, nonalcoholic fatty liver, and liver cirrhosis or fibrosis [81]. Most of them have raised serum levels of the liver enzymes aspartate and alanine transaminase (hypertransaminasemia). In rare cases, untreated CD may lead to fulminant hepatitis and liver failure. A GFD ameliorates the symptoms associated with CD and active screening for CD is recommended in patients suffering from liver diseases with unclear etiology. There were cases of severe liver diseases, where the identification of CD and treatment by a GFD avoided transplantation [82].

A number of studies have demonstrated the association of CD with various types of kidney diseases such as membranous nephropathy, IgA nephropathy, diabetes nephropathy, and chronic kidney disease [83]. A systematic review and meta-analysis confirmed a significantly increased risk of kidney diseases among patients with CD and subgroup analysis showed that the risks were particularly elevated in patients with diabetic nephropathy and IgA nephropathy [84].

These findings may influence the clinical management and primary prevention of kidney diseases in CD patients.

Several studies report the association of CD with accelerated atherosclerosis and cardiovascular diseases, many of which include ischemic diseases like acute myocardial infarction, angina pectoris, ischemic heart disease, and stroke [85,86]. Not-ischemic cardiovascular diseases associated with untreated CD are represented by dilated cardiomyopathy, atrial fibrillation, and myocarditis. Most disorders tend to resolve on a GFD, often in conjunction with the healing of small intestinal villous atrophy. In some cases however, the alterations are irreversible. The findings underscore the need for CD screening and treatment, when cardiovascular issues arise of unknown etiology.

8.7.5 Infections and reproductive disorders

Severe infections are recognized complications of CD. A large population-based cohort study revealed that patients with CD had significantly higher incidence of *Clostridium difficile* infections [87]. This result is consistent with prior findings of higher rates of other infections in CD patients such as tuberculosis and influenza. Moreover, individuals with CD are at increased risk of invasive pneumococcal disease [88]. Therefore CD patients, who have not received the scheduled vaccination series as a child, may be considered for pneumococcal vaccination.

Various nutrient deficiencies may affect sexual hormones, health of the reproductive organs, and the viability of a fetus. It is therefore not surprising that a number of studies in recent decades have associated CD with reproductive complaints. Previous investigations on the association between CD and infertility showed contradictory results. An extract from 12 case-control or cohort design studies from 1966 to 2013 however indicated that untreated CD is in fact a risk for infertility [89]. A meta-analysis, based on a literature search using the keywords "celiac disease", "gluten", and "infertility", demonstrated that CD is more prevalent in both women with all-case infertility (2.3%) and women with unexplained infertility (3.2%) as compared to controls [90]. In men, reduced sperm mobility, abnormal sperm morphology, impotence, and loss of libido were reported by patients with active CD. Altogether, early diagnosis of CD and the implementation of a GFD are keys to the primary prevention of reproductive problems in patients with CD.

Untreated CD may also have implications on gynecological health [91]. Patients are more likely to have menstrual pain, irregular menstrual cycles or no period for 6 or more months and they start menopause earlier [92]. The association between CD and pregnancy complications and adverse birth outcome has been differently judged up to this day. For example, extensive literature research demonstrated that women with undiagnosed CD showed an increased risk of miscarriage, intrauterate growth restriction, low birth weight, and preterm delivery [93]. The risk was significantly reduced by a GFD. In contrast, Lebwohl and coworkers [94] found no evidence that active CD increased risk for intrauterine growth restriction, low birth weight, and preterm birth and a nationwide Danish matched cohort study revealed that the rate of adverse pregnancy outcomes in CD patients and controls was the same [95]. Altogether, more studies are needed to evaluate the influence of active CD and a GFD on pregnancy outcomes.

8.7.6 Neurological disorders

A minor percentage (8–10%) of untreated CD patients shows disturbances of the peripheral or central nervous system and presents neurological symptoms [96]. In such cases, these conditions are often the only manifestations of CD and frequently under-recognized by the medical community. They are among the most common symptoms of CD that are considered atypical and delay diagnosis. Pennisi and coworkers [97] have provided an up-to-date summary of the neurophysiological methods used in CD investigations including electroencephalography, multimodal evoked potentials, and transcranial magnetic stimulation. A part of CD patients, referred to neurological opinion, shows significant brain abnormalities such as less grey matter density and white matter abnormalities in some brain regions, brain atrophy, and destruction of myelin sheets [98]. The precise mechanism of the association to CD is unknown and several, possibly combined factors are in discussion: production of antibodies against components of the nerves and brain, lack of essential nutrients, immunochemical response due to inflammation, opiate-like effects of gluten peptides, and modifications of the intestinal microbiota and their metabolites.

Some authors suggest that antibodies are produced against brain tissue or against nerve components. Preliminary findings demonstrated that anti-GM1 ganglioside antibodies and interleukin 10 may be considered markers of nervous system impairment in the course of CD [99]. It was shown that gliadins bind to GM1 ganglioside and to the GM1-rich small intestinal brush border and the formation of antiganglioside antibodies may be caused on this basis. Beyond that, it was hypothesized that antibodies against gluten are neurotoxic themselves causing severe neurological dysfunctions or are markers for a neurotoxic immunological process [100,101]. Nutrient deficiencies typical of active CD may also lead to various neurological symptoms. Causative malabsorption includes the B vitamin complex, vitamins E and K, amino acids, fatty acids, calcium, magnesium, copper, phosphorus, L-carnitine, and inositol.

Intestinal inflammation caused by CD releases different cytokines and chemokines that may be neurotoxic themselves or may stimulate the brain to produce neuroimmune, neuroendocrine, and neurochemical changes. One of the hypothesized links between the gut and the brain postulates that the increased intestinal permeability, observed in active CD, allows gluten peptides and other antigens to cross the gut barrier, to enter the blood stream, and to cross the blood-brain barrier either causing neuroinflammation or affecting neurotransmission within the nervous system and the endogenous opiate system (see Section 7.5).

Moreover, an imbalanced intestinal microbiota has been suspected to be associated with the onset of neurological complaints. Dysbiosis of the gut microbial function may influence the microbiota-gut-brain axis with the potential to cause behavioral and neurophysical deficits [102].

Among neurological disorders, peripheral neuropathies, cerebellar ataxia, epilepsy, migraine, and cognitive impairment are complications most frequently reported in CD patients [103]. Peripheral neuropathy is a general term for conditions, in which the nerves in the peripheries do not perform properly. Around 10% of people, who have peripheral neuropathies with no obvious cause, actually have CD. The peripheral neuropathy in CD is mainly sensory; motor effects are seen less frequently. The most distant parts of the body such as hands, fingers, feet, toes, nose, and tongue, are affected. There may be numbness, tingling, pressure, and pain or even a lack of perception of pain. An acute treatment consists of pain and/or inflammation medications that should be followed by a strict GFD.

Cerebellar ataxia is a balance disturbance caused by the loss of position and motor coordination. It is one of the most frequent neurological syndromes associated with gluten. Gluten-associated ataxia (GA) may be defined as idiopathic sporadic ataxia with positive markers (AGAs) for gluten sensitization, even in the absence of duodenal enteropathy. GA has a prevalence of 15% amongst all ataxia cases and 40% of all idiopathic sporadic ataxia [104]. Symptoms can include staggering gait, lack of balance, poor coordination, unsteadiness during standing or walking, increased falls, and dysarthric speech. Comparative studies on 31 GA, 48 CD, and 37 nonceliac gluten sensitivity (NCGS) patients indicated that clinical characteristics of GA patients were different to CD patients but very close to NCGS patients [105]. Thus GA patients are better classified within the NCGS group than within the CD group. GA may result from immunological damage of the cerebellum and changes in the cerebellum brain stem and the peripheral nerves. There is no straightforward path to the diagnosis of GA. If no alternative cause for ataxia can be found, a GFD should be maintained for 12–24 months. Stabilization or even improvement due to the diet would be a strong argument that the patient suffers from GA. Continuing the GFD can decrease or stop further progression of symptoms but, if permanent damage has already occurred, the symptoms may not resolve.

Epilepsy is another neuropathy associated with CD. Data on the prevalence of CD cases in people with epilepsy differ from 1% to 8%. A systematic computer-based literature search indicated that epilepsy is 1.8 times more prevalent in patients with CD and CD is over two times more prevalent in patients with epilepsy compared to the general population [106]. Patients with epilepsy of unknown etiology should be investigated for serological markers of CD, as some studies have reported an improvement in patients with epilepsy after going on a GFD. Epilepsy is caused by intermittent, spontaneous electrical activity or discharge in the brain. The clinical picture generally includes a specific trial of symptoms such as occipital calcification and seizures originating from a number of brain locations. Seizure symptoms can include perceptual disturbance prior to seizure, disorientation, photosensitivity, euphoria, aphagia, loss of consciousness, and involuntary movements. Whether or not multiple sclerosis, reported to be associated with elevated AGAs, is part of gluten-mediated disorders, is still in discussion. CD patients are apparently not at increased risk for dementia overall, though subgroup analysis suggested that they may be at increased risk for vascular dementia [107].

CD has been associated with mild to severe migraines and headaches occurring periodically and coming on rapidly. A large cohort study on children and adolescents with recurrent headache demonstrated that CD prevalence was doubled in these patients [108]. A systematic computer-based research, including 40 articles published between 1987 and 2017, revealed that the prevalence of headache amongst those with CD was 26% in adult populations and 18% in pediatric populations [109]. GFD appeared to be an effective management leading to total resolution of headaches in up to 75% of patients. Symptoms may include pain in parts or all of the head such as stabbing, pulsating or throbbing possibly with nausea and vomiting, temporary visual loss or change, and increased sensitivity to noise and smell. They are caused by the constriction of the intracranial arteries. Blood flow abnormalities in the brain show a direct correlation between untreated CD and blood flow alternation.

Much less is known about the transient cognitive impairments ("brain fog") to memory, attention executive function, and the speed of cognitive processing reported by CD patients. The causal links to gluten are not known. The cognitive involvement is variable in its expression, its duration, and its prognosis ranging from transient and reversible subtle involvement to dementia itself. Several mechanisms have been proposed to explain the deleterious influence of CD on cognitive functions: nutritional deficiencies, elevation of circulating cytokine levels due to systemic inflammation, and low brain serotonin levels, for instance [110]. Cognitive impairments are measurable in untreated CD patients by subtle tests and improve with adherence to a GFD [111].

8.7.7 Psychiatric disorders

Numerous studies identified an association between CD and psychological symptoms [112]. Up to one third of adult CD patients was found to have a history of psychiatric disorders. The pathogenesis is thought to be due to immune reactions causing inflammation and damage in the brain and enhancing the production of cytokines and dysfunction in the gut-brain axis. The precise mechanisms underlying these reactions are not known. One of the hypothesized links between the gut and the brain postulates that an increased intestinal permeability allows gluten peptides to cross both gut and blood-brain barriers either causing neuro-inflammation or affecting the endogenous opiate system. Gluten peptides could also target the brain through the microbiota-gut-brain axis, where the vagus nerve is also a key element [113]. Several studies have shown improvement in psychiatric symptoms after the start of a GFD [114].

Anxiety, fatigue, and depression are common psychological complaints in patients with untreated CD and these comorbidities decisively contribute to a lower quality of life [115]. Symptoms of depression include persistent sad, anxious, and "empty" moods, feelings of hopelessness, pessimisms, guilt, and worthlessness as well as a loss of interest in activities that were once enjoyed. It is difficult to determine, whether depression is a separate condition or a result of CD. Some studies that identified an association between CD and depression found that malabsorption of essential nutrients and chronic intestinal inflammation may be responsible. Moreover, the depression might result from years of chronic, painful symptoms and of never feeling well. Three randomized-controlled trials and 10 longitudinal studies demonstrated that a GFD significantly improved pooled depressive symptom scores in patients with gluten-related disorders [116].

Schizophrenia is the psychiatric disorder with the most robust relationship to CD (the so-called bread madness). Schizophrenia is currently thought to be heterogeneous in both etiology and symptomatology and to comprise a range of subtypes. It is a mental disorder affecting approximately 1% of the population worldwide. The symptoms include abnormal social behavior, false beliefs, unclear and confused thinking and speech, auditory hallucinations, and lack of motivation. Difficulties in attention, long-term memory, executive functioning, and speed of processing also commonly occur. A genetic predisposition as well as psychological processes appear to be important contributory factors. The diagnosis is based on observed behavior and the patient's reported experience.

A systematic review and meta-analysis, using literature databases from inception to June 2017, found a significantly higher risk of schizophrenia among patients with CD compared to the general population [117]. Conversely, a study of more than 10,000 patients with schizophrenia showed that these patients were at an increased risk for CD [118] and several studies demonstrated that a GFD may improve schizophrenic symptoms. Opioid-like peptides derived from gluten (see Section 7.5) could be a factor relevant to the development or maintenance of schizophrenia symptoms in at least a subset of patients. The level of AGAs in people with schizophrenia is significantly higher than that observed in the general population indicating that gluten may be involved. The different antibody patterns of CD and schizophrenia however suggest that the mechanisms of the immune responses to gluten in the two conditions are probably different [119].

The spectrum of autism spectrum disorders (ASD) is characterized by deficits in social interactions, delay or lack of language, and restricted interests. These impairments can range from mild to severe. ASD affects approximately 1% of the general population with a male to female ratio of around 3:1. Both genetic and environmental cofactors, underlying the promotion of ASD, are in discussion. The link between gluten and behavioral problems in children with CD is well documented. In a number of young children with untreated CD, an autistic-like behavior can be observed. They appear sad, introverted and unwilling to socialize or communicate even with parents, symptoms that improve on a GFD. There is only minimal research pursuing a link between CD and ASD. Increased inflammatory cytokines, abnormal lymphocyte populations, and immune deregulations are assumed to be possible factors.

Proposed treatments for ASD include the use of various dietary interventions. Findings that a gluten- and casein-free diet appears to be beneficial in subjects with ASD have generated the base for the "opioid excess theory". This theory has been the focus of various investigations, since it partially reflects the symptomatology of the disorder [120]. The gastrointestinal digestion of the milk protein casein and gluten proteins produces small peptides (casomorphins and gliadorphins) that function as exogenous opioids (see Section 7.5). These peptides were identified in the urine of patients with ASD and were shown to bind to opioid receptors. It was therefore speculated that these peptides affect the central nervous system by modulating opioid levels in the brain. According to a detailed review of Buie [121], there is currently insufficient evidence to support instituting a gluten/casein-free diet as a treatment for ASD, although there may be a subgroup of patients, who might benefit from this diet. Large-scale randomized and placebo-controlled trials are needed to demonstrate the efficacy of a dietetic treatment.

Based on the opioid theory, naltrexone, an opioid antagonist, was used in the management of children with ASD. The effects were summarized by a systematic review of 10 relevant studies [122]. Out of 128 children with ASD, who received naltrexone, 77% showed significant improvement in symptoms of irritability and hyperactivity. However, there was not sufficient evidence that it had an impact on core features of autism in the majority of patients. It is likely that only a subgroup of ASD children with abnormal endorphin levels may respond to naltrexone. Beyond opioid antagonists, selected strains of bifidobacteria with high degradative capabilities may be useful as probiotic microorganisms to eliminate the effects of food-derived opioid peptides [123].

The association between CD and attention-deficit/hyperactivity disorders (ADHD) remains unclear up to this day [124]. To examine the prevalence of CD in children and adolescents with ADHD, 102 participants diagnosed with ADHD were examined for CD [125]. The results demonstrated that the prevalence of CD was not higher than in the general population. However, children with ADHD had high levels of AGAs showing that gluten may be involved in the disease. Up to now, there are not enough data to support screening for CD and the introduction of a GFD in children with ADHD unless there are additional indications.

8.7.8 Malignancy and mortality

Untreated CD has been associated with an increased risk of malignancy and mortality, but the strength of this association is contradictory among different studies. A population-based cohort study published in 2009 indicated that the overall risks of malignancies in untreated CD was not elevated compared to the general population during a 20-year follow-up [126]. However, certain malignancies such as non-Hodgkin's lymphoma and esophageal cancer seemed to be significantly associated with untreated CD. The most likely explanation for this increased risk was chronic inflammation. Similarly, a Finnish study demonstrated that the overall incidence ratio of malignant diseases was not significantly increased in untreated CD patient, whereas the special incidence ratios for non-Hodgkin lymphoma, small intestinal cancer, colon cancer, and basal cell carcinoma of the skin were increased [127]. The higher risk to develop a colon cancer was not confirmed by Volta and coworkers [128]. Only 6 out of 1757 individuals with CD (0.34%) developed a colon carcinoma. Another evaluation, using the Dutch nationwide population-based pathology database, showed that the relative risk of T-cell lymphoma, predominantly enteropathy associated T-cell lymphoma, was strongly associated with CD diagnosis [129]. Further CD-associated malignancies were small bowel adenocarcinoma and esophageal squamous cell carcinoma, but absolute risks were relatively low. Other types of lymphomas and gastrointestinal carcinomas were not associated with CD.

To determine, whether mortality in CD is influenced by mucosal healing, 7648 CD patients underwent a follow-up biopsy within 5 years following diagnosis [130]. Persistent villous atrophy was present in 43% of patients. They were not at increased risk of death compared with those with mucosal healing. The risk of mortality may be heightened by several complications of CD such as refractory CD (see below), small bowel carcinoma, and abdominal lymphoma. These complications are quite rare, but they are characterized by a very high mortality [131].

The most feared complication of CD is refractory CD (RCD) [132]. This condition is characterized by persistence or recurrence of villous atrophy and elevated levels of IELs despite a strict GFD for at least 6–12 months. It may also occur in patients, who previously had responded to a long-term GFD. RCD is very rare; for example, a Finnish study revealed a prevalence of 0.31% among diagnosed CD patients and 0.002% in the general population [133]. The diagnosis of RCD is based on specialized investigations of the small intestine (enteroscopy and video capsule endoscopy) and the analysis of IELs (immunohistochemistry and flow cytometry). Early identification of RCD enables early intervention, which results in a reduction of morbidity and mortality. The management of the disease should be undertaken in specialist centers and it requires intensive dietary supervision, strict gluten exclusion, and subsequent reevaluation.

RCD can be subdivided into two different types: Type I is indistinguishable from uncomplicated active CD except in its autonomy toward gluten exposure, whereas type II resembles a low-grade lymphoma characterized by clonal expansion of aberrant IELs [132]. Type I presents IELs having a normal expression of surface markers and the T-cell receptors are polyclonal. In type II, loss of IEL markers CD3 and CD8 and abnormal clonality of the γ/δ receptor of T cells occur. Type II is a far more severe presentation with a 5-year survival prognosis between 44% and 58%. It imposes a serious risk of developing a virtually lethal enteropathy-associated T-cell lymphoma, small intestinal adenocarcinoma, and progressive malabsorption. Patients with RCD II necessarily require evaluation for potential small intestinal cancer.

No curative treatment exists for RCD yet. Corticosteroids may improve clinical symptoms, but the use is limited by systemic side effects. The current treatment of RCD I includes prednisolone in combination with the immunosuppressants azathioprine, mesalamine, or tioguanine [134]. In RCD II, immunosuppressive drugs have no impact on the abnormal clonal IEL population and more aggressive treatments are necessary. Different treatment approaches are currently adopted with various results [135]. They include immunomodulation with azathioprine and steroids, methotrexate, cyclosporine, alemtuzumab, and cladribine or fludarabine sometimes combined with autologous stem cell transplantation.

8.8 Associated diseases

CD is strongly associated with autoimmune diseases and genetically determined syndromes. Autoimmune diseases occur, when the immune system that normally protects the body from foreign agents erroneously sees the body's own proteins and tissues as foreign and attacks them for example by specific antibodies. Like CD, the prevalence of autoimmune diseases has increased significantly over the last 40 years and today about 3% of the general population suffer from one or more autoimmune diseases. In comparison, about 30% of adult CD patients have at least one associated autoimmune disease. The estimated prevalence of CD in individuals with some of the best known autoimmune diseases is summarized in Table 8.6 [136]. All of them are clearly linked

TABLE 8.6 Prevalence of CD associated with autoimmune diseases and genetically determined disorders [136].

Condition	Prevalence (%)
Type 1 diabetes mellitus	4–10
Sjögren's syndrome	4–12
Autoimmune thyroid disease	4–5
Autoimmune hepatitis	3–6
Addison's disease	5
Down syndrome	5–10
Selective IgA deficiency	7

to CD, although some of the associations are stronger than others. Most data on prevalences strongly emphasize the need for routine screening of patients with autoimmune diseases for CD regardless of the presence and absence of symptoms, because late identification can lead to various complications, sometimes even very severe.

CD and autoimmune diseases share important features: They are chronic and multifactorial and have a female-to-male ratio of around 2:1 and common HLA- and non-HLA loci. Major risk factors, associated with the occurrence of autoimmune diseases in CD patients, are the family history of autoimmune diseases, overweight at CD diagnosis, and a delay of CD diagnosis [137]. The close relationship between CD and autoimmune diseases is largely explained by sharing a common genetic background. More than 60% of CD-associated susceptibility gene loci are shared with at least one autoimmune condition suggesting common pathogenic mechanisms. First-degree relatives of individuals with CD are at increased risk for developing CD, but little is known about their risk for autoimmune diseases. A Swedish study found that first-degree relatives of people with CD were at a slightly increased risk of autoimmune diseases (3.9%) compared with controls (3.0%) [138].

Roughly 75% of affected persons have been diagnosed first with an autoimmune disease and then with CD. The risk for CD-associated autoimmune diseases increases with the duration of gluten exposure and the age of patients. For example, the prevalence of autoimmune diseases in CD patients, diagnosed at the age of <2 years, was around 5% and that of CD patients, diagnosed at the age of >20 years, was around 34% [139]. These results indicate that a GFD may be protective against the development of autoimmune diseases, but profound data are still missing.

Type 1 diabetes mellitus (T1D) is one of the most complicated associations with CD, because it requires a lifelong adherence to two restrictive diets and lifestyles (avoidance of both sugar and gluten). T1D is characterized by a lack of insulin, the hormone that enables the body to use and store sugar (glucose). Without insulin, the body is unable to process its most readily available source of energy, namely glucose, provided by food. The symptoms are unexplained weight loss, extreme fatigue, thirst, poor diabetic control, diarrhea, and recurrent episodes of hypoglycemia. T1D and CD share high-risk HLA haplotypes and the prevalence of both diseases has increased in many regions during the past half century [140].

Sjögren's syndrome, encountered with similar frequency in CD patients as T1D, is an autoimmune disease, in which the glands that produce moisture and mucus in the mucous membranes become inflamed and/or destroyed. Glands of the mouth, eyes, nose, lungs, pancreas, and vagina may be affected. Thus Sjögren's syndrome causes extreme discomfort interfering with eyesight, digestion, and sexual function. Autoimmune thyroid (Hashimoto's) disease is characterized by an underactive thyroid gland inflamed by an autoimmune mechanism. The gland cannot produce enough of the necessary thyroid hormone.

Symptoms include fatigue, depression, muscle cramps, and constipation. Patients may also have weight gain, intolerance to cold, and dry thickened skin and scalp.

Several autoimmune hepatitis diseases have been associated with CD; the best known relation is with biliary cirrhosis. This disorder is characterized by an inflammatory process in the liver affecting the bile ducts. The patients have difficulties in excreting cholesterol and bilirubin (a degradation product of red blood cells) leading to an accumulation of bile salts in the liver. The major complaints are itching (pruritus) and development of yellow skin. In some cases, the disease leads to cirrhosis of the liver. Addison's disease is an autoimmune-mediated atrophy of the adrenal glands. Patients do not produce enough of the adrenal hormones like cortisol and aldosterone. Deficiencies of these hormones affect response to stress, immune reactions, cardiovascular functions, and production of sexual hormones. Other autoimmune diseases such as cardiomyopathy, lupus, and rheumatoid arthritis have also considered to be associated with CD.

The causative factors of the comorbidity of CD and autoimmune diseases are unclear. CD is a disease with an autoimmune component (TGAs) and shares different HLA and non-HLA gene loci with autoimmune diseases. The comparison of genetic patterns shows that CD bears a stronger resemblance to organ-specific autoimmune diseases than to inflammatory diseases. The dependence of the prevalence of autoimmune diseases on the age of diagnosis of CD may indicate that the time of gluten exposure is a cofactor for the development of autoimmune diseases (see above). However, the relationship between the treatment of CD and the prevention of autoimmune diseases by a GFD is still debated. The questions whether a GFD may prevent CD patients from contracting autoimmune diseases and whether an autoimmune disease may improve on a GFD should urgently be answered by future research.

The question whether CD belongs to the class of autoimmune diseases is controversially discussed. It is a fact that the immune reactions of CD are directed against the body's own intestinal enzyme TG2 resulting in a production of antibodies against TG2 (TGAs). However, this reaction happens only after oral intake of the exogenous agent gluten and can be inhibited by a GFD. In classical autoimmune diseases, no exogenous factor could be found up to now and autoimmunity cannot be stopped. The proinflammatory Th-1 pathway of the adaptive immune response and the innate response, both involved in the development of CD (see Section 8.9), are directed against gluten peptides and not against the body`s own proteins. The antiinflammatory Th-2 pathway is initiated only by TG2/gluten peptide conjugates and not by TG2 alone and the resulting TGAs are not pathogenic as far as known [141]. This may be confirmed by CD patients, who are seronegative (up to 10% of all diagnosed cases) (see Section 8.6) but can develop full-blown CD symptoms. Considering all these points, CD is not an autoimmune disease. The most decisive difference between CD and autoimmune diseases is that CD symptoms can be abolished by removing the exogenous factor (gluten).

On the other hand, CD shares important features with autoimmune diseases. Both disorders show a number of common HLA- and non-HLA-gene loci suggesting an overlap of genetic mechanisms involved. They are chronic, multifactorial, have female-to-male ratios of around 2:1, produce antibodies (TGAs) against their own tissues or organs and show T-cell mediated destruction of specific cells. Encompassing all aspects, CD should be defined as an immune-mediated food hypersensitivity with an autoimmune component [142]. The association of CD with autoimmune diseases will be an interesting field of future research. For instance, CD might be used as an example to identify exogenous factors that drive autoimmune processes of autoimmune diseases.

Patients with certain genetically determined syndromes also have a higher rate of CD compared to the general population. Examples are Down syndrome and IgA deficiency (Table 8.6). Down syndrome, also known as trisomy 21, is a genetic disorder caused by the presence of all or parts of a third copy of chromosome 21. The disease is associated with growth delays, mild to moderate intellectual disability, and characteristic facial features. Up to 10% of patients with Down syndrome have CD and the level of serum AGAs is frequently elevated. Affected subjects are advised to be screened for CD, because late identification can lead to various complications, sometimes even very severe [143].

IgA deficiency is inherited and associated with differences in chromosomes 18, 14, and 6. Most IgA-deficient individuals are asymptomatic; a minority has a tendency to develop recurrent gastrointestinal infections and disorders, allergies, autoimmune conditions, and malignancies. The presence of selective IgA deficiency, observed in a number of CD patients, can complicate diagnostic blood tests for CD and for this reason, IgG antibodies have to be used for diagnosis (see Section 8.10.1).

Other comorbidities associated with CD are less frequent. Examples are inflammatory bowel diseases including Crohn's disease, ulcerative colitis, herpes zoster, asthma, immune thrombocytopenic purpura, and idiopathic inflammatory myopathies.

8.9 Pathomechanism

Food proteins are usually broken down into small fragments (amino acids, dipeptides, and tripeptides) by gastrointestinal enzymes. However, the high proline content of gluten proteins, combined with the absence of prolyl endopeptidases in the human gastrointestinal tract, makes these proteins resistant to complete digestion so that long-chain peptides reach the intestinal mucosa. Nonetheless, this poor digestion alone is not sufficient to cause CD and there is no known difference between healthy individuals and CD patients in their non-ability to completely digest gluten proteins (see Section 7.3).

The pathological cascade starts during the intestinal epithelial passage. In healthy individuals, undigested gluten peptides mainly pass the enterocytes via

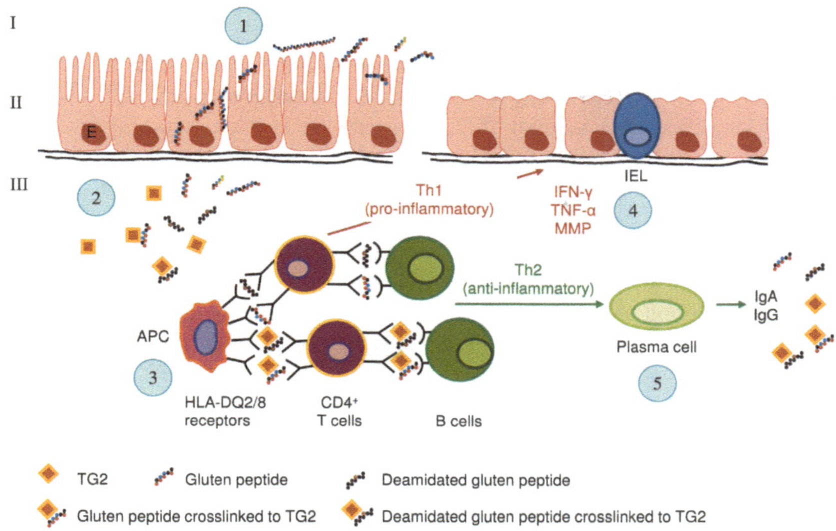

FIG. 8.5 Schematic representation of the adaptive immune response. I = intestinal lumen; II = epithelium; III = lamina propria; APC = antigen-presenting cell; IEL = intraepithelial lymphocytes; IFN = interferon-γ; Ig = immunoglobulin; MMP = matrix metalloproteinase; TG2 = tissue transglutaminase; Th = T helper; TNF = tumor necrosis factor. For steps 1–5, see text.

the transcellular route and are almost totally degraded by the lysosomal enzyme system (see Section 7.4). In CD patients, mucosal permeability is increased, a feature that is typical of CD. Elevated permeability usually precedes the onset of CD and is not just a consequence of intestinal inflammation. Therefore a heightened number of gluten peptides cross the intestinal mucosa and reach the lamina propria. There is evidence that this occurs by both the paracellular route (between the enterocytes) and the transcellular route (through the enterocytes) (step 1 in Fig. 8.5). On the one hand, gluten peptides interact with the chemokine receptor CXCR3 on the luminal side of enterocytes and induce an increased release of zonulin-1 combined with increased paracellular permeability [144]. On the other hand, CD-active peptides were shown to survive the transcellular pathway [145]. Possibly, CD71, a transferrin receptor that is upregulated in active CD helps gluten peptides to escape of lysosomal degradation known as the "Trojan horse" phenomenon [146]. After passing the epithelium via the para- and/or transcellular pathway and reaching the lymphatic tissue, CD-active peptides induce two different immune reactions: the adaptive and the innate responses. Major players in the immune cascade in the lamina propria are TG2, HLA-DQ2/8 heterodimers, T-cell receptors (TCRs), cytokines, and chemokines on the molecular level and antigen-presenting cells (APCs), T cells, B cells, and IELs on the cellular level [147].

8.9.1 Adaptive response

The mechanism of the adaptive immune system is quite well understood (Fig. 8.5) [148]. After arriving at the lamina propria, gluten peptides are specifically modified by intestinal TG2. This enzyme is expressed widely in different human tissues and is involved in several physiological and pathogenic processes. The enzyme usually occurs only intracellular but can be released from intracellular stores upon mechanical stress, inflammation, infection or during apoptosis. TG2 belongs to the ubiquitous family of calcium-dependent transamidation enzymes that catalyze the covalent cross-linking of proteins via an isopeptide bond between the amino acid side chains of lysine and glutamine. These cross-linked products are highly resistant to mechanical challenge and proteolytic degradation and are important for a number of tissues, where such properties are crucial such as skin and hair. They are also necessary for wound healing.

Under certain conditions, specific glutamine residues are deamidated to glutamic acid through the catalytic activity of TG2 instead of being cross-linked. The specificity of TG2 in the deamidation reaction is strongly restricted by the relative positions of proline (P) and hydrophobic amino acids (J) to glutamine (Q) within the amino acid sequences. The sequences QXP and QXXJ (X representing any amino acid) are preferred targets for TG2, whereas the enzyme is not active on QP and QXXP sequences. Thus deamidation by TG2 results in a highly selective introduction of the negative charge of glutamic acid (E) (e.g., EXP and EXXJ) into gluten peptides. Both reaction pathways of TG2 are regarded as important during CD pathogenesis (step 2 in Fig. 8.5): Deamidation of glutamine creates epitopes of gluten peptides that bind to HLA-DQ molecules with increased affinities and transamidation enables cross-linking of TG2 and gluten peptides, which activates the production of autoantibodies against TG2.

Gluten peptides, partly deamidated by TG2, are selectively bound by HLA-DQ2 or -DQ8 heterodimers on the surface of APCs (e.g., dendritic cells, macrophages, and B cells). Key anchor points in the grooves of the heterodimers are at positions p1, p4, p6, p7, and p9 of the involved gluten epitopes [149]. Both HLA-DQ2 and -DQ8 heterodimers contain positively charged pockets with preference for binding negatively charged peptide side chains (deamidated glutamine) mostly in positions p4 and/or p6 (DQ2.5 and DQ2.2) or p1 and p9 (DQ8) of the epitopes, respectively (Table 8.5). The deamidation of specific glutamine residues by TG2 is not an absolute requirement, especially in the case of HLA-DQ8 heterodimers, but it strongly favors peptide binding and participates in the selection of the gluten-sensitive TCR reservoir.

Each of the HLA-DQ2.5, -DQ2.2 and -DQ2.8 heterodimers recognizes different gluten epitopes (Table 8.5) [44,149,150]. As an example, Fig. 8.6 shows schematically that the epitope glia-a2 (PQPELPYPQ) from α-gliadins is linked to HLA-DQ2.5 at positions p1 (P), p4 (E), p6 (P), p7 (Y), and p9 (Q). These linkages are based on hydrogen bonds and ionic bonds. The bound peptides

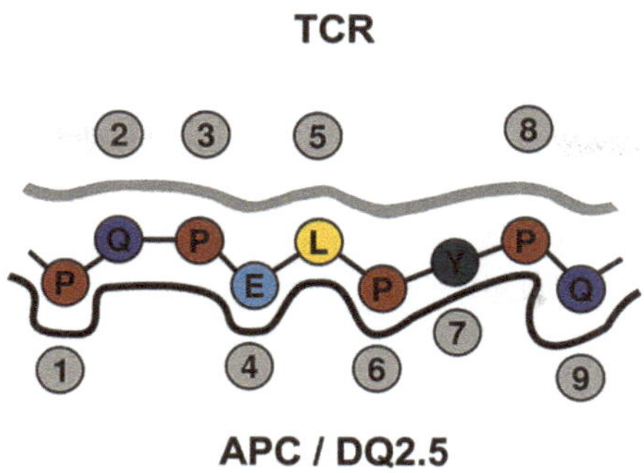

FIG. 8.6 Binding sites of an antigen-presenting cell (APC/DQ2.5) and a T-cell receptor (TCR) for the immunogenic epitope DQ2.5-glia-α2 from α-gliadins.

are then presented to TCRs on the surface of gluten-sensitive T cells (step 3 in Fig. 8.5). The specificity of TCRs for immunodominant gluten epitopes is strongly pronounced: Despite similarity of epitope sequences, most of TCRs respond to either of the epitopes [151]. In case of the epitope shown in Fig. 8.6, positions p2 (Q), p3 (P), p5 (L), and p8 (P) are involved in binding to the corresponding TCR. It appears that the complex formed by APCs, gluten peptides, and TCRs is constructed like a "hot dog", where the immunogenic epitope represents the sausage.

The peptide binding specificities of HLA-DQ2.2 are similar to those of HLA-DQ2.5 but seems to have a preference for binding serine at p3 instead of proline. In case of DQ2.2-restricted epitope glut-L1 from LMW-GS, for instance, binding sites for proline at p1, serine at p3, glutamic acid at p4 and p6, glutamine at p7, and valine at p9 were postulated [150]. The DQ8-restricted peptides have binding specificities that are strikingly different from those of DQ2.5- and DQ2.2-restricted peptides. The DQ8 heterodimer favors the binding of peptides with negatively charged glutamic acid in positions p1 and p9, whilst different amino acid residues occupy the remaining binding positions.

CD4+ T cells, activated by gluten peptides, play a crucial role in the CD pathogenesis and their specificities persist over decades. When the formation of the HLA-DQ/peptide/TCR complex is completed, CD4+ T cells elicit two different immune responses, the proinflammatory Th1 response and the anti-inflammatory Th2 response. In case of Th1, activated CD4+ T cells secrete proinflammatory cytokines and chemokines (e.g., interferon-γ, tumor necrosis factor, and matrix metalloproteinase) leading to mucosal damage (step 4 in Fig. 8.5). In case of Th2, stimulated T cells together with B cells elicit an anti-inflammatory response that produces IgA and IgG serum antibodies to gluten peptides and autoantibodies to the cross-linked gluten peptide/TG2 conjugate (step 5 in Fig. 8.5). These antibodies occur in the intestinal mucosa and the blood of CD patients and can be used as specific indicators to diagnose CD (see Section 8.10.1). It remains to be determined, whether these antibodies contribute to the progression of CD at an intestinal and extraintestinal level.

8.9.2 Innate response

Innate immunity is involved in the pathogenesis of CD in several ways [152]. It collaborates with the adaptive immunity to induce a proinflammatory Th1 response, favor the cytolytic attack on the epithelium, and initiate the increase in IELs, one of the hallmarks of CD (see Section 8.6). IELs represent an abundant and heterogeneous population of T cells located between the epithelial cells at the basolateral side of the epithelium. Activated IELs contribute to cell apoptosis, villous atrophy, and intestinal inflammation and thus damage of epithelial cells. The pathophysiology of cross-talk between enterocytes and IELs and how this interaction is crucial for intestinal integrity and homeostasis are discussed by Vitale and coworkers [153]. A count of more than 30 IELs per 100 enterocytes is characteristic for CD-damaged biopsies and represents "pathologic intraepithelial lymphocytosis" (see Section 8.10.2).

IELs are phenotypically and functionally different from the gluten-sensitive CD4+ T cells that are involved in the adaptive response. The contribution of IELs to villous atrophy in CD is based on the following facts: (1) IELs are the main producers of IFN-γ in active CD; (2) IELs are enriched in cytolytic proteins (perforin, granzymes, and FasL); and (3) the expression of the latter proteins is associated with the increased epithelial apoptosis. IEL infiltration into the epithelium is highly correlated with gluten ingestion and therefore is considered to be the most sensitive histopathological marker of CD. The proliferation of IELs is induced by the proinflammatory cytokine interleukin 15 (IL-15) that is therefore considered to be a central player within the gluten-induced innate immunity [154,155]. IL-15 is produced in excess in the intestine of untreated CD patients and stimulates the accumulation and cytotoxic activity of IELs. In a few cases, IL-15 activates the malignant innate-like IELs in RCD type II (see Section 8.7.8).

Gluten peptides that elicit the innate response are different from those inducing the adaptive response. Gluten peptides, derived from the N-terminal segment of α-gliadins (e.g., peptide α31-43) are commonly believed to induce an innate response in CD [156,157]. Glutamine side chains of these peptides are not deamidated by TG2. Therefore they show only weak or moderate binding to HLA-DQ heterodimers and none of the gluten-sensitive HLA-DQ-restricted T cells recognizes these peptides. They initiate significant toxic effects in *in vivo* (instillation) and organ culture tests and activate IL-15 production only in CD patients but not in healthy controls.

Up to now, only peptides from α-gliadins have been identified as triggering factor of the innate immunity. The fact that these peptides activate the innate response implies that receptors for these peptides should exist. However, these receptors remain elusive, except for implication of the chemokine receptor

CXCR3 on intestinal epithelial cells or the Toll-like receptors in monocytes. Thus more work is needed to delineate the mechanism of action of these peptides.

Amylase trypsin inhibitors (ATIs), present in wheat flour, were suggested as coactivators of the innate system [18] (see Section 10.3). They were identified as strong initiators of the innate response in monocytes, macrophages, and dendritic cells. These activated cells express Toll-like receptors (TLRs), a class of highly conserved membrane-bound molecules that have a principle role in the recognition of microorganisms and thus act in the first line of defense. ATIs engage the TLR4-MD2-CD14 complex, lead to upregulation of maturation markers, and elicit the release of proinflammatory cytokines. These findings define cereal ATIs as novel contributors to CD.

The long pentraxin 3 (PTX3), a marker of the acute-phase inflammatory response, plays an important role in innate immunity and in modulation of the adaptive immune response. The examination of patients with active CD revealed that PTX3 serum levels were significantly enhanced compared with healthy controls and correlated with IgA DGPA levels typical of CD [158]. Moreover, PTX3 serum levels were significantly elevated in patients with Marsh 1–3 type histology compared with Marsh 0 type and healthy controls and thus clearly reflected the histological status of duodenal biopsies from patients with active CD [159]. It was hypothesized that PTX3 is able to modulate the innate response to gliadin and to regulate the adaptive response in CD.

In conclusion, the role of the innate immune response is currently gaining extraordinary importance due to the fact that the underlying mechanism is not fully understood yet. In addition, an unanswered question is how the cross-talk between innate and adaptive immunities is orchestrated. The mutual interdependence of innate and adaptive responses is an outstanding issue that will need to be addressed in future.

8.10 Diagnosis

The precise diagnosis of CD is extremely important for individuals suspected of having CD. False-negative results burden CD patients with permanent impairment of health. False-positive results needlessly force individuals to a strict lifelong GFD, which is a considerable strain not only for the affected persons but also for their families. The diagnosis of CD is complex and tends to require a high level of clinical suspicion. Shannahan and Leffler have reviewed the current strategy to diagnose CD and the newer modalities to assess for associated complications [160]. Due to similar symptoms, CD must be differentiated from other gluten-related disorders (e.g., NCGS and wheat allergy), food intolerances (e.g., lactose intolerance and histamine intolerance), parasitic infections, and chronic bowel diseases (e.g., Crohn's disease, irritable bowel syndrome, and autoimmune enteropathy). The diagnostic approaches have changed during the last decades and recommendations for the diagnostic procedure still disagree. In

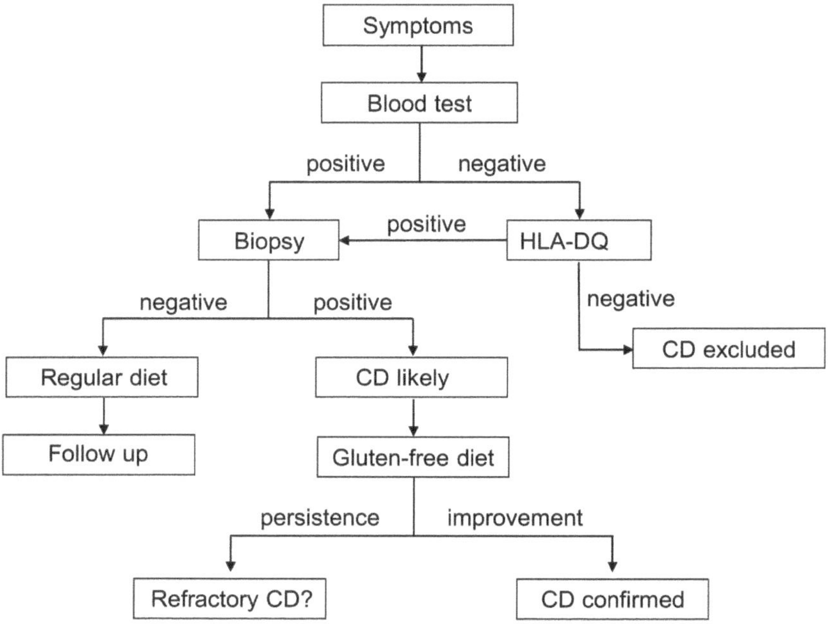

FIG. 8.7 Scheme to diagnose celiac disease in adults.

summary, the diagnostic scheme may consist of (a) clinical history and symptomatology, (b) serology, (c) small intestinal histology, (d) HLA status, and (e) response to a GFD (Fig. 8.7).

In young children with clear symptoms and positive serology (10× the upper limit for normal IgA TGAs), CD diagnosis may be established without histological examination [161]. Since 2012, guidelines of the European Society for Paediatric Gastroenterology Hepatology and Nutrition (ESPGHAN) have made it possible to avoid the biopsy in symptomatic pediatric patients with high levels of IgA TGAs, positivity to IgA antiendomysium antibodies (EMAs), and HLA-DQ2/8 alleles ("triple criteria"). In all other cases, duodenal biopsy is still mandatory to confirm the diagnosis. The intention to avoid an intestinal biopsy has been judged controversially. Pro and contra of a biopsy-avoiding pathway and the contrast between pediatric and adult guidelines have been discussed by Reilly and coworkers [162]. A clinical practice update, including key modalities in the diagnosis of CD in adults as well as in children and adolescents, has been presented by Husby and colleagues [163].

Several recent investigations have verified that biopsy is not always necessary for CD diagnosis. A large international prospective study have demonstrated that symptomatic children with high IgA TGA levels can be accurately diagnosed with CD (positive predictive value >99%) without biopsy analysis [164]. HLA-DQ analysis is not required for accurate diagnosis. Similarly, the diagnosis of CD can be reliably made in a high portion of adults based on serology alone using IgA TGAs [165]. At an antibody cut-off greater than 45 U/mL, the positive predictive value is 100%, a further evidence that biopsy should no longer be considered mandatory for CD diagnosis. Fuchs and colleagues confirmed that CD can safely be diagnosed without biopsy in adults fulfilling the "triple criteria", which enables the number of endoscopies to be reduced by one third [166]. Besides being easier for patients, up to 95% of diagnostic expenses may be spared by omitting the biopsy.

Potential diagnostic problems include patients with positive serology but normal histology, patients with negative serology but abnormal intestinal mucosa, and patients, who do not to respond to a GFD, although serology and histology are positive. It is important to mention that patients should consume normal gluten-containing foods until the end of the diagnostic process. However, a number of patients voluntarily initiate a GFD before diagnostic testing, which can lead to unclear results. In these cases, gluten has to be reintroduced into the diet for some weeks to reproduce serological markers and intestinal damage.

The primary care doctor has an important role at the beginning of the diagnostic procedure. Classical intestinal symptoms such as chronic diarrhea, steatorrhea, and postprandial abdominal pain should prompt the physician to initiate a serological test. Additionally, close relatives of CD patients and persons with autoimmune diseases should be advised by the doctor to be screened for CD. Associations of extraintestinal complaints and CD are more difficult to detect. Therefore medical specialists such as dermatologists, dentists, orthopedists, gynecologists, and neurologists should be aware of the possible extraintestinal manifestations [167]. Asymptomatic patients are often identified through case-finding strategies by screening at risk populations. Delay in diagnosis should be avoided, because it is associated with a greater risk of long-term complications. However, delays in diagnosis of patients without gastrointestinal complaints still remain prolonged with an average delay of 3.5 years [168]. Moreover, many patients have to consult three or more physicians prior to correct diagnosis. Barriers to accurate and timely diagnosis include atypical presentation, lack of physician awareness about current diagnostic criteria, misdiagnosis, and limited access to specialists [169].

8.10.1 Serological tests

CD is characterized by disease-specific serum antibodies in response to gluten exposure including AGAs, DGPAs, EMAs, and TGAs. Immunochemical tests are the methods of choice for antibody determination, because they provide high specificity and sensitivity in diagnosing CD. Most tests are based on the so-called enzyme-linked immunosorbent assay (ELISA) (see Section 12.3.3). The first serological tests were developed in the 1970s using AGAs and were initially based on immunofluorescent staining and then on ELISA. AGA tests seemed very promising, but their specificity was fairly poor, because AGAs was also detected in nonceliac enteropathies as well as in 5–10% of healthy individuals. The problems with AGAs were overcome by the development of tests based on EMAs (detected by indirect immunofluorescence), TGAs (ELISA), and DGPAs (ELISA). A new ELISA test system (tTg-neo), using a purified complex of cross-linked TG2 and gliadin peptides as antigen, has been presented by Porcelli and coworkers [170]. Different assays of tTg-neo (detecting IgA, IgG, and IgA + IgG) were compared with a conventional TGA test [171]. The results showed that serological diagnostic performance, reflection of intestinal damage, and false positivity were better with the tTg-neo test.

Tests with IgA antibodies are usually applied, because antibodies of the IgA class mostly have significantly better specificities compared with IgG class antibodies. In case of selective IgA deficiency that is present in around 7% of CD patients [136], tests against IgG antibodies have to be used, after IgA deficiency has been confirmed by the measurement of total IgA antibodies [172]. IgG TGAs seems to be more reliable markers than IgG DGPAs.

The quality of tests is primarily defined by their sensitivity (percentage of correctly identified CD patients) and specificity (percentage of correctly identified persons without CD). Table 8.7 shows corresponding data resulting from different test systems [173]. The commercially available test kits however can have widely different cut-off levels and overall performance, making the comparison of antibody levels in both individual patients and across populations quite difficult [174]. Numerous studies revealed that the clinical demand for both maximal sensitivity and maximal specificity cannot be achieved with a single test. Therefore the combination of different tests (e.g., IgA TGA and IgG DGPA tests) has been recommended.

Conventional ELISA tests require qualified personnel and laboratory facilities and results are available only after a time delay. Therefore different

TABLE 8.7 Sensitivity and specificity (%) of serological tests [173].

Antibody[a]	Sensitivity (range)	Specificity (range)
IgA AGA	85 (57–100)	90 (47–94)
IgG AGA	85 (42–100)	80 (50–94)
EMA	95 (86–100)	99 (97–100)
IgA TGA	98 (78–100)	98 (90–100)
IgG TGA	70 (45–95)	95 (94–100)
IgA DGPA	88 (74–100)	95 (90–99)
IgG DGPA	80 (63–95)	98 (90–99)

[a] *AGA, antigliadin antibody; DGPA, antideamidated gliadin peptide antibody; EMA, antiendomysium antibody; TGA, antitransglutaminase antibody.*

electrochemical immunosensors ("point-of-care tests"), mainly based on immobilized AGAs, DGPAs or TGAs, were developed for CD diagnosis [175]. The principle of these sensors may comprise working and reference electrodes, binding of the antigen to the surface of the working electrode, and adding of serum antibodies and an analyte-specific second antibody labeled with an enzyme. The reaction of the enzyme with an appropriate substrate initiates a current that can be measured by different electrical methods. The tests can be performed with a drop of blood and allows visual reading after a few minutes. A review by Gupta and coworkers [176], has summarized the progress in immunosensors for diagnosing CD in perspective of the scaled down hardware, multianalyte discovery, and low volume necessity. Results on specificity and sensitivity suggest that immunosensors may be promising alternatives to existing immunochemical tests such as ELISAs in the future. However, performance criteria for specificity, sensitivity, accuracy, handling, time, and cost of analysis should be further established for each system and compared with corresponding ELISA kits [177]. Finally, international collaborative studies should evaluate method performance and clarify, which of the developed sensors is the best candidate for substituting ELISA.

Altogether, serological testing is an extremely valuable tool for CD diagnosis. Standardization of assays and routine reporting of serum IgA concentration should be considered to improve the current diagnostic process. However, it should be kept in mind that the prevalence of seronegative CD accounted for up to 10% of diagnosed cases [172]. Thus seronegativity in individuals does not rule out the possibility of CD.

8.10.2 Intestinal histology

The histological judgment of the small intestinal mucosa is commonly regarded as the gold standard for CD diagnosis. Though villous atrophy is a key diagnostic finding in CD, non-celiac enteropathy is not rare. Many other entities such as peptic duodenitis, inflammatory bowel disease, milk protein allergy, drug-induced duodenitis, tropical sprue, and bacterial overgrowth can present with clinical and/or histologic features similar to CD and has to be ruled out by differential diagnosis. Key histologic mimickers of CD, including the differentiating pictures of biopsies with villous atrophy and epithelial lymphocytosis, have been highlighted by Gibiino and colleagues [178].

Intestinal biopsies are usually taken endoscopically using standard fiberoptic instruments. A review, provided by Lagana and Bhagat [179], has summarized the pathologist's perspective on biopsy adequacy, evaluation, and reporting in light of current knowledge. Four to six specimens should be taken from the second part of the duodenum and two samples from the bulb region (area just under the stomach). The microscopic evaluation of the biopsies should consider the following targets: luminal surface (e.g., check for infectious agents) and morphology of enterocytes (enterocyte thickness, ratio of villous height to crypt

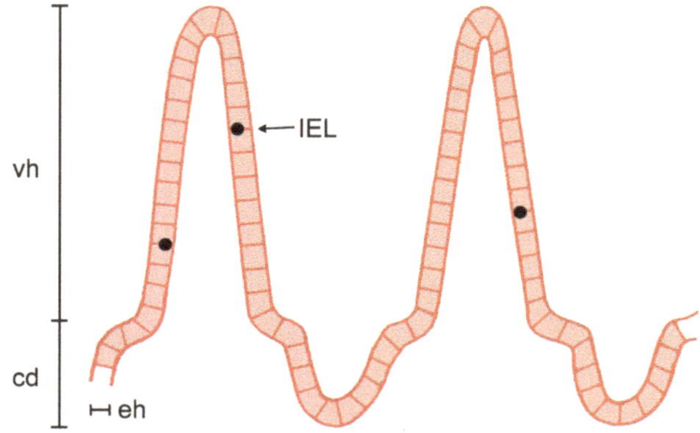

FIG. 8.8 Schematic picture of the intestinal epithelium (cd, crypt depth; eh, enterocyte height; IEL, intraepithelial lymphocyte; vh, villous height).

depth, and crypt hyperplasia) (Fig. 8.8). The increase in IELs, one of the hallmarks of CD, is an additional parameter of CD diagnosis. The number of IELs isolated from biopsies can be measured by flow cytometry, which is particularly useful for the diagnosis of CD in difficult cases [180]. However, the increase of IELs is not a CD-exclusive feature and its diagnostic evidence can only be evaluated in a broader clinical context. Video capsule endoscopy (VCE), introduced for the diagnosis of gastrointestinal diseases several years ago, should be reserved only for CD patients with equivocal diagnosis, alarm symptoms, refractory CD, suspicions of malignancies or for those, who are not willing or unable to undergo upper endoscopy including biopsy [181].

The so-called Marsh-Oberhuber classification (see Section 8.6) is frequently used by pathologists to evaluate the mucosal lesions. Villous to crypt ratio $<3:1$ or total villous atrophy and an IEL count of >30 per 100 enterocytes (Marsh types 2–4) are strong indicators of CD [70]. The Marsh 1 type histology is not specific for CD and can be associated with other immune-mediated disorders, infections or drugs [182]. In the case of equivocal histology, biopsy tissues can be utilized further to measure intestinal EMAs and TGAs. Altogether, there is a move toward a less invasive and perhaps more cost-effective diagnostic approach, but duodenal biopsy remains the gold standard for the majority of CD patients [183].

8.10.3 Genetic tests

Genetic tests, based on HLA-DQ2 and -DQ8 alleles, cannot be used to confirm CD, because they are also common in healthy people ($\approx 30\%$) (see Section 8.3). They should be performed to rule out the existence of CD, since they have a high negative predictive value. If a person suspected having CD is negative for HLA-DQ2/8, CD can be excluded lifelong with a certainty of about 97%. The special role of HLA-DQ typing in CD diagnostic has been reviewed by Lazar-Molnar and Snyder [184]. The genetic analysis is usually not applied routinely in the initial diagnosis of CD but can be useful, where the diagnosis is unclear after serological and histological testing.

For HLA-DQ testing, DNA is extracted from cell or blood samples and analyzed by commercially available assays based on the polymerase chain reaction (PCR). Even if HLA-DQ2.2 is rare, it should be included into testing besides HLA-DQ2.5 and -DQ8. A number of DNA tests, based on conventional PCR, have been developed and are commercially available. However, several comparative studies on the accuracy of single tests revealed high rates of variation. These findings highlight the importance of adherence to standardized guidelines for optimal performance of HLA typing.

New techniques makes HLA typing simple and fast at a relatively low cost. For example, bleed-to-read disposable microsystems for both the genetic and serological analysis of CD markers with amperometric detection demonstrated an excellent correlation with conventional reference methods [185]. Multiplex ligation-dependent probe amplification allows the simultaneous detection of HLA-DQ2.2, -DQ2.5, and -DQ8 alleles. A corresponding assay, validated against a set of 59 samples characterized by conventional techniques, genotyped all samples correctly and thus provided a reliable single-reaction analysis of the three CD risk genes [186].

8.10.4 Effects of a gluten-free diet

The clinical, serological, and optionally histological responses to a GFD should be finally used to confirm CD diagnosis. Most patients show rapid clinical improvement with their symptoms disappearing around 2 weeks after commencing the GFD. CD-specific antibody levels can take from 6 to 12 weeks to normalize. Histological normalization occurs within 3–6 months in children, whereas it might take 2–5 years in adults. However, a minority of CD patients on a GFD shows persistent symptoms or small intestinal damage. The most frequent reason for non-responsiveness is the continued ingestion of gluten, either intentional or unintentional (see Section 12.1). These patients should be monitored in their nutrients uptake by a dietitian for some weeks. Detection of gluten immunogenic peptides in the urine by a lateral flow test with gluten-sensitive antibodies is a simple noninvasive method to monitor continued ingestion of gluten [187]. Other reasons for non-responsiveness are conditions that coexist with CD such as irritable bowel disease, lactose intolerance or gastrointestinal reflux disease. In these cases, a review of the initial diagnosis of CD, a search for concurrent disorders, and a follow-up biopsy are recommended. The most feared cause of failure is refractory CD (see Section 8.7.8).

8.10.5 Mass screening

Since severely underdiagnosed, CD has been suggested appropriate for mass screening. CD fulfills the major criteria for mass screening: high prevalence, high numbers of patients with undetected CD, availability of sensitive and specific tests, available treatment, and protection from complications due to delay in diagnosis. Correspondingly, all individuals regardless of symptoms should undergo serological testing for CD and those, who are positive, should undergo duodenal endoscopy with biopsy. However, widespread screening of the general population is still controversially discussed. Arguments for and against mass screening have been summarized by Aggarwal and colleagues [188].

The arguments against screening include high costs and a considerable high number of false-positive test results, because serological testing is not 100% specific. False-positive results would lead to unnecessary medical procedures and psychological strain on affected individuals. In addition, there are doubts on the benefits of screening for asymptomatic persons. In those cases, it is not clear, if the questionable increase in health status will outweigh the burden of adhering to a GFD. Moreover, adherence to the GFD might be low in a screen-detected asymptomatic patient, who does not feel any apparent effect of a regular diet. The US Prevention Service Task Force concluded that the current evidence is insufficient to assess the balance of benefits and harms of screening for CD in asymptomatic persons [189].

On the other hand, important health problems such as reduced fetal growth and birth weight, delayed growth in height and weight in children, and reduced bone mineral density in both children and adults can be prevented by mass screening [190]. In addition, untreated children with CD may have reduced bone mineral density, lower levels of D vitamin, higher levels of parathyroid hormone, and signs of systemic inflammation compared with children on a GFD and controls [191]. This indicates that children with screening-detected CD clearly benefit from an early diagnosis and treatment. Undiagnosed adult CD patients are more likely to develop osteoporosis, dermatitis herpetiformis, chronic fatigue, thyroitis, and autoimmune diseases, if CD is not detected in time [192]. In patients with autoimmune diseases, known to occur alongside CD (see Section 8.8), screening could have beneficial effects owing to the effectiveness of gluten-free dietary interventions in CD [193].

Compromising the different opinions, screening is recommended for individuals in risk groups such as close relatives of subjects, who have a confirmed diagnosis of CD, and for persons with autoimmune diseases known to be associated with CD ("active case-finding") [194]. A cost-effective case finding policy and new strategies of mass screening could significantly reduce the morbidity and mortality associated with untreated CD [195].

References

[1] Gasbarrini G, Rickards O, Martinez-Labarga C, Pacciani E, Chilleri F, Laterza L, et al. Origin of celiac disease: how old are predisposing haplotypes? World J Gastroenterol 2012;18:5300–4.

[2] Lionetti E, Catassi C. Co-localization of gluten consumption and HLA-DQ2 and -DQ8 genotypes, a clue to the history of celiac disease. Dig Liv Dis 2014;46:1057–63.

[3] Laas MW, Schmitz R, Uhlig HH, Zimmer K-P, Thamm M, Koletzko S. The prevalence of celiac disease in children and adolescents in Germany. Dtsch Arztebl Int 2015;112:553–60.

[4] Singh P, Arora A, Strand TA, Leffler DA, Catassi C, Green PH, et al. Global prevalence of celiac disease: systematic review and meta-analysis. Clin Gastroenterol Hepatol 2018;16. 823–836.e2.
[5] Catassi C, Ratsch IM, Gandolfi L, Pratesi R, Fabiani E, El Asmar R. Why is coeliac disease endemic in the people of the Sahara? Lancet 1999;354:647–8.
[6] Ludvigsson JF, Murray JA. Epidemiology of celiac disease. Gastroenterol Clin North Am 2019;48:1–18.
[7] Rubio-Tapia A, Kyle RA, Kaplan EL, Johnson DR, Page W, Erdtmann F, et al. Increased prevalence and mortality in undiagnosed celiac disease. Gastroenterology 2009;137:88–93.
[8] Lohi S, Mustalahti K, Kaukinen K, Laurila K, Collin P, Rissanen H, et al. Increasing prevalence of coeliac disease overtime. Aliment Pharmacol Ther 2007;26:1217–25.
[9] Almallouhi E, King KS, Patel B, Wi C, Juhn YJ, Murray JA, et al. Increasing incidence and altered presentation in a population-based study of pediatric celiac disease in North America. J Pediatr Gastroenterol Nutr 2017;65:432–7.
[10] Grode L, Bech BH, Jensen TM, Humaidan P, Agerholm IE, Plana-Ripoll O, et al. Prevalence, incidence, and autoimmune comorbidities of celiac disease: a nation-wide, population-based study in Denmark from 1977 to 2016. Eur J Gastroenterol Hepatol 2018;30:83–91.
[11] Shewry PR, Pellny T, Lovegrove A. Is modern wheat bad for health? Nat Plants 2016;2:16097.
[12] Kasarda DD. Can an increase in celiac disease be attributed to an increase in the gluten content of wheat as a consequence of wheat breeding? J Agric Food Chem 2013;61:1155–9.
[13] Pronin D, Scherf K. Wheatscan—Aufklärung der Ursachen für Weizenunverträglichkeiten. Bericht 2017. Leibniz-Institut für Lebensmittel-Systembiologie an der Technischen Universität München; 2017, p. 53–5 ISBN: 978-3-00-058295-0.
[14] Prandi B, Tedeschi T, Folloni S, Galaverna G, Sforza S. Peptides from gluten digestion: a comparison between old and modern wheat varieties. Food Res Int 2017;91:92–102.
[15] Malalgoda M, Meinhardt SW, Simek S. Detection and quantitation of immunogenic epitopes related to celiac disease in historical and modern hard red spring wheat cultivars. Food Chem 2018;264:101–7.
[16] Ribeiro M, Rodriguez-Quijano M, Nunes FM, Carrillo JM, Branlard G, Igrejas G. New insights into wheat toxicity: breeding did not seem to contribute to a prevalence of potential celiac disease's immunostimmulatory epitopes. Food Chem 2016;213:8–18.
[17] Ozuna CV, Barro F. Characterization of gluten proteins and celiac disease-related immunogenic epitopes in the *Triticeae*: cereal domestication and breeding contributed to decrease the content of gliadins and gluten. Mol Breed 2018;38:22.
[18] Yunker Y, Zeissig S, Kim S-J, Barisani D, Wieser H, Leffler DA, et al. Wheat amylase trypsin inhibitors drive intestinal inflammation via activation of Toll-like receptor 4. J Exp Med 2012;209:2395–408.
[19] Dominguez-Bello MG, Costello EK, Contreras M, Magris M, Hidalgo G, Fierer N, Knight R. Delivery mode shapes the acquisition and structure of the initial microbiota across multiple body habitats in newborns. Proc Natl Acad Sci U S A 2010;107:11971–5.
[20] Collin P, Vilppula A, Luostarinen L, Holmes GK, Kaukinen K. Review article: coeliac disease in later life must not be missed. Aliment Pharmacol Ther 2018;47:563–72.
[21] Lebwohl B, Spechler SJ, Wang TC, Green PH, Ludvigsson JF. Use of proton pump inhibitors and subsequent risk of celiac disease. Dig Liver Dis 2014;46:36–40.
[22] Odenwald MA, Turner JR. The intestinal epithelial barrier: a therapeutic target? Nat Rev Gastroenterol Hepatol 2017;14:9–21.
[23] Olivares M, Benitez-Paez A, de Palma G, Capilla A, Nova E, Castillejo G, et al. Increased prevalence of pathogenic bacteria in the gut microbiota of infants at risk of developing celiac disease: the PROFICEL study. Gut Microbes 2018;9:551–8.

[24] Ricano-Ponce I, Wijmenga C, Gutierrez-Achury J. Genetics of celiac disease. Best Pract Res Clin Gastroenterol 2015;29:399–412.
[25] Abadie V, Sollid LM, Barreiro LB, Jabri B. Integration of genetic and immunological insights into a model of celiac disease pathogenesis. Annu Rev Immunol 2011;29:493–525.
[26] Kamycheva E, Goto T, Camargo Jr. CA. Celiac disease autoimmunity is associated with leukocyte telomere shortening in older adults: the U.S. National Health and Nutrition Examination Survey. Exp Gerontol 2017;89:64–8.
[27] Van de Kamer JH, Weijers HA, Dicke WK. Coeliac disease. IV. An investigation into the injurious constituents of wheat in connection with their action on patients with coeliac disease. Acta Pediatr 1953;42:223–31.
[28] Wieser H, Koehler P, Konitzer K. Celiac disease and gluten. multidisciplinary challenges and opportunities. London, Waltham, San Diego: Academic Press, Elsevier; 2014. p. 106–15.
[29] Kilmartin C, Lynch S, Abuzakouk M, Wieser H, Feighery C. Avenin fails to induce a Th1 response in coeliac tissue following in vitro culture. Gut 2003;52:47–52.
[30] Kilmartin C, Wieser H, Abuzakouk M, Kelly J, Jackson J, Feighery C. Intestinal T cell responses to cereal proteins in celiac disease. Dig Dis Sci 2006;51:202–9.
[31] Pinto-Sanchez MI, Causada-Calo N, Bercik P, Ford AC, Murray JA, Armstrong D, et al. Safety of adding oats to a gluten-free diet for patients with celiac disease: systematic review and meta-analysis of clinical and observational studies. Gastroenterology 2017;153:395–409.
[32] Van den Broeck HC, Smulders MJ, Hamer RJ, Gilissen JW, van der Meer IM. Coeliac-safe wheat. A novel wheat to decrease the prevalence and symptoms of coeliac disease. Agro Food Ind Hi-Tech 2011;22:18–21.
[33] Geisslitz S, Wieser H, Scherf KA, Koehler P. Gluten protein composition and aggregation properties as predictors for bread volume of common wheat, spelt, durum wheat, emmer and einkorn. J Cereal Sci 2018;83:204–12.
[34] Gianfrani C, Maglio M, Rotondi-Aufiero V, Camarca A, Vocca I, Iaquinto G, et al. Immunogenicity of monococcum wheat in celiac patients. Am J Clin Nutr 2012;96:1339–45.
[35] Suligoi T, Gregorini A, Colomba M, Ellis HJ, Ciclitira PJ. Evaluation of the safety of ancient strains of wheat in coeliac disease reveals heterogeneous small intestinal T cell responses suggestive of coeliac toxicity. Clin Nutr 2013;32:1043–9.
[36] Zanini B, Villanacci V, de Leo L, Lanzini A. *Triticum monococcum* in patients with celiac disease: a phase II open study on safety of prolonged daily administration. Eur J Nutr 2015;54:1027–9.
[37] Tanner GJ, Blundell MJ, Colgrave ML, Howitt CA. Creation of the first ultra-low gluten barley (*Hordeum vulgare* L.) for coeliac and gluten-intolerant populations. Plant Biotechnol J 2016;14:1139–50.
[38] Mujico JR, Mitea C, Gilissen LJ, de Ru A, van Veelen P, Smulders MJ, Koning F. Natural variation in avenin epitopes among oat varieties: implications for celiac disease. J Cereal Sci 2011;54:8–12.
[39] Catassi C, Fabiani E, Jacono G, D'Agate C, Francavilla R, Biagi F, et al. A prospective, double-blind, placebo-controlled trial to establish a safe gluten threshold for patients with celiac disease. Am J Clin Nutr 2007;85:160–6.
[40] Dewar DH, Amato M, Ellis HJ, Pollock EL, Gonzalez-Cinca N, Wieser H, et al. The toxicity of high molecular weight glutenin subunits of wheat to patients with coeliac disease. Eur J Gastroenterol Hepatol 2006;18:483–91.
[41] Hekkens WTJM. The toxicity of gliadin, a review. In: McNicholl B, McCarthy CF, Fottrell PF, editors. Perspectives in coeliac disease. Lancaster: MTP Press; 1978. p. 3–15.

[42] Tanner GI, Howitt CA, Forrester RI, Campbell PM, Tye-Din JA, Anderson RP. Dissecting the T-cell response to hordeins in coeliac disease can develop barley with reduced immunotoxicity. Aliment Pharmacol Ther 2010;32:1184–91.
[43] Wieser H, Belitz HD, Ashkenazi A. Amino acid sequence of the coeliac active peptide B3142. Z Lebensm Unters Forsch 1984;179:371–6.
[44] Sollid LM, Qiao SW, Anderson RP, Gianfrani C, Koning F. Nomenclature and listing of celiac disease relevant gluten T-cell epitopes restricted by HLA-Q molecules. Immunogenetics 2012;64:455–60.
[45] Schalk K, Lang C, Wieser H, Koehler P, Scherf KA. Quantitation of the immunodominant 33-mer peptide from α-gliadin in wheat flours by liquid chromatography tandem mass spectrometry. Sci Rep 2017;7:45092.
[46] Pes GM, Bibbo S, Dore MP. Coeliac disease: beyond genetic susceptibility and gluten. A narrative review. Ann Med 2019;51:1–16.
[47] Verdu EF, Caminero A. How infection can incite sensitivity to food. Science 2017;356:29–30.
[48] Beyerlein A, Donnachie E, Ziegler AG. Infection in early life and development of celiac disease. Am J Epidemiol 2017;186:1277–80.
[49] Kemppainen KM, Lynch KF, Liu E, Lönnrot M, Simell V, Briese T, et al. Factors that increase risk of celiac disease autoimmunity after a gastrointestinal infection in early life. Clin Gastroenterol Hepatol 2017;15. 694–702.e5.
[50] Lebwohl B, Green PH, Murray JA, Ludvigsson JF. Season of birth in a nationwide cohort of coeliac disease patients. Arch Dis Child 2013;98:48–51.
[51] Krishnareddy S. The microbiome in celiac disease. Gastroenterol Clin North Am 2019;48:115–26.
[52] Verdu EF, Galipeau HJ, Jabri B. Novel players in coeliac disease pathogenesis: role of the gut microbiota. Nat Rev Gastroenterol Hepatol 2015;12:497–506.
[53] Lionetti E, Castellaneta S, Francavilla R, Pulvirenti A, Catassi C. Mode of delivery and risk of celiac disease: risk of celiac disease and age at gluten introduction cohort study. J Pediatr 2017;184:. 81–86.e2.
[54] Koletzko S, Lee HS, Beyerlein A, Aronsson CA, Hummel M, Liu E, et al. Cesarean section on the risk of celiac disease in the offspring: the Teddy study. J Pediatr Gastroenterol Nutr 2018;66:417–24.
[55] Kristensen K, Henriksen L. Cesarean section and disease associated with immune function. J Allergy Clin Immunol 2016;137:587–90.
[56] Namatovu F, Olsson C, Lindkvist M, Myleus A, Högberg U, Ivarsson A, Sandström O. Maternal and perinatal conditions and the risk of developing celiac disease during childhood. BMC Pediatr 2016;16:77.
[57] Bach JF. The hygiene hypothesis in autoimmunity: the role of pathogens and commensals. Nat Rev Immunol 2018;18:105–20.
[58] Kondrashova A, Mustalahti K, Kaukinen K, Viskari H, Volodicheva V, Haapala AM, et al. Lower economic status and inferior hygienic environment may protect against celiac disease. Ann Med 2008;40:223–31.
[59] Croese J, Giacomin P, Navarro S, Clouston A, McCann L, Dougall A, et al. Experimental hookworm infection and gluten microchallenge promote tolerance in celiac disease. J Allergy Clin Immunol 2015;135:508–16.
[60] Lebwohl B, Murray JA, Verdu EF, Crowe SE, Dennis M, Fasano A, et al. Gluten introduction, breastfeeding, and celiac disease: back to the drawing board. Am J Gastroenterol 2016;111:12–4.

[61] Crespo-Escobar P, Mearin ML, Hervas D, Auricchio R, Castillejo G, Gyimesi J, et al. The role of gluten consumption at any early age in celiac disease development: a further analysis of the prospective PreventCD cohort study. Am J Clin Nutr 2017;105:890–6.
[62] Bjarnason I, Scarpignato C, Holmgren E, Olszewski M, Rainsford KD, Lanas A. Mechanisms of damage to the gastrointestinal tract from nonsteroidal anti-inflammatory drugs. Gastroenterology 2018;154:500–14.
[63] Sostres C, Gargallo CJ. Gastroenterological lesions and complications of low-dose aspirin in the gastrointestinal tract. Best Pract Res Clin Gastroenterol 2012;26:141–51.
[64] Meijer C, Shamir R, Szajewska H, Mearin L. Celiac disease prevention. Front Pediatr 2018;6:368.
[65] Schwertz E, Kahlenberg F, Sack U, Richter T, Stern M, Conrad K, et al. Serologic assay based on gliadin-related nonapeptides a highly sensitive and specific diagnostic aid in celiac disease. Clin Chem 2004;50:2370–5.
[66] Ellis HJ, Lozano-Sanchez P, Permudo-Redondo C, Suligoj T, Biagi F, Bianchi PI, et al. Antibodies to wheat high-molecular-weight glutenin subunits in patients with celiac disease. Int Arch Allergy Immunol 2012;159:428–34.
[67] Dieterich W, Ehnis T, Bauer M, Donner P, Volta U, Riecken EO, et al. Identification of tissue transglutaminase as the autoantigen of celiac disease. Nat Med 1997;3:797–801.
[68] Schiepatti A, Sanders DS, Biagi F. Seronegative coeliac disease: clearing the diagnostic dilemma. Curr Opin Gastroenterol 2018;34:154–8.
[69] Dore MP, Pes GM, Dettori I, Villanacci V, Manca A, Realdi G. Clinical and genetic profile of patients with seronegative coeliac disease: the natural history and response to gluten-free diet. BMJ Open Gastroenterol 2017;4:e000159.
[70] Oberhuber G, Granditsch G, Vogelsang H. The histopathology of coeliac disease: time for a standardized report scheme for pathologists. Eur J Gastroenterol Hepatol 1999;11:1185–94.
[71] Jericho H, Sansotta N, Guandalini S. Extraintestinal manifestations of celiac disease: effectiveness of the gluten-free diet. J Pediatr Gastroenterol Nutr 2017;65:75–9.
[72] Do Vale RR, Conci NS, Santana AP, Pereira MB, Menezes NY, Takayasu V, et al. Celiac crisis: an unusual presentation of gluten-sensitive enteropathy. Autops Case Rep 2018;8:e2018027.
[73] Leffler D, Schuppan D, Pallav K, Najarian R, Goldsmith JD, Hansen J, et al. Kinetics of the histological, serological and symptomatic responses to gluten challenge in adults with coeliac disease. Gut 2013;62:996–1004.
[74] Karatepe Hashas AS, Altunel O, Sevinc E, Duru N, Alabay B, Torun YA. The eyes of children with celiac disease. J AAPOS 2017;21:48–51.
[75] Macho VM, Coelho AS, Veloso e Silva DM, de Andrade DJ. Oral manifestations in pediatric patients with coeliac disease—a review article. Open Dent J 2017;11:539–45.
[76] Reunala T, Salmi TT, Hervonen K, Kaukinen K, Collin P. Dermatitis herpetiformis: a common extraintestinal manifestation of coeliac disease. Nutrients 2018;10:E602.
[77] Mansikka E, Hervonen K, Salmi TT, Kautiainen H, Kaukinen K, Collin P, Reunala T. The decreasing prevalence of severe villous atrophy in dermatitis herpetiformis: a 45-year experience in 393 patients. J Clin Gastroenterol 2017;51:235–9.
[78] Rodrigo L, Beteta-Gorriti V, Alvarez N, Gomez de Castro C, de Dios A, Palacios L, Santos-Juanes J. Cutaneous and mucosal manifestations associated with celiac disease. Nutrients 2018;10:E80.
[79] Ungprasert P, Wijarnpreecha K, Kittanamongkolchai W. Psoriasis and risk of celiac disease: a systematic review and meta-analysis. Indian J Dermatol 2017;62:41–6.
[80] Kaplan L, Castelo-Soccio L. When your patient's parent asks: "My child's alopecia areata is not getting better. Should he or she get tested for celiac disease?". Pediatr Dermatol 2018;35:535–8.

[81] Hoffmanova I, Sanchez D, Tuckova L, Tlaskalova-Hogenova H. Celiac disease and liver disorders: from putative pathogenesis to clinical implications. Nutrients 2018;10:E892.

[82] Kaukinen K, Halme L, Collin P, Färkkilä M, Mäki M, Vehmanen P, et al. Celiac disease in patients with severe liver disease: gluten-free diet may reverse hepatic failure. Gastroenterology 2002;122:881–8.

[83] Boonpheng B, Cheungpasitporn W, Wijarnpreecha K. Renal disease in patients with celiac disease. Minerva Med 2018;109:126–40.

[84] Wijarnpreecha K, Thongprayoon C, Panjawatanan P, Thamcharoen N, Pachariyanon D, Nakkala K, et al. Celiac disease and the risk of kidney diseases: a systematic review and meta-analysis. Dig Liver Dis 2016;48:1418–24.

[85] Ciaccio EJ, Lewis SK, Biviano AB, IyerV GH, Green PH. Cardiovascular involvement in celiac disease. World J Cardiol 2017;9:652–66.

[86] Santoro L, de Mattheis G, Fuorlo M, Giupponi B, Martone AM, Landi F, et al. Atherosclerosis and cardiovascular involvement in celiac disease: the role of autoimmunity and inflammation. Eur Rev Med Pharmacol Sci 2017;21:5437–44.

[87] Lebwohl B, Nobel YR, Green PH, Blaser MJ, Ludvigsson JF. Risk of *Clostridium difficile* infection in patients with celiac disease: a population-based study. Am J Gastroenterol 2017;112:1878–84.

[88] Röckert TA, Bonnedahl J, Inghammar M, Egesten A, Kahlmeter G, Naucler P, et al. Coeliac disease and invasive pneumococcal disease: a population-based cohort study. Epidemiol Infect 2017;145:1203–9.

[89] Lasa JS, Zubiaurre I, Soifer LO. Risk of infertility in patients with celiac disease: a meta-analysis of observational studies. Arq Gastroenterol 2014;51:144–50.

[90] Singh P, Arora S, Lal S, Strand TA, Makharia GK. Celiac disease with infertility: a meta-analysis. J Clin Gastroenterol 2016;50:33–9.

[91] Schiepatti A, Sprio E, Sanders DS, Lovati E, Biagi F. Coeliac disease and obstetric and gynaecological disorders: where are we now? Eur J Gastroententerol Hepatol 2019;31:425–33.

[92] Soni S, Badawy SZ. Celiac disease and its effect on human reproduction: a review. J Reprod Med 2010;55:3–8.

[93] Tersigni C, Castellani R, de Waure C, Fattorossi A, de Spirito M, Gasbarrini A, et al. Celiac disease and reproductive disorders: meta-analysis of epidemiologic associations and potential pathogenic mechanisms. Hum Reprod Update 2014;20:582–93.

[94] Lebwohl B, Stephansson O, Green PH, Ludvigsson JF. Mucosal healing in patients with celiac disease and outcomes of pregnancy: a nationwide population-based study. Clin Gastroenterol Hepatol 2015;13:1111–7.

[95] Grode L, Bech BH, Plana-Ripoll O, Bliddal M, Agerholm IE, Humaidan P, Ramlau-Hansen CH. Reproductive life in women with celiac disease: a nationwide, population-based matched cohort study. Hum Reprod, 2018;33:1538–47.

[96] Zis P, Hadjivassiliou M. Treatment of neurological manifestations of gluten sensitivity and coeliac disease. Curr Treat Options Neurol 2019;21:10.

[97] Pennisi M, Bramanti A, Cantone M, Pennisi G, Bella R, Lanza G. Neurophysiology of the "celiac brain": disentangling gut-brain connections. Front Neurosci 2017;11:498.

[98] Currie S, Hadjivassiliou M, Clark MJ, Sanders DS, Wilkinson ID, Griffiths PD, Hoggard N. Should we be `nervous' about coeliac disease? Brain abnormalities in patients with coeliac disease referred for neurological opinion. J Neurol Neurosurg Psychiatry 2012;83:1216–21.

[99] Przybylska-Felus M, Zwolinska-Wcislo M, Piatek-Guziewicz A, Furgula A, Salapa K, Mach T. Concentrations of antiganglioside M1 antibodies, neuron specific enolase, and interleukin 10 as potential markers of autonomic nervous system impairment in celiac disease—preliminary findings. Pol Arch Med Wewn 2016;126:763–71.

[100] Vojdani A, Kharrazian D, Mukherjee PS. The prevalence of antibodies against wheat and milk proteins in blood donors and their contribution to neuroimmune reactivities. Nutrients 2014;6:15–36.
[101] Hadjivassiliou M, Sanders DS, Grünewald RA, Woodrove N, Boscolo S, Aeschlimann D. Gluten sensitivity: from gut to brain. Lancet Neurol 2010;9:318–30.
[102] Oriach CS, Robertson RC, Stanton C, Cryan JF, Dinan DG. Food for thought: the role of nutrition in the microbiota-gut-brain axis. Clin Nutr Exp 2016;6:25–38.
[103] Casella G, Bordo BM, Schalling R, Villanacci V, Salemme M, Di Bella C, et al. Neurological disorders and celiac disease. Minerva Gastroenterol Dietol 2016;62:197–206.
[104] Hadjivassiliou M, Sanders DS, Aeschlimann DP. Gluten-related disorders: gluten ataxia. Dig Dis 2015;33:264–8.
[105] Rodrigo L, Hernandez-Lahoz C, Lauret E, Rodriguez-Pelaez M, Soucek M, Ciccocioppo R, et al. Gluten ataxia is better classified as non-celiac gluten sensitivity than as celiac disease: a comparative clinical study. Immunol Res 2016;64:558–64.
[106] Julian T, Hadjivassiliou M, Zis P. Gluten sensitivity and epilepsy: a systematic review. J Neurol 2019;266:1557–65.
[107] Lebwohl B, Luchsinger JA, Freedberg DE, Green PH, Ludvigsson JF. Risk of dementia in patients with celiac disease: a population-based cohort study. J Alzheimers Dis 2016;49:179–85.
[108] Nenna R, Petrarca L, Verdecchia P, Florio M, Pietropaoli N, Mantrogiorgio G, et al. Celiac disease in a large cohort of children and adolescents with recurrent headache: a retrospective study. Dig Liver Dis 2016;48:495–8.
[109] Zis P, Sarrigiannis PG, Rao DG, Hadjivassiliou M. Quality of life in patients with gluten neuropathy: a case-controlled study. Nutrients 2018;10:E662.
[110] Makhlouf S, Messelmani M, Zaouali J, Mrissa R. Cognitive impairment in celiac disease and non-celiac gluten sensitivity: review of literature on the main cognitive impairments, the imaging and the effect of gluten free diet. Acta Neurol Belg 2018;118:21–7.
[111] Yelland GW. Gluten-induced cognitive impairment ("brain fog") in coeliac disease. J Gastroenterol Hepatol 2017;32(Suppl):190–3.
[112] Cossu G, Carta MG, Contu F, Mela Q, Demelia L, Elli L, Dell'Osso B. Coeliac disease and psychiatric comorbidity: epidemiology, pathophysiological mechanisms, quality-of-life, and gluten-free diet effects. Int Rev Psychiatry 2017;29:489–503.
[113] Forsythe P, Bienenstock J, Kunze WA. Vagal pathways for microbiome-brain-gut axis communication. Adv Exp Med Biol 2014;817:115–33.
[114] Campagna G, Pesce M, Tatangelo R, Rizzulo A, La Fratta I, Grilli A. The progression of coeliac disease: its neurological and psychiatric implications. Nutr Res Rev 2017;30:25–35.
[115] Zingone F, Swift GL, Card TR, Sanders DS, Ludvigsson JF, Bai JC. Psychological morbidity of celiac disease: a review of the literature. United European Gastroenterol J 2015;3:136–45.
[116] Busby E, Bold J, Fellows L, Rostami K. Mood disorders and gluten: It's not all in your mind! A systematic review with meta-analysis. Nutrients 2018;10:. pii: E1708.
[117] Wijarnpreecha K, Jaruvongvanich V, Cheungpasitporn W, Ungprasert P. Association between celiac disease and schizophrenia: a meta-analysis. Eur J Gastroenterol Hepatol 2018;30:442–6.
[118] Chen SJ, Chao YL, Chen CY, Chang CM. Prevalence of autoimmune diseases in in-patients with schizophrenia: nationwide population-based study. Br J Psychiatry 2012;200:374–80.
[119] Samaroo D, Dickerson F, Kasarda DD, Green PH, Briani C, Yolken RH, et al. Novel immune response to gluten in individuals with schizophrenia. Schizophr Res 2010;118:248–55.
[120] Lazaro CP, Ponde MP, Rodrigues LE. Opioid peptides and gastrointestinal symptoms in autism spectrum disorders. Braz J Psychiatr 2016;38:343–6.
[121] Buie T. The relationship of autism and gluten. Clin Ther 2013;35:578–83.

[122] Roy A, Roy M, Deb S, Unwin G, Roy A. Are opioid antagonists effective in attenuating the core symptoms of autism spectrum conditions in children: a systematic review. J Intellect Disabil Res 2015;59:293–306.

[123] Sakurai T, Yamada A, Hashikura N, Odamaki T, Xiao JZ. Degradation of food-derived opioid peptides by bifidobacteria. Benef Microbes 2018;9:675–82.

[124] Erturk E, Wouters S, Imeraj L, Lampo A. Association of ADHD and celiac disease: what is the evidence? A systematic review of the literature. J Atten Disord 2016. pii: 1087054715611493.

[125] Kumperscak HG, Rebec ZK, Sobocan S, Fras VT, Dolinsek J. Prevalence of celiac disease is not increased in ADHD sample. J Atten Disord 2016. pii: 1087054716666953.

[126] Lohi S, Mäki M, Rissanen H, Knekt P, Reunanen A, Kaukinen K. Prognosis of unrecognized coeliac disease as regards mortality: a population-based cohort study. Ann Med 2009;41:508–15.

[127] Ilus T, Kaukinen K, Virta LJ, Pukkala E, Collin P. Incidence of malignancies in diagnosed celiac patients: a population-based estimate. Am J Gastroenterol 2014;109:1471–7.

[128] Volta U, Vincentini O, Quintarelli F, Felli C, Silano M. Low risk of colon cancer in patients with celiac disease. Scand J Gastroenterol 2014;49:564–8.

[129] Van Gils T, Nijeboer P, Overbeek LI, Hauptmann M, Caselijn DA, Bouma G, et al. Risks for lymphoma and gastrointestinal carcinoma in patients with newly diagnosed adult-onset celiac disease: consequences for follow-up: celiac disease, lymphoma and GI carcinoma. United European Gastroenterol J 2018;6:1485–95.

[130] Lebwohl B, Granath F, Ekbom A, Montgomery SM, Murray JA, Rubio-Tapia A, et al. Mucosal healing and mortality in coeliac disease. Aliment Pharmacol Ther 2013;37:332–9.

[131] Biagi F, Gobbi P, Marchese A, Borsotti E, Zingone E, Ciacci C, et al. Low incidence but poor prognosis of complicated coeliac disease: a retrospective multicentre study. Dig Liver Dis 2014;46:227–30.

[132] Malamut G, Cellier C. Refractory celiac disease. Gastroenterol Clin North Am 2019;48:137–44.

[133] Ilus T, Kaukinen K, Virta LJ, Huhtala H, Mäki M, Kurppa K, et al. Refractory coeliac disease in a country with a high prevalence of clinically-diagnosed coeliac disease. Aliment Pharmacol Ther 2014;39:418–25.

[134] Rowinski SA, Christensen E. Epidemiologic and therapeutic aspects of refractory coeliac disease—a systematic review. Dan Med J 2016;63. pii: A5307.

[135] Nasr I, Nasr I, Campling H, Ciclitira PJ. Approach to patients with refractory coeliac disease. F1000Res 2016;5. Faculty Rev-2544.

[136] Kaukinen K, Lindfors K, Collin P, Koskinen O, Mäki M. Coeliac disease—a diagnostic and therapeutic challenge. Clin Chem Lab Med 2010;48:1205–16.

[137] Conti L, Lahner E, Galli G, Esposito G, Carabotti M, Annibale B. Risk factors associated with the occurrence of autoimmune diseases in adult coeliac patients. Gastroenterol Res Pract 2018. 3049286.

[138] Emilsson L, Wijmenga C, Murray JA, Ludvigsson JF. Autoimmune disease in first-degree relatives and spouses of individuals with celiac disease. Clin Gastroenterol Hepatol 2015;13:1271–7.

[139] Ventura A, Magazzu G, Greco L. Duration of exposure to gluten and risk for autoimmune disorders in patients with celiac disease. Gastroenterology 1999;117:297–303.

[140] Verdu EF, Danska JS. Common ground: shared risk factors for type 1 diabetes and celiac disease. Nat Immunol 2018;19:685–95.

[141] Qiao SW, Iversen R, Raki M, Sollid LM. The adaptive immune response in celiac disease. Semin Immunopathol 2012;34:523–40.

[142] Scherf KA, Koehler P, Wieser H. Gluten and wheat sensitivities—an overview. J Cereal Sci 2016;67:2–11.
[143] Pavlovic M, Berenji K, Bukurov M. Screening of celiac disease in Down syndrome—old and new dilemmas. World J Clin Cases 2017;5:264–9.
[144] Fasano A. Leaky gut and autoimmune diseases. Clin Rev Allergy Immunol 2012;42:71–8.
[145] Menard S, Lebreton C, Schumann M, Matysiak-Budnik T, Dugave C, Bouhnik Y, et al. Paracellular versus transcellular intestinal permeability to gliadin peptides in active celiac disease. Am J Pathol 2012;180:608–15.
[146] Matysiak-Budnik T, Moura IC, Acros-Fajardo M, Lebreton C, Menard S, Candalh C, et al. Secretory IgA mediates retrotranscytosis of intact gliadin peptides via the transferrin receptor in celiac disease. J Exp Med 2008;205:143–54.
[147] De Re V, Magris R, Cannizzaro R. New insights into the pathogenesis of celiac disease. Front Med (Lausanne) 2017;4:137.
[148] Sollid LM, Iversen P, Steinsbo O, Qiao SW, Bergseng E, Dorum S, et al. Small bowel, celiac disease and adaptive immunity. Dig Dis 2015;33:115–21.
[149] Du Pre MF, Sollid LM. T-cell and B-cell immunity in celiac disease. Best Pract Res Clin Gastroenterol 2015;29:413–23.
[150] Jabri B, Chen X, Sollid LM. How T cells taste gluten in celiac disease. Nat Struct Mol Biol 2014;21:429–31.
[151] Dahal-Koirala S, Ciacchi L, Petersen J, Risnes LF, Neumann RS, Christophersen A, et al. Discriminative T-cell receptor recognition of highly homologous HLA-DQ2-bound gluten epitopes. J Biol Chem 2019;294:941.
[152] Kim SM, Mayassi T, Jabri B. Innate immunity: actuating the gears of celiac disease pathogenesis. Best Pract Res Clin Gastroenterol 2015;29:425–35.
[153] Vitale S, Picascia S, Gianfrani C. The cross-talk between enterocytes and intraepithelial lymphocytes. Mol Cell Pediatr 2016;3:20.
[154] Abadie V, Jabri B. IL-15: a central regulator of celiac disease immunopathology. Immunol Rev 2014;260:221–34.
[155] Meresse B, Korneychuk N, Malamut G, Cerf-Bensussan N. Interleukin-15, a master piece in the immunological jigsaw of celiac disease. Dig Dis 2015;33:122–30.
[156] Barone MV, Troncone R, Auricchio S. Gliadin peptides as triggers of the proliferative and stress/innate immune response of the celiac small intestinal mucosa. Int J Mol Sci 2014;15:20518–37.
[157] Calvanese L, Nanayakkara M, Aitoro R, Sanseverino M, Tornesello AL, Falcigno L, et al. Structural insights on P31-43, a gliadin peptide able to promote an innate but not an adaptive response in celiac disease. J Peptide Sci 2019;25:e3161.
[158] Assandri R, Monari M, Colombo A, Montanelli A. Pentraxin 3 serum levels in celiac patients: evidences and perspectives. Recent Pat Food Nutr Agric 2014;6:82–92.
[159] Assandri R, Montanelli A. Pentraxin 3 and biopsy status in celiac patients. Gastroenterol Hepatol Bed Bench 2018;11:225–32.
[160] Shannahan S, Leffler DA. Diagnosis and updates in celiac disease. Gastrointest Endosc Clin North Am 2017;27:79–92.
[161] Klapp G, Masip E, Bolonio M, Donut E, Polo B, Ramos D, Ribes-Koninckx C. Celiac disease: the new proposed ESPGHAN diagnostic criteria do work well in a selected population. J Pediatr Gastroenterol Nutr 2013;56:351–6.
[162] Reilly N, Husby S, Sanders DS, Green PH. Coeliac disease: to biopsy or not? Nat Rev Gastroenterol Hepatol 2018;15:60–6.

[163] Husby S, Murray JA, Katzka DA. AGA clinical practice update on diagnosis and monitoring of celiac disease: changing utility of serology and histologic measures: expert review. Gastroenterology 2019;156:885–9.

[164] Werkstetter KJ, Korponay-Szabo IR, Popp A, Villanacci V, Salemme M, Heilig G, et al. Accuracy in diagnosis of celiac disease without biopsies in clinical practice. Gastroenterology 2017;153:924–35.

[165] Holmes GK, Forsyth JM, Knowles S, Seddon H, Hill PG, Austin AS. Coeliac disease: further evidence that biopsy is not always necessary for diagnosis. Eur J Gastroenterol Hepatol 2017;29:640–5.

[166] Fuchs V, Kurppa K, Huhtala H, Laurila K, Mäki M, Collin P, et al. Serology-based criteria for adult coeliac disease have excellent accuracy across the range of pre-test probabilities. Aliment Pharmacol Ther 2019;49:277–84.

[167] Hujoel IA, Reilly NR, Rubio-Tapia A. Celiac disease: clinical features and diagnosis. Gastroenterol Clin North Am 2019;48:19–37.

[168] Paez MA, Gramelspacher AM, Sinacore J, Winterfield L, Venu M. Delay in diagnosis of celiac disease in patients without gastrointestinal complaints. Am J Med 2017;130:1318–23.

[169] Cichewicz AB, Mearns ES, Taylor A, Boulanger T, Gerber M, Leffler DA, et al. Diagnosis and treatment patterns in celiac disease. Dig Dis Sci 2019;64:2095–106.

[170] Porcelli B, Ferretti F, Vindigni C, Terzuoli L. Assessment of a test for the screening and diagnosis of celiac disease. J Clin Lab Anal 2016;30:65–70.

[171] Lerner A, Ramesh A, Matthias T. Serological diagnosis of celiac disease: new biomarkers. Gastroenterol Clin North Am 2019;48:307–17.

[172] Lebwohl B, Rubio-Tapia A, Assiri A, Newland C, Guandalini S. Diagnosis of celiac disease. Gastrointest Endosc Clin N Am 2012;22:661–77.

[173] Leffler DA, Schuppan D. Update on serologic testing in celiac disease. Am J Gastroenterol 2010;105:2520–4.

[174] Adriaanse M, Leffler DA. Serum markers in the clinical management of celiac disease. Dig Dis 2015;33:236–43.

[175] Scherf KA, Koehler P, Wieser H. Electrochemical immunosensors for the diagnosis of celiac disease. Adv Chem Engineering Sci 2015;5:83–95.

[176] Gupta S, Kaushal A, Kumar A, Kumar D. Recent advances in biosensors for diagnosis of celiac disease: a review. Biotechnol Bioeng 2019;116:444–51.

[177] Scherf KA, Ciccocioppo R, Pohanka M, Rimarova K, Opatrilova R, Rodrigo L, et al. Biosensors for the diagnosis of celiac disease: current status and future perspectives. Mol Biotechnol 2016;58:381–92.

[178] Gibiino G, Lopetuso L, Ricci R, Gasbarrini A, Cammarota G. Coeliac disease under a microscope: histological diagnostic features and confounding factors. Comput Biol Med 2019;104:335–8.

[179] Lagana SM, Bhagat G. Biopsy diagnosis of celiac disease: the pathologist's perspective in light of recent advances. Gastroenterol Clin North Am 2019;48:39–51.

[180] Valle J, Morgado JM, Ruiz-Martin J, Guardiola A, Lopes-Nogueras M, Garcia-Vela A, et al. Flow cytometry of duodenal intraepithelial lymphocytes improves diagnosis of celiac disease in difficult cases. United European Gastroenterol J 2017;5:819–26.

[181] Lewis SK, Semrad CE. Capsule endoscopy and enteroscopy in celiac disease. Gastroenterol Clin North Am 2019;48:73–84.

[182] Tortora R, Capone P, Imperatore N, de Stefano G, Gerbino N, Leo M, et al. Predictive value of "Marsh 1" type histology in subjects with suspected celiac disease. Scand J Gastroenterol 2014;49:801–6.

[183] Lau MS, Sanders DS. Optimizing the diagnosis of celiac disease. Curr Opin Gastroenterol 2017;33:173–80.
[184] Lazar-Molnar E, Snyder M. The role of human leukocyte antigen in celiac disease diagnostics. Clin Lab Med 2018;38:655–68.
[185] Ortiz M, Joda H, Höth J, Boni V, Katakis I, Klemm R, et al. Bleed-to-read disposable microsystems for the genetic and serological analysis of celiac disease markers with amperometric detection. Electrophoresis 2015;36:1920–6.
[186] Vijzelaar R, van der Zwan E, van Gammeren A, Yilmaz R, Verheul A, van Hoogstraten I, et al. Rapid detection of the three celiac disease risk genotypes HLA-DQ2.2, HLA-DQ2.5, and HLA-DQ8 by multiplex ligation-dependent probe amplification. Genet Test Mol Biomarkers 2016;20:158–61.
[187] Moreno ML, Cebolla A, Munoz-Suano A, Carrillo-Carrion C, Comino C, Pizarro A, Leon F, et al. Detection of gluten immunogenic peptides in the urine of patients with coeliac disease reveals transgressions in the gluten-free diet and incomplete mucosal healing. Gut 2017;66:250–7.
[188] Aggarwal S, Lebwohl B, Green PH. Screening for celiac disease in average-risk and high-risk populations. Therap Adv Gastroenterol 2012;5:37–47.
[189] Bibbins-Domingo K, Grossman DC, Curry SJ, Barry MJ, Davidson KW, Doubeni CA, et al. Screening for celiac disease: US Prevention Services Task Force Statement. JAMA 2017;317:1252–7.
[190] Mearin ML. The prevention of coeliac disease. Best Pract Res Clin Gastroenterol 2015;29:493–501.
[191] Björck S, Brundin C, Karlsson M, Agardh D. Reduced bone mineral density in children with screening-detected celiac disease. J Pediatr Gastroenterol Nutr 2017;65:526–32.
[192] Hujoel IA, van Dyke CT, Brantner T, Larson J, King KS, Sharma A, et al. Natural history and clinical detection of undiagnosed coeliac disease in a North American community. Aliment Pharmacol Ther 2018;47:1358–66.
[193] Lundin KE, Wijmenga C. Coeliac disease and autoimmune disease—genetic overlap and screening. Nat Rev Gastroenterol Hepatol 2015;12:507–15.
[194] Ludvigsson JF, Card TR, Kaukinen K, Bai J, Zingone F, Sanders DS, Murray J. Screening for celiac disease in the general population and in high-risk groups. United European Gastroenterol J 2015;3:106–20.
[195] Catassi C, Gatti S, Lionetti E. World perspective and celiac disease epidemiology. Dig Dis 2015;33:141–6.

Chapter 9

Wheat allergy

Allergenicity may be defined as the ability of antigens (allergens) to induce humoral and cellular immune responses that can produce allergen-specific IgE antibodies and T-helper 2 cytokines and result in clinical symptoms. Allergies affect up to 25% of the population; they are more common in industrialized countries than in developing countries and more common in urban populations than in rural populations. Allergies are strongly dependent on genetics: Allergic parents are more likely to have allergic children and monozygotic twins have the same allergy in 70% and non-identical twins in 40% of the cases. Among environmental factors, chronic stress, a high hygiene standard, and a changed intestinal microbiota may contribute to the increasing development of allergies. Most allergens are proteins or glycoproteins (proteins linked with saccharides) that usually are water-soluble, relatively stable to heat, acid, and proteolytic enzymes and have molecular masses between 10,000 and 70,000. The most common sources of allergens include pollen, foods, insect stings, and medications, in parts causing severe reactions. This chapter is concerned with food allergies in general and the wheat-specific types of dietary allergy, wheat-dependent exercise-induced anaphylaxis, respiratory allergy, and contact urticaria.

9.1 Food allergies

Food allergies are an important part of the spectrum of allergies [1]. They have to be differentiated from food intolerances (e.g., lactose intolerance) that are not considered to have an immunological basis. Food allergies likely affect nearly 5% of adults and 8% of children with growing evidence of an increase in prevalence. The clinical manifestations of food allergies are characterized by symptoms on the skin, the gastrointestinal tract, and the respiratory tract. To protect sensitive consumers from harmful allergic symptoms, the "General Standard for the Labelling of Prepacked Foods" of the Codex Alimentarius (CODEX STAN 1-1985) states that the following foods and ingredients that contain proteins, known to cause allergies and other types of hypersensitivities, shall always be declared (the top eight food allergens) [2]: Cereals containing gluten (i.e., wheat, rye, barley, oats, spelt or their hybridized strains or products of these), crustaceans, eggs, fish, peanuts, soybeans, milk, and tree nuts.

The general approach to diagnose food allergies includes medical history, skin tests, blood tests, elimination diets, and provocation tests [3]. The patient's

Wheat—An Exceptional Crop. https://doi.org/10.1016/B978-0-12-821715-3.00009-5

history might clarify the causal food, the ingested quantities, the time course of reactions, and various cofactors. If a food is suspected to be allergenic, the patient will be instructed to modify her/his diet and to avoid the food for some weeks. In case of improving symptoms, the patient may then be challenged by reintroducing the potential allergenic food to see, if symptoms are reproducible.

Skin tests are most frequently used to find out the culprit allergen and food, respectively. A small amount of the potential allergen is introduced into the patient's skin (inner forearm or back), which provokes a small controlled allergic response seen in the form of a rash or hives within less than 1 h. A couple of different techniques can be used to apply the allergen: (1) pricking the skin with a needle or pin (skin prick test); (2) deep scratching of the skin with the blunt bottom of a lancet (skin scratch test); (3) injection underneath the dermis with a hypodermic syringe (intradermic test; (4) superficial scraping of the skin with the bovel of a needle (skin scrape test); and (5) application of a patch that contains the allergen to the skin (patch test). Skin prick tests are most commonly used for the diagnosis of allergies induced by oral food intake. In cases of contact urticaria, the patch test is recommended; positive results are obtained after 24–48 h. The patch test may also be useful to identify foods that are responsible for non-IgE-mediated allergy.

Blood tests for allergy are quick and simple and can be utilized even for very young children. They are based on the fact that IgE-mediated allergic reactions are linked to the release of IgE antibodies and histamine as well as basophil proliferation. The quantitation of food-specific serum IgE antibodies has been widely used in the diagnosis of food allergies. Significantly increased IgE levels (>0.80 kU/L in the case of wheat allergy) are correlated with the likelihood of allergy. Most commercially available tests are based on fluorescence immunoassays. The determination of total serum IgE antibody concentrations is less valuable than the determination of those antibodies specific for the potential allergen.

The basophil activation test measures the specific reaction of basophilic cells of peripheral blood against allergens. Blood samples and the suspected allergen are incubated to initiate the degranulation process. Afterwards, specific protein markers for degranulation (e.g., CD63), expressed on the basophil membrane, are labeled with a fluorescent antibody. The numbers of labeled cells are then quantitated by flow cytometry. Another blood test is the histamine release test. Whole blood is mixed with the suspected allergen in a glass fiber-coated well. Histamine, released by degranulation of basophils, is bound to the glass fiber matrix. After a washing step, histamine is dissolved from the matrix and measured by means of commercially available test kits.

The oral provocation test is still considered as the gold standard to confirm the diagnosis of allergy, because the positive predictive values of skin or blood tests are less than 75%, notably in adults due to cross-reactivity with grass pollen. Another difficulty with both skin and blood tests is that individuals can be sensitized to a food allergen but may not have any symptom. The challenge procedure is not yet fully standardized and may be performed as follows: The suspected allergen is introduced orally or through inhalation. The starting dose for oral provocation is different for single allergenic foods; 100 mg have been recommended for wheat-containing food [4]. The dose may be doubled every 15–30 min. The top dose administered should be the normal daily intake. Symptoms most often occur 3–15 min after intake. However, allergologists appear to avoid oral provocation tests because of risks, costs, and time. It is a procedure of over 3–4 h and even longer considering the time of preparation. Moreover, the provocation test occasionally produces serious symptoms such as anaphylaxis and therefore requires intensive monitoring and experienced personnel.

The primary treatment of food allergies is to avoid the culprit food. If the avoidance of the allergenic food may result in nutritional deficiency, dietary supplementation is necessary. Several medications are helpful to prevent the action of allergic mediators and degranulation processes. They include antihistamines, glucocorticoids, mast cell stabilizers, and antileukotriene agents. The prompt administration of epinephrine is essential for food-induced anaphylaxis. Among future therapeutic options, the oral, sublingual, and epicutaneous immunotherapy is the mostly investigated approach. Doses of the allergenic food protein are given in gradually increasing amounts toward a maintenance dose. The aim of the treatment is to restore or induce a tolerant state. However, more investigations are still needed to assess efficacy and safety of this approach.

Wheat allergy (WA) can be caused by exposure to wheat-based products through ingestion, inhalation or skin/mucosal contact [5]. Several studies have indicated that WA represents approximately 11–25% of all food allergies and affects about 1% of the general population. Wheat-based products are staple foods worldwide and due to its valuable functional properties, wheat flour is used as an ingredient in a surprisingly high percentage of processed foods (see Chapter 4), which highlights the problems for those suffering from WA. Moreover, many workers in the milling, baking, and food processing professions are exposed to wheat flour. Based on different targets, symptoms, and pathogenesis, WA can be divided into wheat food allergy, wheat-dependent exercise-induced anaphylaxis, respiratory allergy, and contact urticaria; they are discussed in the following sections.

9.2 Wheat food allergy

The most commonly mentioned dietary allergies to cereals are related to wheat and to a lesser extent to rice, rye, and barley. WA, caused by the intake of wheat-containing foods (dietary WA), is roughly divided into IgE-mediated, non-IgE-mediated, and mixed immune responses [6]. IgE-mediated WA (IgE-WA) is subdivided into immediate onset reactions (immediate in time) and immediate plus late-phase reactions (in which the immediate onset symptoms are followed by prolonged or ongoing symptoms). IgE-WA usually develops during early infancy or childhood with an estimated prevalence of around 4%. It is frequently outgrown between the ages of 1–8 years, because immune tolerance increases when progressing in age. Therefore IgE-WA is less common in adolescents and adults. The prevalence estimates for Europe depend on the assessment method used and range from 0.1% (positive oral challenge) to 3.6% (self-reported prevalence) [7]. Symptoms typically develop within a few hours after wheat ingestion, but delayed symptoms may also appear. Around 1–2 g of wheat proteins, corresponding to around 15–25 g of bread, are usually needed to induce symptoms in most of sensitized children; a minor group reacts to smaller amounts (< 10 mg of protein) [8]. In the adult allergic population, protein amounts of up to 6 g have been reported to cause the reaction.

The clinical manifestations appearing after ingestion are similar to those of other food allergies. They can occur in the mouth, throat, nose, and eyes (swelling, itching, and angioedema), the skin (rash, hives, swelling, itching, urticaria, and atopic dermatitis), the respiratory tract (wheezing, impaired breathing, and bronchial obstruction) and/or the intestinal tract (cramps, nausea, bloating, abdominal pain, and diarrhea). Even anaphylaxis has been described in relation to IgE-WA. The intestinal manifestations of IgE-WA are similar to those of celiac disease but do not include mucosal damage. The mechanism of IgE-WA includes two phases: (1) sensitization by the initial exposure to the allergen and (2) immune reaction upon reexposure to the allergen (see Section 7.1). Activated mast cells and basophils start degranulation and release inflammatory mediators such as histamine and heparin.

As the diagnostic accuracies of skin prick tests and serum IgE assays have shown unsatisfactory results, the diagnosis of IgE-WA is currently based almost exclusively on the results of oral wheat challenge and wheat elimination tests. Treatment of IgE-WA is usually based on the avoidance of wheat products and severe immune reactions can be treated with antihistamines and corticosteroids. Oral immunotherapy appears to be a promising therapeutic alternative, while sublingual or epicutaneous immunotherapy may have less efficacy [9]. A multicenter, placebo-controlled clinical trial, including 46 patients with WA aged 4–22 years, revealed that oral immunotherapy with vital wheat gluten induced desensitization in about one half of the subjects after 1 year of treatment [10]. Another study, including 100 children and adolescents with WA aged 6–18 years, showed that wheat oral immunotherapy was modestly successful (57%) but was associated with frequent moderate or severe adverse events [11].

Non-IgE-WA is well-known in children under the age of three and is less recognized in adults, but there is an increased body of evidence that they may be underrecognized. Non-IgE-WA is typically delayed in onset, primarily affects the esophagogastrointestinal mucosa, and provokes two main clinical conditions: eosinophilic esophagitis and eosinophilic gastritis [12]. Eosinophil esophagitis is a systemic disease and is associated with severe inflammation of the esophagus displaying high levels of intraepithelial eosinophils [13]. Typical symptoms are reflux, vomiting, dysphagia, cough, and chest pain. Eosinophilic gastritis is a less severe disease characterized by abdominal complaints such as pain, bloating, and nausea. Diagnosis of both diseases is made by esophagogastroduodenoscopy assessing the number of eosinophils in corresponding biopsies. The treatment relies on eliminating wheat by a gluten-free diet (see Chapter 12) and acute symptoms are usually treated with corticosteroids [14]. The pathomechanism of non-IgE-WA is currently not well known; it is believed to be T-cell-mediated and characterized by high levels of interleukin-13 and tumor necrosis factor-α as drivers of intestinal epithelial damage and eosinophil infiltration. Mixed WA involves both IgE- and non-IgE-mediated immune responses and is of less importance.

Numerous wheat proteins have been identified as related to WA [15,16]. Most of them are associated with storage functions (gluten proteins), metabolic functions (e.g., enzymes), and protective functions (e.g., enzyme inhibitors). The broad spectrum of allergenic proteins may vary in induced symptoms and affected populations. A number of clinical studies have shown that both gliadins and glutenins are reactive to IgE antibodies in sera from WA patients. Strong IgE binding epitopes mainly include amino acid sequences of the repetitive segments (see Section 8.4.1) such as QFPQQQFPQQ (ω5-gliadins), QQSFPLQPQQ (ω1,2-gliadins), VQQQQFPGQQ (α-gliadins), QQLPQPQQP (γ-gliadins), and SQQQPPF (low-molecular-weight glutenin subunits) (one-letter code of amino acids see Table 2.3). Apart from gluten proteins, a number of studies identified non-gluten proteins within the albumin/globulin fraction as reactive proteins. α-Amylase inhibitor subunits appear to be the major allergens; the precise assignment of reactive subunits however varies between studies [15]. This may reflect differences between the patients studied or between the analytical approaches. Other non-gluten proteins that bound to IgE from patients with WA are lipid-transfer proteins, germ agglutinin, peroxidase, serpin, β-amylase, α-amylase/trypsin inhibitors, and thioredoxin.

A small number of patients exhibit WA, when eating foods that contain hydrolyzed wheat protein (HWP), although they can eat normal bread, pasta, and pastries [17]. HWP is the water-soluble product obtained from wheat gluten after different procedures of acid or alkaline partial hydrolysis. Due to their foaming and emulsifying properties, HWP preparations are used as food additives, e.g., in sausages and soaps. The processes leading to deamidation of wheat proteins, the clinical features of the patients suffering from this allergy, and the possible sensitizing mechanism have been discussed by Tranquet and coworkers [18]. The IgE-binding epitopes of HWP are derived from those repetitive amino acid sequences, in which glutamine residues (Q) of gluten proteins are in parts modified into glutamic acid residues (E) by partial hydrolysis (e.g., epitope QPEEPFPE derived from QPQQPFPQ). According to a report of the Japanese Society of Allergy, 2026 cases with WA caused by HWP were registered in Japan in 2013. Most patients had likely been sensitized to HWP through the use of HWP-supplemented soap.

9.3 Wheat-dependent exercise-induced anaphylaxis

Wheat-dependent exercise-induced anaphylaxis (WDEIA) is a special type of WA, because wheat intake alone does not induce symptoms [19]. Allergic reactions are elicited only with an additional triggering cofactor. The most prevalent cofactor is physical activity, e.g., jogging, cycling, or dancing, but milder exercise like walking or gardening may also be sufficient. Symptoms of WDEIA are usually elicited by physical activity one to four hours after wheat intake. Acetylsalicylic acid (aspirin) and other non-steroidal antiinflammatory drugs (NSAIDs) (e.g., indomethacin, ibuprofen, and diclofenac), alcohol consumption, and even the patient's general condition (e.g., stress or infections) have also been described as cofactors [20]. Combined with wheat consumption, they can induce symptoms without physical activity. A combination of two cofactors (e.g., exercise and aspirin) increases the risk of reactions.

WDEIA is a rare disease with an estimated prevalence below 0.1% and occurs more frequently in adults than in children. Most cases have been reported in Far East (mainly Japan) and Europe. Typical symptoms of WDEIA are pruritus, local and generalized urticaria, dyspnea, angioedema, upper respiratory tract obstruction, hypotension, and gastrointestinal symptoms. In severe cases, bronchospasm, vascular collapse, and live-threatening anaphylactic shock have been observed. The frequency of WDEIA episodes varies from patient to patient and ranges from a singular episode to multiple events.

Diagnosis of WDEIA is challenging and significant delays in diagnosis are common. For example, a recent multicenter evaluation of 132 patients with NCGS revealed significant delays in diagnosis ranging from 1 to 5 years (40% of patients) and more than 5 years (29%) [21]. Results of skin tests and determination of wheat-specific IgE serum antibodies may be negative despite a clear anamnesis. The detection of specific IgE antibodies to ω5-gliadins may have good sensitivity and specificity, but oral challenge tests with gluten plus exercise or aspirin are the gold standard to confirm the diagnosis [22], even if their success depends on the amount of gluten administered, intensity of exercise, and individual thresholds [23]. Recent studies, including 71 patients with a case history of WDEIA, demonstrated that an oral challenge test with gluten at rest and in combination with treadmill exercise is a safe confirmatory test for WDEIA [24].

The most reliable prophylaxis is a strict limitation of wheat ingestion before exercise or a limitation of exercise after wheat ingestion. NSAIDs and excessive alcohol consumption should also be avoided. A strict gluten-free diet is recommended in cases of patients with low-threshold sensitivity and severe reactions and in cases of highly demanding professions (e.g., police and military personnel). The acute treatment of severe symptoms involves application of epinephrine or antihistamines.

WDEIA is an IgE-mediated disorder triggered by gluten proteins/peptides. Mast cell degranulation and histamine release are the key features. The role of the cofactors underlying the pathomechanism however remains poorly understood. Several theories have been proposed; for example, exercise, NSAIDs, and other augmenting cofactors might increase gastrointestinal permeability and osmolality, redistribute blood flow, and lower the threshold for mast cell degranulation [25]. Further studies are required to find out the precise mechanism.

Among wheat proteins, ω5-gliadins with four dominant IgE-binding epitopes (QQIPQQQ, QQFPQQQ, QQSPEQQ, and QQSPQQQ) and high-molecular-weight glutenin subunits with three main epitopes (QQPGQ, QQPGQGQQ, and QQSGQGQ) have been reported as the major allergens [26]. However, other types of gluten proteins such as α-gliadins, γ-gliadins, and low-molecular-weight glutenin subunits can also sensitize WDEIA patients. The allergenic epitopes are localized in the repetitive domains of gluten proteins (see Section 2.2.1). The amount of ingested wheat proteins, necessary for provoking WDEIA, has not been well characterized; different studies reported a wide range from 100 mg to 25 g.

A special form of WDEIA is an allergic reaction caused by epicutaneous sensitization with cosmetics (e.g., soaps, shampoos, or creams) containing HWP (see Section 9.2) followed by exercise. The predominant symptom is angioedema on the eyelids (Fig. 9.1) and even anaphylactic shock occasionally develops. Conventional WDEIA and WDEIA caused by HWP diverge in symptoms, basophil tests, and the specificity of IgE antibodies; α-, γ-, and ω1,2-gliadins are the major allergenic proteins of HWP-mediated WDEIA [27].

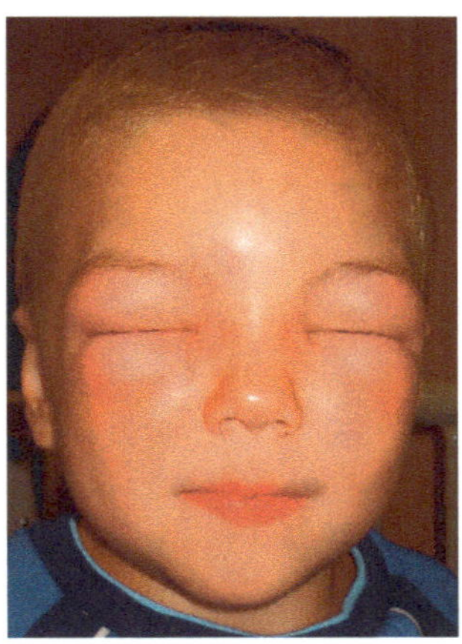

FIG. 9.1 Angioedema on the eyelids of a WDEIA patient.

9.4 Respiratory allergy

Allergenic materials that trigger respiratory allergies are airborne particles such as dust and pollen. Symptoms arise in body areas that are in contact with air, mainly nose, mouth, and lungs. Inhaled allergens reach the mucous layers of affected areas unmodified, i.e., they are not digested via the gastrointestinal tract as in dietary allergies. Baker's asthma and allergic rhinitis are well characterized allergic responses to the inhalation of flours and dusts from wheat and related cereals (rye and barley). They affect around 1% of the population in industrialized countries and are among the most common occupational diseases mainly infesting millers, baker, and confectioners.

Baker's asthma has been recognized since the time of the Roman Empire, when slaves working in bakeries protected themselves by using cloth as a primitive respirator to cover their faces, because their breathing suffered from inhaling flour [28]. Symptoms include shortness of breath, tightness of the chest, nasal irritation, coughing, wheezing, and increased production of mucus in the lungs. Allergic rhinitis is an immune-mediated inflammation of the nasal mucosa and usually precedes asthma. Symptoms are sneezing, pruritus, intense mucus production, obstruction of the nose, and red eyes; conjunctivitis and skin symptoms may also occur. Usually, symptoms of respiratory allergies resolve, if exposure to the offending allergens is avoided.

More than 100 individual wheat proteins that are recognized by IgE antibodies from patients' sera have been identified until now. A selection of allergenic proteins and their activities are shown in Table 9.1 [15]. Similar to dietary WA, they comprise both gluten and non-gluten proteins. Active gluten proteins

TABLE 9.1 Allergens in baker's asthma and their reactivity [15].

Protein	Number of patients with positive reactions
ω-Gliadins	15/24–21/24
α-, γ-Gliadins	12/24–24/24
Total glutenins	7/15–23/24
Monomeric amylase/trypsin inhibitors	9/31–11/11
Dimeric amylase/trypsin inhibitors	4/8–5/31
Tetrameric amylase/trypsin inhibitors	7/31–14/31
Glycerinaldehyde 3-phosphate dehydrogenase	8/10
Triosephosphate isomerase	5/10
Serpin	4/10

include total gliadins, ω-, α-, and γ-gliadins and total glutenins. The group of α-amylase inhibitors contains the most important allergens among non-gluten proteins. Furthermore, a number of other proteins such as serpin, triosephosphate isomerase, germ agglutinin, peroxidase, and lipid transfer proteins, have been shown to react with IgE. Non-flour enzymes that are added to flour to improve the baking performance (e.g., fungal α-amylase, xylanase, and cellulase) have also been reported to induce allergic reactions.

The diagnosis is usually based on skin prick tests, specific serum IgE antibody tests, and inhalative challenge tests. Wheat, rye or barley flours as well as wheat α-amylase are used as reference allergenic agents. Whole flour extracts are still the best option for specific IgE-determination, but single protein components might help to discriminate between baker's asthma, dietary WA, and pollen allergy [29]. For prophylaxis, the exposition to cereal flours and dusts has to be avoided, even if this involves a change of occupation for bakers and millers. Acute treatment is achieved with antihistamines. A long-term specific immunotherapy with subcutaneous injections may be possible but needs further evaluation.

9.5 Contact urticaria and contact dermatitis

Similar to baker's asthma and allergic rhinitis, predisposed employees dealing with wheat are most commonly afflicted by skin allergies such as contact urticaria and contact dermatitis. Both disorders are based on an IgE-mediated immune reaction that appears on the skin following contact with an eliciting substance. The immune system of the affected patient has been previously sensitized to the allergenic agent, which can occur through exposure on the skin and the mucous membranes of the respiratory or gastrointestinal tracts. The symptoms of contact urticaria are localized wheal-and-flare reactions that appear within 10–30 min after allergen exposure and fade away within hours. Most common symptoms are hives and a raised, burning and/or pruritic swelling of the skin. Contact dermatitis is accompanied by the appearance of large, burning and itchy rashes, blisters, and wheals that take several days to weeks to heal.

Hands, arms, and the face are affected most often by skin allergies, but beside the contact area, other organs may be involved leading to conjunctivitis, rhinitis, asthmatic attack or even anaphylactic shock. Most cases are work-related; frequently affected occupations are millers, bakers, and flour handlers. Once symptoms have developed, topical corticosteroids or emollients help relieve pruritus, blisters, erythema, and eczema. Afterwards, the exposure to flours of wheat and related cereals has to be avoided with protective measures such as gloves or barrier creams or complete avoidance of contact.

Both gluten and non-gluten proteins, similar to those causing dietary WA and baker's asthma (see Table 9.1), are involved in the immune reactions. More recently, contact allergies have been described to be triggered by HWP-containing cosmetics (see Section 9.2). The patients are likely to have been sensitized to

HWP primarily through percutaneous and/or rhino-conjunctival routes after using HWP-supplemented products. Most patients have characteristic features of facial angioedema, particularly eyelid edema.

The initial step toward diagnosis includes the patient's thorough history of the allergy. Afterwards, testing should follow a step-wise approach comprising open challenge and skin prick, scratch or patch tests (see Section 9.1). The open test is performed by using a solution of the suspected allergen and spreading it over a 3×3 cm area of the ventral forearm for immediate reactions or the back for delayed reactions [30]. The reaction (e.g., follicular erythema followed by a wheal or wheal-and-flare response) is measured at 20, 40, and 60 min.

References

[1] Muraro A, Werfel T, Hoffmann-Sommergruber K, Roberts G, Beyer K, Bindslev-Jensen C, et al. EAACI food allergy and anaphylaxis guidelines: diagnosis and management of food allergy. Allergy 2014;69:1008–25.

[2] CODEX STAN 1—1985. Codex Alimentarius Commission; 2010. Amendment 7.

[3] Sicherer SH, Sampson HA. Food allergy: epidemiology, pathogenesis, diagnosis, and treatment. J Allergy Clin Immunol 2014;133:291–307.

[4] Bindslev-Jensen C, Ballmer-Weber BK, Bengtsson U, Blanco C, Ebner C, Hourihane J, et al. Standardization of food challenges in patients with immediate reactions to food—position paper from the European Academy of Allergology and Clinical Immunology. Allergy 2004;59:690–7.

[5] Quirce S, Boyano-Martinez T, Diaz-Perales A. Clinical presentation, allergens, and management of wheat allergy. Expert Rev Clin Immunol 2016;12:563–72.

[6] Pasha I, Saeed F, Sultan MT, Batool R, Aziz M, Ahmed W. Wheat allergy and intolerance: recent updates and perspectives. Crit Rev Food Sci Nutr 2016;56:13–24.

[7] Nwaru BI, Hickstein L, Panesar SS, Roberts G, Muraro A, Sheikh A. Prevalence of common food allergies in Europe: a systematic review and meta-analysis. Allergy 2014;69:992–1007.

[8] Hischenhuber C, Crevel R, Jarry B, Mäki M, Moneret-Vautrin DA, Romano A, et al. Review article: safe amounts of gluten for patients with wheat allergy or coeliac disease. Aliment Pharmacol Ther 2006;23:559–75.

[9] Pacham P, Vichyanond P. Immunotherapy for IgE-mediated wheat allergy. Hum Vaccin Immunother 2017;13:2462–6.

[10] Nowak-Wegrzyn A, Wood RA, Nadeau KC, Pongracic JA, Henning AK, Lindblad RW, et al. Multicenter, randomized, double-blind, placebo-controlled clinical trial of vital wheat gluten oral immunotherapy. J Allergy Clin Immunol 2019;143. 651–661.e9.

[11] Kulmala P, Pelkonen AS, Kuitunen M, Paassilta M, Remes S, Schultz R, et al. Wheat oral immunotherapy was moderately successful but was associated with very frequent adverse events in children aged 6-18 years. Acta Pediatr 2018;107:861–70.

[12] Cianferoni A. Wheat allergy: diagnosis and management. J Asthma Allergy 2016;9:13–25.

[13] Vinit C, Dieme A, Courbage S, Dehaine C, Dufeu CM, Jacquemot S, et al. Eosinophilic esophagitis: pathophysiology, diagnosis, and management. Arch Pediatr 2019;26:182–90.

[14] Turnbull JL, Adams HN, Gorard DA. Review article: the diagnosis and management of food allergy and food intolerances. Aliment Pharmacol Ther 2015;41:3–25.

[15] Tatham AS, Shewry PR. Allergens to wheat and related cereals. Clin Exp Allergy 2008;38:1712–26.

[16] Juhasz A, Belova T, Florides CG, Maulis C, Fischer I, Gell G, et al. Genome mapping of seedborne allergens and immunoresponsive proteins in wheat. Sci Adv 2018;4. eaar8602.
[17] Nakamura M, Yagami A, Hara K, Sano A, Kobayashi T, Aihara M, et al. A new reliable method for detecting specific IgE antibodies in the patients with immediate type wheat allergy due to hydrolyzed wheat protein: correlation of its titer and clinical severity. Allergol Int 2014;63:243–9.
[18] Tranquet O, Larre C, Denery-Papini S. Allergic reactions to hydrolysed wheat proteins: clinical aspects and molecular structures of the allergens involved. Crit Rev Food Sci Nutr 2018;21:1–10.
[19] Scherf KA, Brockow K, Biedermann T, Koehler P, Wieser H. Wheat-dependent exercise-induced anaphylaxis. Clin Exp Allergy 2016;46:10–20.
[20] Christensen MJ, Eller E, Mortz CG, Brockow K, Bindslev-Jensen C. Wheat-dependent cofactor-augmented anaphylaxis: a prospective study of exercise, aspirin, and alcohol efficacy as cofactors. J Allergy Clin Immunol Pract 2019;7:114–21.
[21] Kennard L, Thomas I, Rutkowski K, Azzu V, Yong PF, Kasternow B, et al. A multicenter evaluation of diagnosis and management of omega-5 gliadin allergy (also known as wheat-dependent exercise-induced anaphylaxis) in 132 adults. J Allergy Clin Immunolog Pract 2018;6:1892–7.
[22] Brockow K, Kneissl D, Valentini L, Grosber M, Kugler C, Werich M, et al. Using a gluten oral food challenge protocol to improve diagnosis of wheat-dependent exercise-induced anaphylaxis. J Allergy Clin Immunol 2015;135:977–84.
[23] Kohno K, Matsuo H, Takahashi H, Niihara H, Chinuki Y, Kaneko S, et al. Serum gliadin monitoring extracts patients with false negative results in challenging tests for the diagnosis of wheat-dependent exercise-induced anaphylaxis. Allergol Int 2013;62:229–38.
[24] Christensen MJ, Eller E, Mortz CG, Brockow K, Bindslev-Jensen C. Exercise lowers threshold and increases severity, but wheat-dependent, exercise-induced anaphylaxis can be elicited at rest. J Allergy Clin Immunol Pract 2018;6:514–20.
[25] Wölbing F, Fischer J, Köberle M, Kaesler S, Biedermann T. About the role and underlying mechanismens of cofactors in anaphylaxis. Allergy 2013;68:1085–92.
[26] Matsuo H, Morita E, Tatham AS, Morimoto K, Horikawa T, Osuna H, et al. Identification of the IgE-binding epitope in ω-5 gliadin, a major allergen in wheat-dependent exercise-induced anaphylaxis. J Biol Chem 2004;279:12135–40.
[27] Yokooji T, Kurihara S, Murakami T, Chinuki Y, Takahashi H, Morita E, et al. Characterization of causative allergens for wheat-dependent exercise-induced anaphylaxis sensitized with hydrolyzed wheat proteins in facial soap. Allergol Int 2013;62:435–45.
[28] Brisman J. Baker's asthma. Occup Environ Med 2002;59:498–502.
[29] Raulf M. Allergen component analysis as a tool in the diagnosis and management of occupational allergy. Mol Immunol 2018;100:21–7.
[30] Bhatia R, Alikhan A, Maibach HJ. Contact urticaria: present scenario. Indian J Dermatol 2009;54:264–8.

Chapter 10

Non-celiac gluten sensitivity

Non-celiac gluten sensitivity (NCGS) may be defined by the following criteria: a clinical entity induced by the ingestion of wheat (gluten) leading to intestinal and/or extraintestinal symptoms that resolve once the wheat (gluten)-containing foodstuff is eliminated from the diet and when celiac disease (CD), wheat allergy (WA), and irritable bowel syndrome (IBS) have been ruled out [1]. In literature, some other terms have been used for this disorder, for example, gluten sensitivity, gluten hypersensitivity, non-celiac wheat sensitivity, and non-celiac gluten intolerance. NCGS of adults was first described in the early 1980s and the first cases of NCGS in children were reported in 2012. Since that time, NCGS has been accepted as a wheat hypersensitivity of its own. The evidence for NCGS and its surrounding controversies and uncertainties have been discussed by Aziz and coworkers [2]. Unlike CD and WA, NCGS is still an unclear entity and currently underrecognized and undertreated. It can be regarded as a syndrome rather than a unique disorder due to the wide spectrum of intestinal and extraintestinal symptoms. During the last decade, NCGS has gained considerable recognition in the medical community and general public and the number of patients diagnosed with NCGS as well as publications on this topic have increased substantially [3]. The following sections are devoted to the general characteristics, clinical features, pathogenesis, triggering factors, and diagnosis of this "new" wheat (gluten)-related disorder. The treatment of NCGS and other wheat-mediated disorders is described in Chapter 12.

10.1 General characteristics

Among wheat hypersensitivities, NCGS is least well-defined, because the triggering factors have not been unambiguously identified, specific biomarkers are missing, and the diagnosis relies on exclusion of other wheat-related disorders. The clinical differentiation of NCGS from other wheat hypersensitivities like CD and WA is a major problem due to common features: They show similar symptoms, wheat proteins (gluten) are the triggering factor, and a wheat (gluten)-free diet is administered for treatment. The shared and different characteristics of CD, WA, and NCGS are summarized in Table 10.1 [4].

The overall prevalence of NCGS in the general population is not exactly known, which reflects the uncertainty of diagnostic results very well. Many NCGS cases are self-reported and not always supported by clinical evidence making prevalence difficult to evaluate. Thus the prevalence in the general

Wheat—An Exceptional Crop. https://doi.org/10.1016/B978-0-12-821715-3.00010-1

TABLE 10.1 Characteristics of wheat-dependent disorders [4].[a]

Characteristics	CD	WA	NCGS
Prevalence	≈1%	≈1%	0.6–6%
Symptoms	Intestinal/ extraintestinal	Intestinal/ extraintestinal	Intestinal/ extraintestinal
Onset of symptoms	Days-weeks	Minutes-hours	Hours-days
Complications	Long-term	Short-term	Long-term
Comorbidities	Yes	No	Yes
Genetics	HLA-DQ2/8	not HLA-DQ2/8	not HLA-DQ2/8
Immunity	Adaptive/innate	Allergic	Innate?
Serology	IgA/IgG	IgE	IgG?
Anti-TG2 antibodies	Yes	No	No
Histology (Marsh)	2–4	0	0–1
Triggering proteins	Gluten	Gluten/non-gluten	Gluten/non-gluten
Treatment	Gluten-free	Wheat-free	Wheat-/gluten-free or -reduced

[a] *CD, celiac disease; HLA, human leukocyte antigen; Ig, immunoglobulin; NCGS, non-celiac gluten sensitivity; TG, tissue transglutaminase; WA, wheat allergy.*

population has been estimated differently from slightly lower (0.6%) to much higher (6%) than that of CD (≈1%) [5]. NCGS occurs in adults with a mean age onset of 40 years. The female gender is more prone to NCGS with a female/male ratio of around 5:1. While HLA-DQ2 or HLA-DQ8 haplotypes are present in almost all CD patients, these genes occur in around 50% of NCGS patients, a percentage still higher compared to the general population (≈30%) [6]. In contrast to CD, NCGS does not have a strong hereditary basis.

10.2 Clinical features

Symptoms of NCGS usually occur within hours or days after ingestion of gluten-containing foods and disappear rapidly after elimination of these foods from the diet. Whether NCGS is a permanent or a transient condition, is not yet clear. The heterogeneous collective of NCGS patients is composed of a number of subgroups and each demonstrates different clinical and pathologic manifestations. Frequent symptoms, presented by adult patients, comprise both intestinal and extraintestinal complaints [7]. Predominant intestinal symptoms are abdominal pain, bloating, epigastric pain, nausea, and chronic diarrhea with weight loss [8]. Tiredness, headache, anxiety, foggy mind, muscle and joint pain, eczema, and leg or arm numbness are extraintestinal symptoms typical of NCGS [9]. Children with NCGS mainly have intestinal symptoms such as abdominal pain and chronic diarrhea without weight loss. Less frequently they present with extraintestinal manifestations; tiredness and attention deficit hyperactivity are the most common symptoms.

In recent years, several studies found a link between NCGS and neuropsychiatric manifestations including ataxia, neuropathy, epilepsy, depression, autism spectrum disorder, and schizophrenia. Gluten ataxia (GA) has been traditionally associated with CD (see Section 8.7.6). Comparative clinicopathological investigations however revealed that patients with GA are better classified within the NCGS group than within the CD group [10]. A comparative study evaluated the neurological manifestations of NCGS patients ($n=334$) and CD patients ($n=228$) [11]. The results demonstrated that most common symptoms were cerebellar ataxia, peripheral neuropathy, and encephalopathy. Patients equally responded to a gluten-free diet (GFD) suggestive of common pathophysiological mechanisms. However, additional studies are urgently required to clarify the role of NCGS regarding the wide spectrum of neurological and psychiatric symptoms. In contrast to CD, NCGS is not associated with malabsorption or nutritional deficiencies and any increased risk for autoimmune diseases or malignancy.

10.3 Pathogenesis and triggering factors

NCGS represents a disorder with poorly recognized pathogenesis. Serum IgA and IgG antibodies, typical of CD [antibodies against endomysium (EMAs), transglutaminase (TGAs), and deamidated gliadin peptide (DGPAs)] as well as IgE antibodies typical of WA, are negative. However, IgG antigliadin antibodies (AGAs) may be positive in 40–60% of cases, whereas healthy people develop these antibodies only with the frequency of 2–8%. Regarding intestinal permeability, some studies have reported normal permeability and others indicated increased permeability in HLA-DQ2/DQ8-positive patients [12, 13]. Patients with NCGS have low levels of zonulin-1 and claudin-1, whereas the expression of claudin-4 is upregulated. The intestinal mucosa of NCGS patients is usually normal (Marsh type 0). The number of intraepithelial lymphocytes (IELs) in the superficial epithelium of the intestinal villi may be slightly increased suggesting the existence of a low grade mucosal inflammation (Marsh type 1). The infiltration of duodenal lamina propria with eosinophils and activation of circulating basophils is increased in NCGS patients [14].

Several studies indicated that NCGS is associated with an increased expression of Toll-like receptors (TLR2 and TLR4), a reduction of the T-regulatory cell marker FoxP3, and an increase of IELs but no increase in cytokines related to the adaptive immunity. These findings suggest an important role of the innate immune system without involvement of the adaptive immune response. However, the detection of IgG AGAs in around 50% of NCGS patients provides evidence that adaptive immunity might also contribute to NCGS development.

Because symptoms of NCGS frequently disappear on a GFD, gluten and particularly gluten proteins have been considered a precipitating factor similar as in CD. For example, studies of 34 adult NCGS patients showed that 16 g of gluten per day induced a rapid onset of gastrointestinal symptoms and tiredness within the first week of a 6-week intervention [15]. The important role of gluten has been confirmed by a randomized, double-blind placebo-controlled (DBPC) cross-over trial performed in 61 NCGS patients [16]. However, not all patients showed an evident worsening of symptoms after gluten ingestion. This could imply that other wheat constituents apart from gluten could be additionally responsible for NCGS. Some reports have even indicated that gluten is not the cause of NCGS. Recent studies raised the possibility that beside gluten proteins, fermentable oligo-, di-, and monosaccharides and polyols (FODMAPs), amylase/trypsin inhibitors (ATIs), and wheat germ agglutinins (WGAs) might contribute to the symptoms of NCGS [17]. These new findings indicate that the term gluten sensitivity is probably too restrictive.

Gliadins and glutenins, the major components of gluten, are commonly considered main triggers of NCGS. However, it is still unclear whether all types of gluten proteins (see Section 2.2) are precipitating factors, which amino acid sequences are responsible for harmful biochemical reactions, and which pathomechanism underlies the development of the disease.

FODMAPs (see Section 2.1) have recently experienced much attention regarding their role in NCGS symptomatology [18]. The physiological effects of dietary FODMAPs are well-known. While they are resistant to digestion and absorption in the small intestine, their rapid fermentation by microorganisms in the colon induces the production of gas like hydrogen, carbon dioxide, and methane and causes luminal distension and intestinal complaints. Using a randomized, DBPC cross-over design, Biesiekierski and coworkers found no evidence of specific or dose-dependent effects of gluten in patients with NCGS placed on a low FODMAP diet [19]. They therefore concluded that gluten might not be a specific trigger of intestinal symptoms, once dietary FODMAPs are reduced. However, FODMAPs may be ruled out as a main trigger of NCGS because of the following reasons [1]: (1) FODMAP intolerance is usually characterized by a quick onset (minutes to a few hours) after ingestion of the triggering foods, whereas NCGS has a slower onset (hours to a few days). (2) NCGS patients resolve symptoms after they eliminate gluten-containing foods despite continuing to ingest FODMAPs from other sources such as fruits and vegetables. (3) FODMAPs cause mild wheat sensitivities at most limited to intestinal symptoms and cannot explain extraintestinal symptoms that frequently occur in NCGS.

ATIs have been qualified as prominent triggers of innate immunity of both CD and NCGS. They were also identified as adjuvants of preexisting inflammatory adaptive immune responses such as allergic airway inflammation [20]. ATIs are protective proteins in wheat and other cereals that inhibit enzymes (amylase and trypsin) of common parasites such as mealworms and meal bugs, but they might also have a regulatory effect during seed maturation. Wheat ATIs are a family of up to 17 homologous proteins and exist either as monomers, as non-covalently linked homodimers or as tetramers [21]. The monomers have molecular masses of around 15,000 and variable amino acid sequences but conserved steric structures characterized by α-helices and five intrachain disulfide bonds. Fig. 10.1 shows the three-dimensional structure of ATI 0.19 modeled with SWISS-MODEL [23].

ATIs are encoded by the B and D subgenomes of wheat; thus they are missing in the diploid species einkorn. ATIs are salt-soluble proteins located in the endosperm and account for around 2–4% of the total protein of common wheat grains. Thus wheat-based foods may have levels of ATIs that are as much as 100-fold higher than in most wheat-free foods [24]. Accordingly, an adult, who consumes an average of 200g of wheat-based food per day, will ingest around 0.2–0.4g of ATIs. ATIs resist heat and proteolytic digestion in the gastrointestinal tract maintaining their activity throughout food processing, oral ingestion, and intestinal passage. In contrast to human gastrointestinal enzymes, peptidases of selected Lactobacillus strains were found to degrade ATIs and experiments with gluten-sensitized mice expressing HLA-DQ8 demonstrated that these strains reduced intestinal dysfunction and inflammatory effects induced by gluten-containing diet [25].

In vitro and *in vivo* studies have suggested that wheat ATIs induce innate immune responses that involve the activation of innate immune cells (dendritic cells, macrophages, and monocytes) via TLR4, while intestinal epithelium cells are non-responsive [26]. ATIs 0.19 and CM3 from common wheat have the

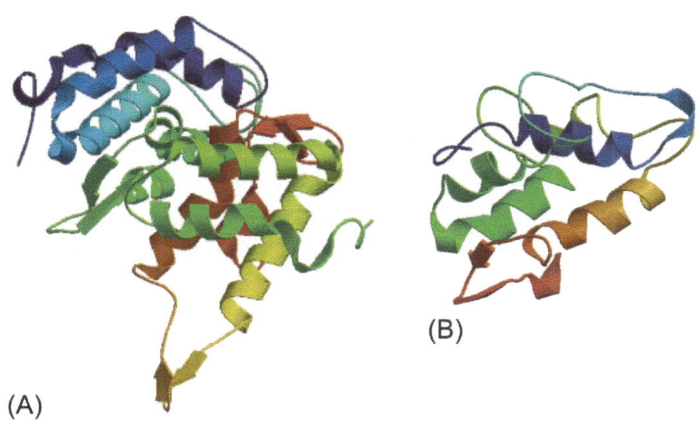

FIG. 10.1 Three-dimensional structure of the α-amylase/trypsin inhibitor 0.19 (A) and CM3 (B) modeled with SWISS MODEL [22].

highest bioactivity on monocytes expressing TLR4, whereas ATI extracts from commercially available emmer and spelt flours only have 30–70% of the activity of common wheat [24]. A targeted liquid chromatography/mass spectrometry analysis, developed to quantitate the major ATIs in eight varieties of the five main wheat species each, revealed that overall ATI contents were slightly higher in spelt and emmer (4.1–6.3 mg/g flour) compared to common wheat and durum wheat (3.2–5.5 mg/g flour) [27]. The quantitative data confirmed the absence or presence of very low amounts of ATIs in einkorn.

The activation of the TLR4-MD2-CD14 complex by ATIs is dose-dependent and results in release of the proinflammatory cytokines and chemokines interleukin-8, tumor necrosis factor-α, and C-C motif chemokine ligand 2. These findings define wheat ATIs as novel contributors to NCGS. They appear to play a role in promoting other immune-mediated diseases within and outside the gastrointestinal tract. Gluten-free cereals display no or little TLR4 stimulating activity [28].

WGAs have been enumerated as another potential trigger of NCGS. They are enriched in the germ of wheat grains; their amount in whole grain flour ranges between 30 and 50 mg/kg [29]. WGAs belong to the class of lectins, proteins that specifically bind the carbohydrate moieties of glycoproteins. WGAs are heterodimers with a molecular mass of around 40,000. Their function is to protect wheat from bacteria, yeasts, and insects by binding to N-acetylglucosamine and sialic acid groups of certain glycoproteins on the cell membrane surface leading to agglutination. WGAs are very stable proteins and resistant to the effects of heat and digestive enzymes. They increase intestinal permeability and stimulate directly the release of several proinflammatory cytokines from monocytes and macrophages through the binding to glycoconjugates [30]. However, there are no human data on the effect of WGAs on NCGS development.

10.4 Diagnosis

Due to the lack of awareness and clear diagnostic criteria, NCGS was underdiagnosed for many years. The major problem in NCGS diagnosis is the differentiation from other wheat-related diseases like CD [22] and IBS [31]. In general, patients should avoid self-diagnosis and should not withdraw gluten-containing foods on their own, before NCGS has been verified in a referral center. Currently, there are no laboratory biomarkers specific for NCGS. The only positive requirement to diagnose NCGS is the presence of troubles caused by wheat ingestion and their disappearance on a GFD [5]. Therefore NCGS diagnosis is primarily based on exclusion criteria, i.e., other food hypersensitivities with similar symptoms, particularly intestinal complaints, have to be ruled out before diagnosing a patient as NCGS-positive.

Regarding differentiation from other wheat-related disorders, non-inflammatory hypersensitivities to carbohydrates, mainly FODMAPs, can usually be excluded clinically or by simple tests (e.g., hydrogen breath test). WA can be diagnosed with conventional IgE and skin testing and/or confocal laser endoscopy. However, non-IgE-mediated WA may be difficult to distinguish from NCGS. The presence of CD-specific serum antibodies (e.g., EMAs and TGAs) supports the diagnosis of CD and excludes NCGS. Duodenal biopsy is also important to rule out the underlying presence of seronegative CD. When corresponding symptoms are existent, Kabbani and colleagues proposed the following diagnostic algorithm to differentiate CD from NCGS [32]: Subjects with negative CD serology (IgA TGAs or IgA/IgG DGPAs are missing) on a regular diet are unlikely to have CD. Those with negative serology, who also lack clinical evidence of malabsorption and CD risk factors, are highly likely to have NCGS. Those with equivocal serology should undergo HLA typing to determine the need for biopsy.

Patients with IBS are identified by the so-called Rome III diagnostic criteria (see Chapter 11). However, there is a remarkable overlap between IBS and NCGS, because numerous patients, who fulfilled the criteria for IBS, were found to be sensitive to gluten and their symptoms improved on a GFD [33]. Assessment of IgG AGAs can be useful to confirm NCGS and should be integrated in the overall management of the patient [34]. A basophil activation test with wheat has also shown a high sensitivity for NCGS and may be an additional tool for diagnosis [14].

Once other disorders have been excluded, the best available final test for NCGS is the "food sensitivity test" by a gluten challenge after a period of treatment with a GFD [35, 36]. The gluten challenge is an important diagnostic tool for all gluten-related disorders, but there is great variability in regards to deciding, which patients should be challenged, what amount of gluten should be used, what the challenge duration should be, when and where the challenge should occur, and sometimes, why to conduct a challenge [37]. Moreover, there is a need for a standardized procedure to establish the diagnosis [38].

The 3rd International Expert Meeting on Gluten Related Disorders, held in Salerno (Italy) in 2014, reached a consensus on how the diagnosis of NCGS should be confirmed [35]. The recommended procedure should include two steps: (1) assessment of the clinical response to a GFD after having been on a normal gluten-containing diet and (2) the reintroduction of gluten after the treatment with GFD and measurement of the clinical response. When a patient with suspected NCGS is already on GFD, only step 2 should be started. NCGS can be excluded in subjects failing to show symptomatic improvement in step 1. Step 2 is required to confirm the diagnosis of NCGS: a definitive demonstration of NCGS can be counted, if the symptoms reappear upon gluten reintroduction. Since the improvement after a GFD can be induced by a placebo effect [39], a DBPC crossover challenge is highly recommended to confirm NCGS. This means that neither the patients undergoing challenge nor the medical personnel know about the intake of gluten or placebo. Both preparations (gluten and placebo) must be undistinguishable in look, texture, taste, and flavor. The proposed challenge procedure for instance involves 4 weeks of strict GFD, 1 week of gluten or placebo challenge (recommended amount: 8 g gluten per day), 1 week wash-out on a strict GFD and after crossover, 1 week of placebo or gluten challenge [35]. During the past years, these guidelines have supported a number of clinicians to reach a firm and positive diagnosis of adult NCGS patients. For the first time, a randomized DBPC gluten challenge with crossover was used to evaluate the prevalence of NCGS in pediatric subjects ($n=28$), who had gastrointestinal symptoms associated with gluten ingestion but were not affected by CD and WA [40]. Eleven children tested positive, while NCGS was ruled out in >60% of cases showing the significance of DBPC gluten challenge for confirming NCGS. However, this procedure is difficult to undertake in routine clinical practice and many patients refuse to introduce gluten into their diet because of quickly recurring symptoms and signs of discomfort.

A number of different materials, for example wheat-containing daily meals, wheat bread, cookies, muffins, udon noodles, or vital gluten, have been used for oral challenge. However, the overall composition and the amount of wheat proteins, used in these materials, mostly were not determined and thus validity and comparability of results were impeded. Already in 2004, the European Academy of Allergy and Clinical Immunology demanded that challenge tests should be carried out with standardized foods or food components guaranteeing the safety of patients, practical feasibility, and the comparability of results between different study centers [41]. However, there has not been any standard method for the preparation of the gluten vehicle and placebo for carrying out challenges for NCGS assessment. To fill this gap, two materials, commercially available gluten and a food-grade gluten hydrolysate ("Pepgluten"), were extensively characterized [42]. Pepgluten was prepared from vital gluten (baking ingredient) by incubation with pepsin (dietary supplement) in a mixture of acetic acid essence (for salads and meals) and mineral water at pH 3 for 120min. The freeze-dried preparation contained 70.7% crude protein ($N \times 5.7$), 10.4% starch, 5.9% water, 4.7% lipids, 4.1% dietary fiber, and 1.1% ash. The contents of gluten equivalents and ATIs were 56.5 and 2.2g per 100g, respectively. The slightly acidic, dough-like smell and bitter taste of Pepgluten could be completely camouflaged in a mixture of multivitamin juice and bitter lemon, grapefruit juice or vegetable and fruit smoothies. Thus Pepgluten meets the criteria for placebo-controlled challenges (active and placebo materials were identical regarding appearance, taste, smell, and texture [35]) and is appropriate as a standard preparation for the oral challenge and clinical investigations to study wheat hypersensitivities.

After the diagnosis of NCGS has been confirmed, patients are advised to start with a GFD (see Section 12.1). In contrast to CD, where are strict GFD has to be maintained, patients with NCGS could adapt a more liberal diet reducing the gluten intake by approximately 90% [24]. Clinical manifestations of NCGS are usually ameliorated or resolved within days to weeks of commencing a GFD. Because it is not known, whether NCGS is a permanent or transient condition, periodic reintroduction of gluten is advised. The efficacy of a gluten-digesting enzyme mixture, derived from *Aspergillus* and *Penicillium* strains and papaya, in improving the symptoms of NCGS was tested by a randomized, single-blind, placebo-controlled, crossover clinical trial [43]. Thirty NCGS patients on a strict GFD were divided in two groups. One group was challenged with gluten (3 × 1.5 g daily) and the enzyme mixture (3 × 250 mg daily) for 2 weeks and the other group was gluten-challenged with placebo (without enzyme) for 2 weeks followed by crossover. Symptom questionnaire index scores indicated that the symptoms of NCGS were improved by the enzyme mixture. There were significantly decreases in the scores of incomplete evacuation feeling, bloating, and headaches. The enzyme mixture can be expected to be a useful dietary supplement to decrease the symptoms of NCGS patients without adverse events.

10.5 Conclusions

NCGS is now accepted as a distinct clinical entity outside of other wheat-related disorders. However, the scientific understanding of NCGS is still restricted and may be in a stage, where CD research was in the 1970s. Exact information on potential triggers, genetic components, and pathomechanisms are missing. Many questions remain unanswered. These include how frequent NCGS is, how it can be reliably diagnosed, is the elimination of dietary gluten alone sufficient for the treatment, and how it can be appropriately managed. Future research is urgently needed to generate more knowledge regarding etiology, mechanism, and biomarkers of NCGS, a hypersensitivity that is widely accepted but has only few certainties and many unresolved issues.

References

[1] Fasano A, Sapone A, Zevallos V, Schuppan D. Nonceliac gluten and wheat sensitivity. Gastroenterology 2015;148:1195–204.

[2] Aziz I, Hadjivassiliou M, Sanders DS. The spectrum of noncoeliac gluten sensitivity. Nat Rev Gastroenterol Hepatol 2015;12:516–26.

[3] Volta U, Caio G, Karunaratne TB, Alaedini A, de Giorgo R. Non-coeliac gluten/wheat sensitivity: advances in knowledge and relevant questions. Expert Rev Gastroenterol Hepatol 2017;11:9–18.

[4] Scherf KA, Koehler P, Wieser H. Gluten and wheat sensitivity—an overview. J Cereal Sci 2016;67:2–11.

[5] Mansueto P, Seidita A, D'Alcamo A, Carroccio A. Non-celiac gluten sensitivity: literature review. J Am Coll Nutr 2014;33:39–54.

[6] Bizzaro N, Tozzoli R, Villalta D, Fabris M, Tonutti E. Cutting-edge issues in celiac disease and in gluten intolerance. Clin Rev Allergy Immunol 2012;42:279–87.

[7] Volta U, de Giorgio R, Caio G, Uhde M, Manfredini R, Alaedini A. Nonceliac wheat sensitivity: an immune-mediated condition with systemic manifestations. Gastroenterol Clin North Am 2019;48:165–82.

[8] Czaja-Bulsa G. Non coeliac gluten sensitivity—a new disease with gluten intolerance. Clin Nutr 2015;34:189–94.

[9] Losurdo G, Principi M, Iannone A, Amoruso A, Ierardi E, di Leo A, Barone M. Extra-intestinal manifestations of non-celiac gluten sensitivity: an expanding paradigm. World J Gastroenterol 2018;24:1521–30.

[10] Rodrigo L, Hernandez-Lahoz C, Lauret E, Rodriguez-Pelaez M, Soucek M, Ciccocioppo R, et al. Gluten ataxia is better classified as non-celiac gluten sensitivity than as celiac disease: a comparative clinical study. Immunol Res 2016;64:558–64.

[11] Hadjivassiliou M, Rao DG, Grinewald RA, Aeschlimann DP, Sarrigianis PG, Hoggaed N, et al. Neurological dysfunction in coeliac diseas and non-coeliac gluten sensitivity. Am J Gastroenterol 2016;111:561–7.

[12] Meijer CR, Shamir CR, Mearin ML. Coeliac disease and noncoeliac gluten sensitivity. J Pediatr Gastroenterol Nutr 2015;60:429–32.

[13] Sapone A, Lammers KM, Casolaro V, Cammarota M, Giuliano MT, de Rosa M, et al. Divergence of gut permeability and mucosal immune gene expression in two gluten-associated conditions: celiac disease and gluten sensitivity. BMC Med 2011;9:23.

[14] Carroccio A, Mansueto P, Jacono G, Soresi M, D'Alcamo A, Cavataio F, et al. Non-celiac wheat sensitivity diagnosed by double-blind placebo-controlled challenge: exploring a new clinical entity. Am J Gastroenterol 2012;107:1898–906.

[15] Biesiekierski JR, Newnham ED, Irving PM, Barrett JS, Haines M, Doecke JD, et al. Gluten causes gastrointestinal symptoms in subjects without celiac disease: a double-blind randomized, placebo-controlled trial. Am J Gastroenterol 2011;106:508–14.

[16] Di Sabatino A, Volta U, Salvatore C, Biancheri P, Caio G, de Giorgio R, et al. Small amounts of gluten in subjects with suspected nonceliac gluten sensitivity: a randomized, double-blind, placebo-controlled, cross-over trial. Clin Gastroenterol Hepatol 2015;13. 1604-12.e3.

[17] Gibson PR, Muir JG, Newnham ED. Other dietary confounders: FODMAPS et al. Dig Dis 2015;33:269–76.

[18] Priyanka P, Gayam S, Kupec JT. The role of low fermentable oligosaccharides, disaccharides, monosaccharides, and polyol diet in nonceliac gluten sensitivity. Gastroenterol Res Pract 2018;1561476.

[19] Biesiekierski JR, Peters SL, Newnham ED, Rosella O, Muir JG, Gibson PR. No effects of gluten in patients with self-reported non-celiac gluten sensitivity after dietary reduction of fermentable, poorly absorbed short-chain carbohydrates. Gastroenterology 2013;145:320–8.

[20] Zevallos VF, Raker VK, Maxeiner J, Scholtes P, Steinbrink K, Schuppan D. Dietary wheat amylase trypsin inhibitors exacerbate murine allergic airway inflammation. Eur J Nutr 2019;58:1507–14.

[21] Schuppan D, Zevallos V. Wheat amylase trypsin inhibitors as nutritional activators of innate immunity. Dig Dis 2015;33:260–3.

[22] Leonard MM, Sapone A, Catassi C, Fasano A. Celiac disease and nonceliac gluten sensitivity: a review. JAMA 2017;318:647–56.

[23] Waterhouse A, Bertoni M, Bienert S, Studer G, Tauriello G, Gumienny R, et al. SWISS-MODEL: homology modelling of protein structures and complexes. Nucleic Acids Res 2018;46:W296–303.

[24] Zevallos VF, Raker V, Tenzer S, Jimenez-Calvente C, Ashfaq-Khan M, Rüssel N, et al. Nutritional wheat amylase-trypsin inhibitors promote intestinal inflammation via activation on myeloid cells. Gastroenterology 2017;152. 1100-13.e12.

[25] Caminero A, McCarville JL, Zevallos VF, Pigrau M, Yu XB, Jury J, et al. Lactobacilli degrade wheat amylase trypsin inhibitors to reduce intestinal dysfunction induced by immunogenic wheat proteins. Gastroenterology 2019;156:2266–80.

[26] Junker Y, Zeissig S, Kim SJ, Barisani D, Wieser H, Leffler DA, et al. Wheat amylase trypsin inhibitors drive intestinal inflammation via activation of Toll-like receptor 4. J Exp Med 2012;209:395–408.

[27] Geisslitz S, Ludwig C, Scherf KA, Koehler P. Targeted LC-MS/MS reveals similar contents of α-amylase/trypsin inhibitors as putative triggers of nonceliac gluten sensitivity in all wheat species except einkorn. J Agric Food Chem 2018;66:12395–403.
[28] Schuppan D, Pickert G, Ashfaq-Khan M, Zevallos V. Non-celiac wheat sensitivity: differential diagnosis, triggers and implications. Best Pract Res Clin Gastroenterol 2015;29:469–76.
[29] De Punder K, Pruimboom L. The dietary intake of wheat and other cereal grains and their role in inflammation. Nutrients 2013;5:771–87.
[30] Dalla-Pellegrina C, Perbellini O, Scupoli MT, Tomelleri C, Zanetti C, Zoccatelli G, et al. Effects of wheat germ agglutinin on human gastrointestinal epithelium: insights from an experimental model of immune/epithelial cell interaction. Toxicol Appl Pharmacol 2009;237: 146–53.
[31] Catassi C, Alaedini A, Bojarski C, Bonaz B, Bouma G, Carroccio A, et al. The overlapping area of non-celiac gluten sensitivity (NCGS) and wheat-sensitive irritable bowel syndrome (IBS): an update. Nutrients 2017;9. pii: E1268.
[32] Kabbani TA, Vanga RR, Leffler DA, Villafuerte-Galvez J, Pallav K, Hansen J, et al. Celiac disease or non-celiac gluten sensitivity? An approach to clinical differential diagnosis. Am J Gastroenterol 2014;109:741–6.
[33] Shahbazkhani B, Sadeghi A, Malekzadeh R, Khatavi F, Etemadi M, Kalantri E, et al. Non-celiac gluten sensitivity has narrowed the spectrum of irritable bowel syndrome: a double-blind randomized placebo-controlled trial. Nutrients 2015;7:4542–54.
[34] Infantino M, Manfredi M, Meacci F, Grossi V, Severino M, Benucci M, et al. Diagnostic accuracy of anti-gliadin antibodies in non celiac gluten sensitivity (NCGS) patients: a dual statistical approach. Clin Chim Acta 2015;451:135–41.
[35] Catassi C, Elli L, Bonaz B, Bouma G, Carroccio A, Castillejo G, et al. Diagnosis of non-celiac gluten sensitivity (NCGS): the Salerno experts' criteria. Nutrients 2015;7:4966–77.
[36] Branchi F, Aziz I, Conte D, Sanders DS. Noncoeliac gluten sensitivity: a diagnostic dilemma. Curr Opin Clin Nutr Metab Care 2015;18:508–14.
[37] Bascunan KA, Roncoroni L, Branchi F, Doneda L, Scricciolo A, Ferretti F, et al. The 5 Ws of a gluten challenge for gluten-related disorders. Nutr Rev 2018;76:79–87.
[38] Dale HF, Biesiekierski JR, Lied GA. Non-coeliac gluten sensitivity and the spectrum of gluten-related disorders: an update overview. Nutr Res Rev 2018;16:1–10.
[39] Dale HF, Hatlebakk JG, Hovdenak N, Ystad SO, Lied GA. The effect of a controlled gluten challenge in a group of patients with suspected non-coeliac gluten sensitivity: a randomized, double-blind placebo-controlled challenge. Neurogastroenterol Motil 2018;30:e13332.
[40] Francavilla R, Cristofori F, Verzillo L, Gentile A, Castellaneta S, Polloni C, et al. Randomized double-blind placebo-controlled crossover trial for the diagnosis of non-celiac gluten sensitivity in children. Am J Gastroenterol 2018;113:421–30.
[41] Bindslev-Jensen C, Ballmer-Weber BK, Bengtsson U, Blanco C, Ebner C, Hourihane J, et al. Standardization of food challenges in patients with immediate reactions to foods—position paper from the European Academy of Allergology and Clinical Immunology. Allergy 2004;59:690–7.
[42] Wieser H, Scherf KA. Preparation of a defined gluten hydrolysate for diagnosis and clinical investigations of wheat hypersensitivities. Nutrients 2018;10. pii: E1411.
[43] Ido H, Matsubara H, Kuroda M, Takahashi A, Kojima Y, Koikeda S, Sasaki M. Combination of gluten-digesting enzymes improved symptoms of non-celiac gluten sensitivity: a randomized single-blind, placebo-controlled crossover study. Clin Transl Gastroenterol 2018;9:181.

Chapter 11

Irritable bowel syndrome

Irritable bowel syndrome (IBS) is one of the most frequently diagnosed functional gastrointestinal disorders [1]. Its prevalence in the Western adult population has been estimated in a range from 5% to 20%. IBS is not a single disease but rather a symptom cluster resulting from diverse pathologies [2]. Typical intestinal symptoms are similar to those of celiac disease (CD), nonceliac gluten sensitivity (NCGS), small intestinal bacterial overgrowth, and lactose intolerance including abdominal discomfort and pain, gas, bloating, and diarrhea changing with constipation [3]. The Rome IV criteria categorize IBS by the most predominant presenting symptom: diarrhea (IBS-D), constipation (IBS-C), mixed (IBS-M), or unspecified (IBS-U) [4]. Additionally, IBS is often associated with other functional digestive and nondigestive disorders.

Roughly two-thirds of IBS patients perceive their symptoms as being related to specific meals, notably foods rich in carbohydrates, while experiencing an improvement during fasting. Wheat has been regarded as a culprit of symptom generation in a substantial subset of IBS patients [5,6]. Using strict criteria as recommended for IBS studies, about one third of patients with IBS-D or IBS-M are wheat sensitive with a similar proportion of these types [7]. Both endomicroscopic features and histopathological abnormalities, occurring after wheat was administered into the duodenal mucosa, provided a morphological basis to the existence of wheat sensitivity in some cases of IBS [8]. Some types of IBS, especially diarrhea-predominant cases, showed symptomatic improvement on a gluten-free diet (GFD) [9]. A clinical trial in IBS-D patients ($n=27$) confirmed the close relationship between the intake of wheat-containing food and IBS development [10]. Withdrawal of wheat led to a significant improvement of intestinal symptoms, while wheat-containing food challenge was followed by the recurrence of symptoms.

Previously, people, who did not have any other cause for their symptoms, were diagnosed with IBS. Today, the identification of IBS patients is no longer based on the exclusion of other diseases but on a positive diagnosis using symptom-related criteria, the so-called Rome III diagnostic criteria. They require recurrent abdominal pain or discomfort at least 3 days a month in the past 3 months associated with two or more of the following points [11]: (1) improvement with defecation; (2) onset associated with a change in frequency of stools; and (3) onset with a change in form (appearance) of stools. However, the diagnostic procedure performs only moderately well

Wheat—An Exceptional Crop. https://doi.org/10.1016/B978-0-12-821715-3.00011-3

and misdiagnosis is common. In particular, differential diagnosis to exclude symptom-related diseases (e.g., NCGS and non-IgE-mediated wheat allergy) is often difficult. A placebo-controlled oral challenge test with a gluten-containing preparation (see Section 10.3) may be a valuable help to identify IBS-D patients. A novel, noninvasive approach by the analysis of a urinary volatile organic compound (1,3,5,7-cyclooctadecaen) enables the differentiation of IBS-D from CD [12].

Screening for CD in pediatric and adult IBS patients have revealed a significantly higher CD prevalence compared to the general population [13,14]. IBS patients ($n = 48$), were more likely to exhibit positive levels of antibodies against deamidated gliadin peptides and transglutaminase 2 than healthy controls ($n = 20$) [15]. They (50%) had more often than controls (20%) the predisposing genetic factors (HLA-DQ2/8) underlying the development of CD. A case-control study, including 400 patients with IBS according to the Rome III criteria, demonstrated that up to 5.2% of subjects were positive for at least one of the CD-specific antibodies and 2.5% had biopsy-confirmed CD [16]. The IBS-D subtype had the highest prevalence for serological positivity (12%). Therefore screening for CD especially in IBS-D patients appears to be a reasonable strategy.

A substantial proportion of patients with IBS may have NCGS, because IBS and NCGS can coexist independently from each other without necessarily sharing a common pathophysiological basis [17,18]. The incomplete knowledge of the pathogenesis of IBS and NCGS, lack of data on their real epidemiology, and the absence of a gold standard for their diagnosis make the overall picture difficult to understand [19]. Further research is required to help distinguish between IBS and NCGS.

The pathomechanism of IBS is complicated and not clearly identified. The drawbacks are the heterogeneous clinical manifestations and the absence of structural abnormalities and biomarkers. Factors that seem to be important for the development of IBS include alterations of the intestinal microbiota, intestinal bacterial overgrowth, intestinal permeability, intestinal immune function, visceral sensation, gut-brain interactions, and psycho-social status [20]. Genetic, immune, environmental, inflammatory, neurological, and psychological factors in addition to visceral hypersensitivity can all play an important role. Regarding altered intestinal permeability, there is no uniformity of data on this issue. A case-control study revealed that both IBS-D patients and controls had significantly lower values for biomarkers of increased permeability (e.g., urinary lactulose/mannitol ratio) than patients with active CD [21]. The results allowed the categorization of IBS-D patients into two subtypes, either with normal or increased intestinal permeability. This finding might be useful to assess a better evaluation of the heterogeneous IBS-D population. Elevated intestinal permeability in IBS-D patients is additionally reflected by increased markers of permeability such as myosin light chain kinase activity and claudin-15 expression [10].

There is no agreement on a gold standard for the treatment of IBS. Strategies are based on the nature and severity of the symptoms, the degree of functional impairment of the bowel habits, and the presence of psycho-social comorbidity. Traditional pharmaceutical agents may be indicated for constipation, diarrhea or abdominal pain. There is no special diet for IBS patients, but small and frequent meals are recommended and high-fat food, dairy products, high amounts of carbohydrates, caffeine, and alcohol should be avoided. In many cases, patients can associate the ingestion of certain foods with their complaints; an improvement may be achieved after the restriction of these foods. Diets, restricted in their contents of FODMAPs (see Section 2.1), have been used for a long time to manage patients with IBS and should be the first dietary approach [22]. For example, the use of sourdough cultures in bread-making was shown to reduce the quantities of FODMAPs (mostly fructans) resulting in bread products that were well tolerated by patients with IBS [23]. However, the drastic reduction of FODMAP intake could have negative consequences on the nutritional state and the guidance of an experienced dietitian should help ensure both success and overall nutritional adequacy [24]. A number of studies support the role of a GFD as specific dietary intervention in IBS. In particular, IBS-D patients should be advised to test a GFD for several months.

In conclusion, the pathogenesis, diagnosis, and treatment of IBS will remain subjects of much ongoing research. Well-structured multicenter studies are urgently needed to get insight into the pathomechanism of this disease. Although the relationship between food ingestion and symptom generation has long been established, the understanding of dietary triggers such as wheat and its components is still poorly defined and awaits further investigations. The identification and validation of specific biomarkers will be essential to identify IBS in diagnostically difficult cases. Moreover, further well-structured studies are needed to improve the management of IBS and to discover appropriate therapeutic targets.

References

[1] Rej A, Sanders DS. Gluten-free diet and its 'cousins' in irritable bowel syndrome. Nutrients 2018;10:E1727.

[2] Chey WD, Kurlander J, Eswaran S. Irritable bowel syndrome: a clinical review. JAMA 2015;313:949–58.

[3] Borghini R, Donato G, Alvaro D, Picarelli A. New insights in IBS-like disorders: Pandora's box has been opened; a review. Gastroenterol Hepatol Bed Bench 2017;10:79–89.

[4] Drossmann DA. Functional gastrointestinal disorders: history, pathophysiology, clinical features and Rome IV criteria. Gastroenterology 2016;150:1262–79.

[5] De Giorgio R, Volta U, Gibson PR. Sensitivity to wheat, gluten and FODMAPs in IBS: facts or fiction? Gut 2016;65:169–78.

[6] Volta U, Pinto-Sanchez MI, Boschetti E, Caio G, de Giorgio R, Verdu EF. Dietary triggers in irritable bowel syndrome: is there a role for gluten? J Neurogastroenterol Motil 2016;22:547–57.

[7] Barmeyer C, Schuhmann M, Meyer T, Zielinski C, Zuberbier T, Siegmund B, et al. Long-term response to gluten-free diet as evidence for non-celiac wheat sensitivity in one third of patients with diarrhea-dominant and mixed-type irritable bowel syndrome. Int J Colorectal Dis 2017;32:29–39.
[8] Fritscher-Ravens A, Schuppan D, Ellrichmann M, Schoch S, Röcken C, Brasch J, et al. Confocal endomicroscopy shows food-associated changes in the intestinal mucosa of patients with irritable bowel syndrome. Gastroenterology 2014;147:1012–20.
[9] Zanwar VG, Pawar SV, Gambhire PA, Jain SS, Surude RG, Shah VB, et al. Symptomatic improvement with gluten restriction in irritable bowel syndrome: a prospective, randomized, double blinded placebo controlled trial. Intest Res 2016;14:343–50.
[10] Wu RL, Vazquez-Roque MI, Carlson P, Burton D, Grover M, Camillere M, et al. Gluten-induced symptoms in diarrhea-predominant irritable bowel syndrome are associated with increased myosin light chain kinase activity and claudin-15 expression. Lab Invest 2017;97:14–23.
[11] Longstreth GF, Thompson WG, Chey WD, Houghton LA, Mearin F, Spiller RC. Functional bowel disorders. Gastroenterology 2006;130:1480–91.
[12] Arasaradnam RP, Westenbrink E, McFarlane MJ, Harbord R, Chambers S, O'Connelli N, et al. Differentiating coeliac disease from irritable bowel syndrome by urinary volatile organic compound analysis—a pilot study. PLoS One 2014;9:e107312.
[13] Cristofori F, Fontana C, Magista A, Capriati T, Indrio F, Castellaneta L, et al. Increased prevalence of celiac disease among pediatric patients with irritable bowel syndrome: a 6-year prospective cohort study. JAMA Pediatr 2014;168:555–60.
[14] Irvine AJ, Chey WD, Ford AC. Screening for celiac disease in irritable bowel syndrome: an updated systematic review and meta-analysis. Am J Gastroenterol 2017;112:65–76.
[15] Domzal-Magrowska D, Kowalski MK, Szczesniak P, Bulska M, Orszulak-Michalak D, Malecka-Panas E. The prevalence of celiac disease in patients with irritable bowel syndrome and its subtype. Prz Gastroenterol 2016;11:276–81.
[16] Sanchez-Vargas LA, Thomas-Dupont P, Torres-Aguilera M, Azamar-Jacome AA, Ramirez-Ceervanes KL, Aedo-Garces MR, et al. Prevalence of celiac disease and related antibodies in patients diagnosed with irritable bowel syndrome according to the Rome III criteria. A case-control study. Neurogastroenterol Motil 2016;28:994–1000.
[17] Makharia A, Catassi C, Makharia GK. The overlap between irritable bowel syndrome and non- celiac gluten sensitivity: a clinical dilemma. Nutrients 2015;7:10417–26.
[18] Rej A, Sanders DS. The overlap of irritable bowel syndrome and noncoeliac gluten sensitivity. Curr Opin Gastroenterol 2019;35:199–205.
[19] Soares RL. Irritable bowel syndrome, food intolerance and non-celiac gluten sensitivity. A new clinical challenge. Arq Gastroenterol 2018;55:417–22.
[20] Bellini M, Gambaccini D, Stasi C, Urbano MT, Marchi S, Usai-Satta P. Irritable bowel syndrome: a disease still searching for pathogenesis, diagnosis and therapy. World J Gastroenterol 2014;20:8807–20.
[21] Linsalata M, Riezzo G, D'Attoma B, Clemente C, Orlando A, Russo F. Noninvasive biomarkers of gut barrier function identify two subtypes of patients suffering from diarrhea predominant-IBS: a case-control study. BMC Gastroenterol 2018;18:167.
[22] Mansueto P, Seidita A, D'Alcamo A, Carroccio A. Role of FODMAPs in patients with irritable bowel syndrome. Nutr Clin Pract 2015;30:665–82.
[23] Muir JG, Varney JE, Ajamian M, Gibson PR. Gluten-free and low-FODMAP sourdoughs for patients with coeliac disease and irritable bowel syndrome: a clinical perspective. Int J Food Microbiol 2019;290:237–46.
[24] Catassi G, Lionetti E, Gatti S, Catassi C. The low FODMAP diet: many question marks for a catchy acronym. Nutrients 2017;9. pii: E292.

Chapter 12

Treatment of dietary wheat hypersensitivities

Permanent adherence to a gluten-free diet (GFD) is the current effective treatment of hypersensitivities caused by wheat ingestion. Patients with celiac disease (CD) and related disorders (dermatitis herpetiformis and gluten ataxia), dietary wheat allergy (WA), and nonceliac gluten sensitivity (NCGS) have to avoid or at least reduce the intake of products that contain wheat. In most cases, also rye and barley have to be excluded, whereas oats appear to be harmless for most patients. Traditional cereal-based gluten-containing foodstuffs like bread, pasta, and beer have to be replaced by corresponding surrogates made from raw materials that do not contain gluten. These gluten-free foods are dietetic foods for special medical purposes and the allowed level of gluten underlies national and international regulations. The accurate quantitation of gluten contaminations, potentially present in foods labeled gluten-free, is essential for patients, industry, and food control authorities. Numerous analytical methods have been developed during the past decades. Most of them comprise the extraction of gluten proteins and their quantitative determination by immunochemical methods and use of a suitable protein reference for calibration. Patients are frequently unsatisfied with dietetic gluten-free products: They are expensive, often have poor quality, and are not available everywhere. The patients' most pronounced desire is therefore the development of medications that would allow them to eat conventional gluten-containing foods. A number of novel strategies for the prevention and treatment of wheat hypersensitivities have been developed and clinical trials to test medications are already ongoing. In this chapter, basic principles underlying a GFD, production and quality of dietetic gluten-free products, gluten analysis, and potential alternative treatments are discussed.

12.1 Gluten-free diet

Adopting a GFD has become increasingly popular in the Western world, because wheat consumption has been held responsible for a variety of health risks. Pseudoscientific books and the mainstream press have fueled the impression that wheat grains make people sick, stupid, fat, and addicted (see Chapter 7). As a consequence, the consumption of gluten-free foods has significantly increased over the last decades. Millions of people voluntarily abstain from

Wheat—An Exceptional Crop. https://doi.org/10.1016/B978-0-12-821715-3.00012-5

wheat-containing foodstuffs without any medical advice. Individuals may restrict gluten from their diets for a variety of reasons such as improvement of gastrointestinal and other symptoms as well as the perception that gluten is potentially harmful and thus restriction represents a healthy lifestyle [1]. Palmieri and coworkers [2] have answered some questions about the appropriateness of follow this trend such as how safe are the current gluten-free products, what are the benefits and side effects of a GFD, and what are clinical conditions that might benefit from gluten avoidance. However, emerging evidence shows that wheat avoidance may be associated with adverse health effects due to the deficiencies of essential nutrients such as dietary fiber, minerals, and vitamins (see Section 12.4). Taken together, there is no scientific evidence for eliminating wheat from the diet for the vast majority (90–95%) of the population [3]. The GFD is a vital necessity only for CD patients and persons affected by other wheat hypersensitivities such as WA, NCGS, and diarrhea-predominant irritable bowel syndrome (IBS-D).

Patients, diagnosed with wheat hypersensitivities, have to avoid or at least reduce the intake of products that contain wheat and in most cases, also rye and barley have to be excluded. Oats appear to be harmless for most patients, however only pure oats should be introduced into the diet. Patients, who decide to eat oat products, should be monitored for evidence of adverse reactions. A strict lifelong adherence to a GFD is currently the only treatment of CD and related disorders such as dermatitis herpetiformis and gluten ataxia. The current state of research, regarding the definition of a strict GFD in terms of the gluten thresholds considered tolerable as well as the indication for a GFD in the absence of symptoms or histological abnormalities, has been summarized by Itzlinger and coworkers [4]. As a rule, a strict GFD is currently indicated in all cases of symptomatic CD. The vast majority of patients, who start a GFD, can feel an improvement of intestinal complaints within few days or weeks after initiation. A GFD has also been recommended for asymptomatic patients but not for patients with potential CD, since they do not tend to develop villous atrophy [5]. For other forms of hypersensitivities like dietary WA and NCGS, the reduction of wheat intake may be sufficient. Even though the overall aim of a GFD is to achieve healing and to maintain health of affected individuals, distinguishing the different conditions is important for long-term therapy [6].

Patients with CD have to maintain a strict GFD and the total daily intake of gluten should not be beyond 20 mg/day, which approximately corresponds to one hundredth of a slice of bread. CD patients should be followed-up closely for dietary adherence, nutritional deficiencies, and the development of possible comorbidities. They should be referred to a health-care team, including a physician and an expert dietician. Celiac societies can provide patients with accurate and useful information. To avoid symptoms in dietary WA, the limitation of daily wheat intake in a range of 1–2 g of wheat proteins, corresponding to around 15–25 g of bread, is usually sufficient. Patients with wheat-dependent exercise-induced anaphylaxis (WDEIA) are advised to limit wheat ingestion within the temporal proximity of physical activity and intake of nonsteroidal antiinflammatory drugs. Patients with NCGS may adapt a more liberal diet than CD patients by reducing the normal gluten intake to approximately 10%. They require regular reassessment for gluten tolerance allowing with time the reintroduction of increasing amounts of gluten. Apart from GFD, a low FODMAP diet may lead to a significant improvement of symptoms in NCGS [7]. Because some studies support the role of a GFD as specific dietary intervention in IBS, in particular IBS-D patients should be advised to test a GFD for several months.

12.2 Gluten-free products

The GFD is based on products that are naturally free from wheat or gluten and on dietetic gluten-free products, i.e., alternatives to traditional gluten-containing foods. Patients with dietary wheat hypersensitivities are allowed to eat a wide range of foods gluten-free by nature such as meat, fish, dairy products, vegetables, nuts, and fruits. However, they should be aware of numerous composite foods that contain "hidden" sources of gluten. Hidden gluten may be found in numerous foods, cosmetics, and medications (Table 12.1). Thus unintentional gluten intake can amount to up to 1000 mg/day. Patients are therefore advised to check the ingredients labeled on prepacked foods or to get information about unpacked foods. At home, gluten-free foods should always be stored and handled separately from gluten-containing foods. In cases, where separate areas are not available, gluten-free meals should be prepared before other meals. Eating at restaurants, workplaces, and other peoples' home remains a risk for unintentional gluten exposure.

Traditional cereal-based gluten-containing goods, for instance bread and other baked products, pasta, and beer, have to be replaced by corresponding surrogates. These are made from cereals that do not contain gluten such as rice, corn, sorghum, and millet or pseudocereals such as amaranth, buckwheat, and quinoa [8]. Recent advances in the formulation of cereal-based gluten-free products by utilizing alternate ingredients and processing techniques have been summarized by Rai and colleagues [9]. The manufacture of gluten-free baked goods and paste products is still a challenge to food technologists, because substituting gluten functionality such a cohesivity, viscosity, elasticity, and gas-holding ability (see Section 5.2.3) is extremely difficult. Replacing gluten in gluten-free products requires utilizing a complex mix of gluten-free flours or starches, proteins, hydrocolloids, and other additives in an attempt to mimic gluten's multifunctional roles [10]. The main goal is to create products similar to the gluten-containing counterparts currently on the market. However, the quality of gluten-free products still tends to be unsatisfactory. In particular, commercial gluten-free bread has been considered poor in quality due to low volume, small-pored unelastic crumb, pale and soft crust, off-flavor, and rapid onset of staling.

TABLE 12.1 Possible sources of hidden gluten in food and nonfood products.

Foods	
Baked beans	Licorice
Blue cheese	Maltodextrin
Bouillon	Marinades
Breading	Meat loaf
Broth	Nuts
Candy	Puddings
Cereal binding	Rice mixes
Chocolate	Roux
Color (artificial caramel)	Salad dressing
Communion wafer	Sauces
Dry roasted nuts	Sausage
Flavoring	Seasonings
French fries	Self-basting poultry
Gravy	Soups
Herbal teas	Soy sauce
Ice cream	Stuffing
Imitation bacon	Thickeners
Imitation seafood	Vegetarian burger
Nonfood	
Cremes	Shampoos
Lipsticks	Soaps
Medications	Toothpaste
Postage stamps	

Gluten-free bread is the most important gluten-free product and plays a crucial role in the nutrition of people suffering from wheat hypersensitivities. Usually, starch-containing flours from safe sources (e.g., corn, rice, buckwheat, or amaranth) or starches (e.g., from maize, potato, cassava, or rice) have to be used as base material in the production of gluten-free bread. Starch is important, because it absorbs up to 45% water during dough preparation and acts as inner filler in the continuous matrix of dough and bread. Pure wheat starch is appreciated as ingredient for gluten-free baked goods due to its favorable properties (high water absorption, high dough density, and increased bread volume). Products that contain wheat starch or starch derivatives (e.g., maltodextrin and glucose syrup) and meet the Codex Alimentarius Standard (see below) were shown to be safe for CD patients [11].

Because gluten provides about 6–7% protein in standard wheat bread, the supplementation of gluten-free basic material with other protein sources may be necessary. Dairy ingredients such as caseinates or skim milk, soy products, and egg or corn protein are frequently applied alternatives to increase the protein content. To mimic the water absorbance capacity and viscosity of gluten, several hydrocolloids are in use. Examples are guar gum, xanthan gum, sodium alginate, carrageenan, hydroxypropyl methylcellulose, and carboxymethyl cellulose. These compounds act as water binders, improve dough properties and bread texture, and slow down staling. The addition of different emulsifiers (e.g., lecithin or diacetyl tartaric acid ester of mono- and diglycerides) and enzymes (e.g., lipases, oxidases or transglutaminases) may additionally help enhancing dough and bread quality. To promote dietary fiber uptake, gluten-free products can be enriched with psyllium husk or cellulose for instance.

The process of gluten-free bread-making distinctly differs from that of standard wheat bread (see Section 4.2.1). The complex mixture of ingredients and the lack of gluten make the actions quite difficult. Gluten-free doughs are much less cohesive and elastic compared to wheat doughs. They are highly smooth, sticky, and difficult to handle. They require shorter mixing and proofing time. The diminished functionality of gluten-free doughs can be improved through a number of treatments including acid/base deamidation, cross-linking by oxidizing agents and transglutaminase or high-pressure treatment. Proofed gluten-free doughs are usually placed into a pan, because they are not self-supporting, and are baked shorter than standard wheat doughs. Bread volumes are mostly smaller, crumbs firmer, and crusts softer. The use of sourdough may represent an alternative to increase bread quality. Despite numerous past efforts to improve bread properties, ongoing attempts on optimized formulations and technologies are still necessary.

Foods labeled gluten-free are dietetic foods and underlie international and national regulations. The "Codex Standard for Foods for Special Dietary Use for Persons Intolerant to Gluten" (CODEX STAN 118—1979) [12] defines dietetic gluten-free foods as:

(a) Consisting of or made only from one or more ingredients, which do not contain wheat (i.e., all Triticum species, such as durum wheat, spelt, and khorasan wheat, which is also marketed under different trademarks such as kamut), rye, barley, oats or their cross-bred varieties, and the gluten content level does not exceed 20 mg/kg in total based on the food as sold or distributed to the consumer.

(b) Consisting of one or more ingredients from wheat (i.e., all *Triticum* species, such as durum wheat, spelt, and kamut), rye, barley, oats or their cross-bred varieties, which have been specially processed to remove gluten, and the gluten level does not exceed 20 mg/kg in total, based on the food as sold or distributed to the consumer.

The Crossed Grain symbol is the internationally recognized lable for gluten-free foods.

Foods specially processed to reduce gluten content may contain above 20 up to 100 mg/kg in total. Decisions on the marketing of these products may be determined at the national level. Regarding oats, the following footnote was added: *Oats can be tolerated by most but not all people, who are intolerant to gluten. Therefore, the allowance of oats that are not contaminated with wheat, rye or barley in foods covered by this standard, may be determined at the national level.* Since commercial oat products are frequently contaminated with wheat, rye or barley, CD patients are advised to buy only oat products that are explicitly labeled gluten-free.

Dietetic gluten-free foods are no longer a niche products and are now widely available not only in health food stores but also in supermarkets and particularly online. The GFD currently appears to be an increasing "fad" diet as a treatment for all kinds of medical disorders. Its popularity has steadily increased during the last decade and is now the trendiest alimentary habit in the USA and other countries. In parallel, the production of gluten-free foodstuffs is now a multibillion-(US) dollar industry. The worldwide market is expected to increase to around 24 billion dollars by the year 2020.

Nevertheless, numerous people with wheat hypersensitivities are still unsatisfied with dietetic gluten-free foods, because they are much more expensive than regular food products [13]. They often are poor in flavor, taste, texture, and mouthfeel compared to their gluten-containing counterparts. In particular, the replacement of wheat bread and barley beer is one of the most critical aspects of a GFD. Despite the improved quality of gluten-free breads in the past years, most products on the market are still described as low quality products and the nutritional value of a large number of them is poor (see Section 12.4). The nutritional profile of gluten-free products, currently available on the market, and the possible relationship thereof with the nutritional status of CD patients on a GFD have been discussed by Melini and Melini [14]. There is still a gap between science and market especially regarding the approaches to improve the diversity of gluten-free bread and its nutritional quality [15].

Numerous scientific efforts have been made to remove gluten from gluten-containing materials if possible without deteriorating product quality. Products, where the techno-functionality of wheat gluten (e.g., elasticity, cohesivity, and viscosity) is not needed, are specifically suitable for gluten removal. In most cases, special enzymes were used to degrade (detoxify) gluten proteins into nonimmunogenic fragments. Examples are peptidases from different microorganisms (bacteria and fungi) or germinated cereals. Novel procedures that enable the production of gluten-free products from wheat, rye, and barley have been summarized by Scherf and coworkers [16]. They include flour, starch, bran, different baked goods, and pasta from wheat, sourdough and bread from rye as well as fermented beverages from barley like malt beer and beer. All these products were gluten-free according to the Codex Alimentarius and had properties equal to corresponding gluten-containing products. It can be expected that producers of dietetic gluten-free foods will transform these novel approaches into products ready for the market, so that the assortment of these special foodstuffs will be distinctly extended.

12.3 Gluten analysis

The detection and accurate quantitation of gluten contaminations, possibly present in foods labeled gluten-free, is essential for food producers and food control to guarantee the safety for patients with gluten hypersensitivities. To provide some guidance, Codex Alimentarius Standard 118-199 [12] includes a number of general requirements for methods of gluten analysis. It states that the quantitative determination of gluten in foods and food ingredients shall be based on an immunologic method or other methods providing at least equal sensitivity and specificity. The antibody used should react with cereal protein fractions that are toxic or immunogenic for persons intolerant to gluten and should not cross-react with other constituents. The methods should be validated, calibrated against a certified reference material if available, and the detection limit should be 10 mg gluten/kg or below.

Numerous reports on gluten detection and quantitation have been published during the past decades without reaching a worldwide acceptance of a general analytical procedure that should involve the following steps: (1) complete extraction of gluten proteins from the food, (2) application of a suitable protein reference for calibration, and (3) accurate quantitation of extracted gluten proteins [17].

12.3.1 Extraction

The most frequent solvents, used for the extraction of gluten proteins, are aqueous alcohols; Codex Alimentarius recommends 60% ethanol. However, only the prolamins of nonheated products can be completely dissolved. For an extraction of total gluten proteins (prolamins plus glutelins) from both nonheated and heated products, aqueous alcohols (e.g., 60% ethanol or 50% propanol), containing a reducing agent (e.g., 2-mercaptoethanol or tris(2-carboxyethyl)-phosphine) and a disaggregating agent (e.g., guanidine or sodium dodecylsulfate), are most suitable [18]. All components of the extraction solvent need to be compatible with the analytical method. Additional steps may be necessary such as defatting (products with more than 10% fat) or removal of polyphenols.

12.3.2 Reference material

Various protein references, mainly wheat prolamins (gliadins) and wheat gluten, have been produced by different research groups and companies. These references were isolated from different sources by variable procedures. In consequence, protein contents and compositions were scarcely comparable and resulted in divergent calibration curves, when the same analytical method was applied. To date, the best characterized and widely used reference material is the so-called Prolamin Working Group (PWG) gliadin standard, a highly purified gliadin preparation isolated from 28 European wheat cultivars [19]. However, PWG gliadin represents only the prolamin fraction of wheat. Commonly accepted references, representing the glutelin fraction or total gluten proteins from wheat, rye, and barley, are still missing. Reference material for total gluten proteins would allow better calibration, comparability, and harmonization of analytical results. Therefore an international working group started an initiative to prepare a new reference material based on wheat flours that represent typical gluten protein compositions. Among 27 international wheat cultivars the following five cultivars were finally selected according to qualitative and quantitative selection criteria: Akteur and Mv Magvas from Europe, Carberry from North America, Yumai from Asia, and Yitpi from Australia [20]. The flour blend of these cultivars will be used for further work.

One disadvantage of using flours as reference material is their limited shelf life. In contrast to flours, gluten protein isolates like PWG-gliadin can be stored for decades at deep-freezing temperatures without significant changes. Therefore a comprehensive strategy to isolate gluten protein fractions and types from wheat, rye, barley, and oat flours was developed by Schalk and coworkers [21]. The fractions and single types were fully characterized using high-performance column chromatography (HPLC), sodium dodecylsulfate polyacrylamide gel electrophoresis (SDS-PAGE), N-terminal sequencing, and liquid chromatography/mass spectrometry (LC/MS). The analytical methods confirmed that all gluten protein types could be reproducibly isolated in high purity and were suitable to be used as reference material, e.g., for calibration of immunochemical or LC/MS methods.

12.3.3 Immunochemical methods

At the moment, immunochemical assays are the method of choice for gluten analysis and have been recommended by the Codex Alimentarius and commercially offered by different companies. They are based on monoclonal or polyclonal antibodies that have been produced by immunization of mammals such as mice, rabbits or goats after injection of the antigen (gluten proteins). So-called enzyme-linked immunosorbent assays (ELISAs) are currently used most often to quantitate gluten. Two different ELISA systems, sandwich and competitive, are currently on the market (Fig. 12.1). The sandwich ELISA needs two binding sites of the antigen and is suitable only for large antigens

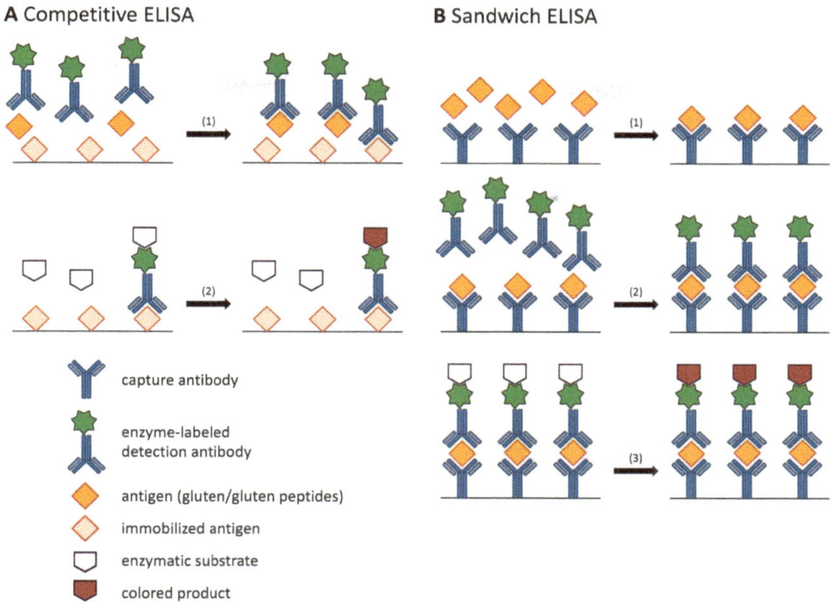

FIG. 12.1 Principle of ELISA: (A) competitive ELISA and (B) sandwich ELISA.

(proteins). The competitive ELISA needs only one binding site of the antigen and is suitable for both proteins and small antigens (peptides). Calibration curves, created with a reference protein or peptide, enable the quantitation of sample antigen in both systems.

In sandwich ELISA, the gluten-specific antibodies are immobilized onto the wells of a microtiter plate and incubated with an aliquot of the extract containing the antigen to be determined, which results in the formation of an antibody/antigen complex (step 1). After washing, antibodies labeled with an indicator enzyme are added and bound to a second binding site of the antigen (step 2). After washing, the substrate of the enzyme is added and transformed into a colored end product (step 3) that is measured by spectrophotometry. The obtained value is directly proportional to the gluten concentration. In competitive ELISA, antigens to be determined are immobilized onto the microtiter plate. A limited and constant quantity of the enzyme-labeled antibody and an aliquot of the antigen extract are added so that free and immobilized antigens compete for the limited enzyme-labeled antibody binding site (step 1). The larger the quantity of added sample antigen, the smaller the quantity of the labeled antibody bound to the immobilized antigen will be. After removal of unbound antibodies, antigens, and antigen/antibody complexes by washing, the enzymatic substrate is added and transformed into a colored product (step 2) that is measured spectrophotometrically. The obtained value is inversely proportional to the gluten concentration.

Numerous ELISA variants have been developed by different research groups and companies and some of them are now commercially available. The most frequently employed test kits use the R5, G12, and α20 monoclonal antibodies, all of which recognize glutamine- and proline-rich sequence epitopes typical of gluten proteins (see Section 2.2.1). ELISA tests require no sophisticated laboratory equipment, provide results after 2–4 h, and are easy to perform, suitable for routine analysis, and often cheaper than other techniques. Unfortunately, there are also some disadvantages, because each test kit has its specific features such as different extraction procedures, reference materials, and antibodies. It is, therefore, not surprising that the comparison of results from different test kits for the same sample shows systematic deviations [22,23]. The following examples may impressively demonstrate this fact.

The analysis of prolamin and glutelin fractions from wheat, rye, and barley with five different sandwich ELISA test kits resulted in partially different responses [24]. Whereas wheat prolamins (gliadins) were detected quite accurately by all antibodies, highly variable results were obtained for rye and barley prolamins and rye glutelins. The largest discrepancies were found for wheat and barley glutelins. Further studies of the five ELISA test kits targeted the specificities and sensitivities to different gluten protein types occurring in the gliadin and glutenin (wheat), secalin (rye), and hordein (barley) fractions [25]. The results showed high variability of results depending on the type of cereal, the gluten protein type, and the antibody used. In another study, three sandwich and two competitive ELISA kits exhibited significant differences in gluten quantities, when gluten fractions and their hydrolysates from wheat, rye, and barley were analyzed [26]. Kit-dependent variations were determined by grain source, antibody specificity, gluten proteins, and the level of their hydrolysis. The gluten contents of 30 wheat starches (14 declared as gluten-free) and one defined assay control were analyzed with seven commercial ELISA test kits using the respective recommended extraction protocol for wheat starch samples [27]. The comparative analysis of gluten yielded up to six significantly different results per wheat starch. Similarly, the determination of gluten in different cultivars of common wheat, spelt, durum wheat, emmer, and einkorn, using two monoclonal antibody kits and a polyclonal antibody kit, revealed clear quantitative differences between kits [28]. The use of a common calibrator reduced the between-assay variability to some extent, but variation due to differences in selectivity of the assays was unaffected [23].

Further problems were found in determining the gluten content of products with different matrices. A set of 14 ELISA kits for gluten detection was used to analyze gluten content in a series of relevant food matrices varying in complexity [22]. The results demonstrated that there was no single ELISA method that could accurately detect and quantify gluten in all different matrices. Additional difficulties were generated by modifications during food processing, for example by Maillard reactions that impaired the detection of gluten in heated or extruded products like bread and pasta. In fermented products such as beer and soy sauce, partial hydrolysis of gluten proteins made their accurate detection even more difficult [29]. In summary, despite the merits of commercially available ELISAs, they still have their limitations of accurate gluten determination and show substantial variations due to different reference materials, extraction methods, and antibodies with variable specificities and sensitivities. Therefore, the need for a generally applicable analytical procedure remains.

The performance of ELISA can be modified and simplified by rapid methods, for example, by lateral flow devices and dipstick systems [30]. The main applications are swab tests of potentially contaminated surfaces and checks for gluten contamination of raw material or processed material. The statistical evaluation of a collaborative study with 18 participants showed that an R5 immunochromatographic dipstick system was capable of detecting gluten in corn flour and heat-processed corn products below the threshold of 20 mg/kg [31].

Electrochemical immunosensors seem to be among the most promising alternative analytical tools [32]. Most constructions are either based on antibody/antigen or aptamer/antigen interactions. For example, a competitive disposable amperometric immunosensor, based on gliadin-functionalized carbon/nanogold screen-printed electrodes, allowed the rapid and sensitive determination of gliadin in food samples [33]. The analysis of different flours revealed limit of detection (LOD) and limit of quantitation (LOQ) of 8 and 22 ng gliadin/mL ethanol extract, respectively. The combination of the quartz crystal microbalance technology with the so-called photonic immobilization enabled the development of a sensitive and reliable biosensor that had a LOD of 4 mg gliadin/kg food [34]. A newly developed label-free impedimetric aptasensor for gliadin revealed linearity in the range of 5–1000 μg gliadin/L and a LOD of 5 μg gliadin/L extract [35]. Measurement of gliadin in gluten-containing and gluten-free products showed a good agreement with the results obtained with the R5 ELISA method. A portable, handheld gluten detection device, the Nima™ sensor, is now available for consumers wishing to determine, if gluten is present in food. The sensor integrates food processing, gluten detection, result interpretation, and data transmission in a portable device [36]. The gluten sensor performed reliably in the detection of gluten in foods having ≥ 20 mg/kg of added gluten with only 6% of failures and worked with greater reliability, as the gluten levels increased. In summary, studies of the last decade have demonstrated that immunosensors are reliable, user-friendly, and cost-effective with short operation times. However, collaborative studies should clarify, whether the developed sensors are capable of substituting conventional ELISAs.

The glutelin fractions of gluten-containing cereals are insufficiently targeted by most ELISA systems, although both prolamins and glutelins contain immunogenic epitopes (see Section 8.4.4). To account for this bias, the determined prolamin content is usually multiplied by a factor of two to obtain the gluten content assuming a prolamin/glutelin ratio of one [12]. This factor tends to introduce an element of uncertainty, because the true prolamin/glutelin ratios may differ from 0.2 (wheat starch) to 13.9 (einkorn flour) [37]. For improvement,

ELISA tests are currently completed with antibodies directed against additional immunogenic gluten proteins, in particular glutenins [38]. At present, the performance of a novel gluten ELISA system, comprising the R5 antibody in combination with two new antibodies against high- and low-molecular-weight glutenin subunits, are evaluated by an international collaborative study.

12.3.4 Nonimmunochemical methods

In the past years, a number of nonimmunochemical methods for gluten quantitation have been established. Examples are genomic-based methods and different forms of MS combined with chromatographic or electrophoretic methods [17]. Genomic-based methods target specific DNA or RNA indicating the presence of gluten-containing cereals, but they do not react with proteins themselves. The advantage is that DNA/RNA detection is more sensitive by several orders of magnitude compared to protein detection, because specific DNA/RNA fragments can be amplified by real-time polymerase chain reaction (qPCR) or real-time loop-mediated isothermal amplification (qLAMP) [39]. A number of qPCR systems have been developed for the differential detection of wheat, rye, barley, and oats in foods. The comparison of qPCR and ELISA test results, presented in the literature over the course of 6 years, revealed a good overall performance of both methods [17]. Therefore, qPCR can be seen as complementary to ELISA and has its strength in highly sensitive screening for the presence of gluten-containing cereals.

Apart from PCR, literature presents a variety of MS systems to detect and quantify gluten. The typical workflow includes the extraction of gluten proteins, their enzymatic partial hydrolysis, and the analysis of selected marker peptides, derived from gluten proteins, by LC-MS/MS. For absolute quantitation of marker peptides, targeted LC-MS/MS, using stable isotope labeled internal peptide standards, is the method of choice. A number of recent studies applied this method for gluten determination in different materials. For example, a targeted LC-MS/MS analysis was used to evaluate wheat contamination of oats [40]. The developed method enabled the quantitation of two wheat-specific gluten peptides in different wheat flour/oat flour mixtures. Using these marker peptides, 10 mg peptides/kg was the lowest concentration, at which the wheat contaminant could be confidently detected in spiked oat flour, which corresponded to approximately 1 mg wheat gluten/kg flour. Proteomic profiling of grains from 16 cereal samples, utilizing SDS-PAGE, Western blotting, and LC-MS/MS, revealed that four wheat marker peptides were reproducible detected with a variation of 22–26% in 14 commercially available wheat cultivars [41]. Based on these four peptides, 15 mg/kg of wheat contamination in soy flour, spiked with wheat flour, were reliable detected. In another study, targeted LC-MS/MS was used to quantitate 16 wheat marker peptides derived from all gliadin and glutenin types [42]. The specific peptide yields enabled the conversion of peptide into protein type concentration and the sum of all determined protein type concentrations was expressed as gluten content. This method was applied to quantitate gluten in wheat starches and compared to R5 ELISA, which resulted in a strong correlation between LC-MS/MS and ELISA. Similarly, seven barley, seven rye, and three oat marker peptides, derived from all types of hordein, secalin, and avenin fractions, were used to determine gluten concentrations in samples from the brewing process and sourdough fermentation by means of targeted LC-MS/MS [43]. The first quantitative method, including all gluten-containing cereals into one LC-MS/MS run, was based on one marker peptide each for rye, barley, and oats and two peptides for wheat [44]. Quantitative parameters were achieved using mixtures of the gluten-containing flours and rice flour resulting in LOQ from 2 to 18 mg peptide/kg flour. In conclusion, gluten quantitation, based on marker peptides, may be a promising alternative to existing immunochemical tests like ELISAs, but drawback is the need for expensive equipment and expertise that are only available in research institutes or specialized laboratories.

12.4 Life on a gluten-free diet

GFD is the only safe treatment for CD and related diseases (dermatitis herpetiformis and gluten ataxia) and is also indicated in dietary WA, severe cases of WDEIA, NCGS, and wheat sensitive IBS. In particular, a lifelong strict GFD is difficult to sustain and can severely impact patients in their quality of life (QOL). A deeper understanding of disease-specific characteristics, including daily management, participation in food-related activities, strategies, and supporting or hindering factors in everyday life, is necessary. Higher costs and restricted availability of dietetic gluten-free foods, their poorer quality of flavor, taste, texture and mouthfeel, and the dilemmas experienced, when eating out, travelling, and socializing with friends, can lead to a serious impairment of the QQL of the patients [45]. Especially patients with CD have reported doing things they enjoy less because of their restrictions, are more worried about their health, and have frequent feelings of frustration. The additional diagnosis of diabetes has an even more pronounced negative effect on QQL. A systematic review with meta-analysis demonstrated that CD patients, strictly adhering to a GFD, have a better health-related QOL compared with nonstrictly adhering patients [46].

Compliance to the GFD among CD patients is reported to be in a range from 45% to 90% depending on definition and method of assessment [47]. Intentional nonadherence to the GFD has been found to be less frequent than inadvertent lapses that are caused by gluten contaminations. For instance, a study in the USA revealed that 16 out of 78 foods labeled "gluten-free" contained gluten levels of more than 20 mg/kg ranging from 20.3 to 60.3 mg/kg [48]. Another study estimated that CD patients on a GFD may be inadvertently exposed to 100–400 mg gluten/day [49]. Thus numerous individuals, who believe they are following a strict GFD, are subjected to an ongoing gluten consumption.

Nonadherence to the GFD is highest among ethnic minorities, adolescents, and adults diagnosed in childhood. Reasons for nonadherence include lack of knowledge about the disease, unclear labeling of foods, and difficulties in identifying gluten-free foods and drinks when dining out. Higher costs of gluten-free food are a further barrier to adherence. Moreover, some patients in remission incorrectly believe that gluten can be tolerated, when gluten-containing food is occasionally eaten. A new and noninvasive method to monitor GFD compliance was presented by Moreno and coworkers [50]. Concentrated urine samples of patients were analyzed for the presence and quantities of immunogenic gluten peptides with a gluten-specific ELISA. The method enabled the detection of peptides in positive correlations with the amount of gluten intake as little as 25 mg.

Follow-up of patients, especially of those diagnosed for CD, is necessary to make an objective assessment of the patient's response, to ensure strict adherence to a GFD and nutritional adequacy, and to improve QOL [51]. After diagnosis, patients should be evaluated at regular intervals by a health-care team including a physician and a registered dietitian, who have expertise in the disease and GFD. Detailed follow-up and monitoring plans for adult CD patients have been presented by Kelly and colleagues [52]. The transition period from adolescence to young adulthood is considered a fragile and high-risk period for intentional and unintentional gluten intake [53]. Adolescent patients with CD should gradually assume exclusive responsibility for their care, although parental support is still important. For the correct evaluation of children at follow-up, it is advisable to assess compliance, nutrition, comorbidities, and possible complications once a year at the referral center [54]. Celiac societies help CD patients by providing accurate and useful information and recommend local support groups with fellow patients, who are already familiar with a GFD.

A strict GFD will result in symptomatic, serologic, and histological remission in most patients [55]. However, a GFD does not always guarantee adequate nutrient intake and some nutritional deficiencies have been described after treatment with a long-term GFD [56,57]. Several studies report an unbalanced intake of carbohydrates and fat as well as a limited supply with certain essential nutrients. For example, the intake of dietary fiber, vitamins (folates, vitamins D, and B_{12}) and minerals (iron, zinc, magnesium, and calcium) may be insufficient [58]. Monitoring the nutritional state by blood tests and the use of appropriate gluten-free supplementation should be integral components of gluten-free nutrition. Dietetic gluten-free products often have high levels of lipids, sugars, and salt to improve food palatability and consistency [59]. In consequence, subjects on a GFD frequently show an excessive consumption of hypercaloric and hyperlipidic foods to compensate dietetic restrictions. Thus a considerable part of CD patients actually has a higher body mass index after following a GFD [60]. GFD may therefore have a negative impact on cardiometabolic risk factor like obesity and atherosclerosis [61]. For this reason, patients are advised to increase their consumption of naturally gluten-free food rich in fiber and vitamins such as vegetables, fruits, legumes, and nuts.

12.5 Alternative therapies

The conventional dietary treatment of wheat-dependent hypersensitivities is a big challenge for the patients, in particular, when a strict lifelong GFD is essential. Therefore patients most often wish has been to have access to a pill or vaccine that would allow them to eat normal gluten-containing foods. Especially on the part of CD patients, there is an urgent need to develop safe and effective alternatives that might replace the GFD. Owing to the enormous increase of research, a number of strategies to eliminate the harmful effects of gluten have been developed [62,63]. Most of them are targeting the CD pathomechanism and consist of interventions in the digestive process and permeability of the intestinal mucosa, inhibition of CD-related tissue transglutaminase and cytokines as well as blockage of receptors. Clinical trials are already ongoing and results appear promising. The first medications to become available most likely will be marketed as a supplement to the GFD rather than a substitute. In the following, trials that are in clinical testing are shortly described.

An attractive approach for all individuals, suffering from wheat hypersensitivities, is based on the oral treatment with special enzymes (peptidases) that degrade harmful proteins and peptides to nontoxic and nonimmunogenic fragments. The research has mainly focused on peptidases from bacteria, fungi, and germinating cereals. A common feature of these peptidases is their ability to hydrolyze proline-rich peptides usually resistant to gastrointestinal peptidases. A major advantage of most of these enzymes is that they work in the stomach and in the lumen of the small intestine and do not themselves take part in harmful side effects to the host. Prolyl endopeptidases (PEPs) from bacteria and fungi have been shown to be the most promising candidates.

An enzyme preparation named ALV003, consisting of a bacterial PEP from *Sphingomonas capsulata* and a peptidase from green barley (EP-B2), was found to be highly effective, safe, and well tolerated in clinical trials [64,65]. Among fungal peptidases, the PEP from *Aspergillus niger* (AN-PEP) has been shown to degrade gluten proteins and peptides much faster than bacterial peptidases. Clinical trials on AN-PEP are widely advanced [66] and a corresponding preparation (named Tolerase G) is now available in the USA. This preparation is not yet intended to replace a GFD but to protect the consumer from unintentional intake of gluten via contaminated foods. It is important to bear in mind that the meal composition determines the amount of AN-PEP needed for gluten elimination [67]. The combination of AN-PEP with dipeptidyl peptidase IV (named STAN1) is also able to detoxify moderate amounts of gluten [68] and is currently in clinical trials with CD patients. Additionally, lactobacilli and specific intestinal and oral bacteria are potent sources of gluten-degrading enzymes (glutenases) that may contribute to commercialization of a novel therapy [69].

Another novel therapeutic option is to prevent the passage of harmful proteins and peptides through the intestinal layer. Zonulin has been found to increase the paracellular mucosal permeability of CD patients (see Section 7.4). Several zonulin antagonists have been shown to prevent the passage of immunogenic peptides through the intestinal enterocytes. One of the specific agents is larazotide acetate (named AT-1001) [70]. Clinical trials demonstrated that the application of AT-1001 was well tolerated and reduced inflammatory cytokines and gastrointestinal symptoms in CD patients [71]. A phase IIb trial on 342 CD patients, who had symptoms despite being on a GFD, showed consistent improvement with larazotide (0.5 mg three times a day) [72].

With the aim of inducing a "tolerogenic" response in gluten-sensitive individuals, a desensitizing vaccine (named NexVax2) has been developed [73]. The vaccine is based on a mixture of three selected immunogenic peptides from α- and γ-gliadins and hordeins. Subcutaneous immunization of patients was well tolerated without any serious adverse effect on patients [74]. Findings, based on two randomized, double-blind, placebo-controlled phase 1 studies, supported the continued development of this potential therapeutic vaccine for CD [75].

Infestation with the hookworm *Necator americanus* has been suggested as a possible treatment of CD [76]. Twelve adult CD patients on a GFD were inoculated with *Necator americanus* larvae and challenged with increasing amounts of gluten over 26 weeks (3 g daily for 2 weeks at most). Two subjects were withdrawn after 2 weeks. The clinical evaluation of the remaining patients revealed that hookworm infection promoted tolerance to gluten and stabilized or improved all tested indices of gluten toxicity (e.g., TGA level and Marsh score). Other strategies for CD treatment such as inhibition of tissue transglutaminase, modulation of inflammation, and HLA-DQ blocking are in a preclinical state.

In summary, current clinical investigations on agents for gluten detoxification and treatment of gluten-mediated disorders have shown promising results. However, studies with longer duration periods and greater significance will be needed to assess both the efficacy and long-term safety. An important point to address is the amount of gluten that can be detoxified *in vivo*. Moreover, there is currently no widely accepted measure of disease activity used in clinical trials. A systematic review including 286 studies indicated that in cases of CD quantitative morphometric analysis of the small intestinal mucosa had better reliability and responsiveness compared with qualitative scales like patient-reported outcome or symptom diary [77]. As the development of drugs move out of the proof-of-concept stage to phase 3 confirmatory trials, they need to be evaluated by endpoints that are tailored to the drug, the disease, and the target patient population [78]. Careful evaluation and reporting of outcomes, including patient's clinical features, serology, histology, and quality of life, will increase transparency and comparability of therapeutic trials and will benefit patients, healthcare, and the pharmaceutical industry [79]. Altogether, one must carefully weigh the risks, benefits, and costs of alternative treatments in comparison to the GFD, a therapy without side effects and no clinical complications.

References

[1] Niland B, Cash BD. Health benefits and adverse effects of a gluten-free diet in non-celiac disease patients. Gastroenterol Hepatol 2018;14:82–91.

[2] Palmieri B, Vadala M, Laurino C. A review of the gluten-free diet in non-celiac patients: beliefs, truths, advantages and disadvantages. Minerva Gastroenterol Dietol 2019;65:153–62.

[3] Shewry PR, Hey SJ. Do we need to worry about eating wheat? Nutr Bull 2016;41:6–13.

[4] Itzlinger A, Branchi F, Elli L, Schumann M. Gluten-free diet in celiac disease—forever and for all? Nutrients 2018;10. pii: E1796.

[5] Volta U, Caio G, Giancola F, Rhoden KJ, Ruggeri E, Boschetti E, et al. Features and progression of potential celiac disease in adults. Clin Gastroenterol Hepatol 2016;14:686–93.

[6] Leonard MM, Sapone A, Catassi C, Fasano A. Celiac disease and nonceliac gluten sensitivity: a review. JAMA 2017;318:647–56.

[7] Dieterich W, Schuppan D, Schink M, Schwappacher R, Wirtz S, Aqaimy A, et al. Influence of low FODMAP and gluten-free diets on disease activity and intestinal microbiota in patients with non-celiac gluten sensitivity. Clin Nutr 2019;38:697–707.

[8] Rosell CM, Barro F, Sousa C, Mena MC. Cereals for developing gluten-free products and analytical tools for gluten detection. J Cereal Sci 2014;59:354–64.

[9] Rai S, Kaur A, Chopra CS. Gluten-free products for celiac susceptible people. Front Nutr 2018;5:116.

[10] O'Shea N, Arendt E, Gallagher E. State of the art in gluten-free research. J Food Sci 2014;79:R1067–76.

[11] Kaukinen K, Salmi T, Collin P, Huhtala H, Kärjä-Lahdensuu T, Mäki M. Clinical trial: gluten microchallenge with wheat-based starch hydrolysates in coeliac disease patients—a randomized, double-blind, placebo-controlled study to evaluate safety. Aliment Pharmacol Ther 2008;28:1240–8.

[12] Codex Standard 118-179. Codex standard for foods for special dietary use for persons intolerant to gluten. Codex Alimentarius Commission; 2015. revision 2008, amendment.

[13] Fry L, Madden AM, Fallaize R. An investigation into the nutritional composition and cost of gluten-free versus regular food products in the UK. J Hum Nutr Diet 2018;31:108–20.

[14] Melini V, Melini F. Gluten-free diet: gaps and needs for a healthier diet. Nutrients 2019;11:170.

[15] Santos FG, Aguiar EV, Capriles VD. Analysis of ingredient and nutritional labeling of commercially available gluten-free bread in Brazil. Int J Food Sci Nutr 2019;8:1–8.

[16] Scherf KA, Wieser H, Koehler P. Novel approaches for enzymatic gluten degradation to create high-quality gluten-free products. Food Res Int 2018;110:62–73.

[17] Scherf KA, Poms RE. Recent developments in analytical methods for tracing gluten. J Cereal Sci 2016;67:112–22.

[18] Garcia E, Llorente M, Hernando A, Kieffer R, Wieser H, Mendez E. Development of a general procedure for complete extraction of gliadins for heat processed and unheated foods. Eur J Gastroenterol Hepatol 2005;17:529–39.

[19] Van Eckert R, Berghofer E, Ciclitira PJ, Chirdo F, Denery-Papini S, Ellis HJ, et al. Towards a new gliadin reference material—isolation and characterisation. J Cereal Sci 2006;43:331–41.

[20] Hajas L, Scherf KA, Török K, Bugyi Z, Schall E, Poms RE, et al. Variation in protein composition among wheat (*Triticum aestivum* L.) cultivars to identify cultivars suitable as reference material for wheat gluten analysis. Food Chem 2018;267:387–94.

[21] Schalk K, Lexhaller B, Koehler P, Scherf KA. Isolation and characterization of gluten protein types from wheat, rye, barley and oats for use as reference material. PLoS One 2017;12:e0172819.

[22] Bruins Slot ID, Bremer MG, van der Fels-Klerx I, Hamer RJ. Evaluating the performance of gluten ELISA test kits: the numbers do not tell the tale. Cereal Chem 2015;92:513–21.
[23] Rzychon M, Brohee M, Cordeiro F, Haraszi R, Ulberth F, O'Connor G. The feasibility of harmonizing gluten ELISA measurements. Food Chem 2017;234:144–54.
[24] Lexhaller B, Tompos C, Scherf KA. Comparative analysis of prolamin and glutelin fractions from wheat, rye, and barley with five sandwich ELISA test kits. Anal Bioanal Chem 2016;408:6093–104.
[25] Lexhaller B, Tompos C, Scherf KA. Fundamental study on reactivities of gluten protein types from wheat, rye and barley with five sandwich ELISA test kits. Food Chem 2017;237:320–30.
[26] Rallabhandi P, Sharma GM, Pereira M, Williams KM. Immunological characterization of the gluten fractions and their hydrolysates from wheat, rye and barley. J Agric Food Chem 2015;63:1825–32.
[27] Scherf KA. Gluten analysis of wheat starches with seven commercial ELISA test kits—up to six different values. Food Anal Meth 2017;10:234–46.
[28] Schopf M, Scherf KA. Wheat cultivar and species influence variability of gluten ELISA analyses based on polyclonal and monoclonal antibodies R5 and G12. J Cereal Sci 2018;83:32–41.
[29] Colgrave ML, Byrne K, Howitt CA. Liquid chromatography-mass spectrometry analysis reveals hydrolyzed gluten in beers crafted to remove gluten. J Agric Food Chem 2017;65:9715–25.
[30] Scherf KA, Koehler P, AACCI Approved Methods, Protein and Enzyme MethodsTechnical Committee Collaborative Study. Detection of gluten in corn flour and corn-based foods by R5 qualitative test method. Cereal Foods World 2017;62:156–60.
[31] Lacorn M, Scherf K, Uhlig S, Weiss T. Determination of gluten in processed and nonprocessed corn products by qualitative R5 immunochromatographic dipstick: collaborative study, first action 2015.16. J AOAC Int 2016;99:730–7.
[32] Scherf KA, Koehler P, Wieser H. Electrochemical immunosensors for the diagnosis of celiac disease. Adv Chem Eng Sci 2015;5:83–95.
[33] Manfredi A, Gianetto M, Mattarozzi M, Constantini M, Mucchino C, Careri M. Competive immunosensor based on gliadin immobilization on disposable carbon-nanogold screen-printed electrodes for rapid determination of celiotoxic prolamins. Anal Bioanal Chem 2016;408:7289–98.
[34] Funari R, Terracciano I, Della Ventura B, Ricci S, Cardi T, D'Agostino N, Velotta R. Label-free detection of gliadin in food by quartz crystal microbalance-based immunosensor. J Agric Food Chem 2017;65:1281–9.
[35] Malvano F, Albanese D, Pilloton R, Di Matteo M. A new label-free impedimetric aptasensor for gluten detection. Food Control 2017;79:200–6.
[36] Zhang J, Portela SB, Horrell JB, Leung A, Weitmann DR, Artiuch JB, et al. An integrated, accurate, rapid, and economical handheld consumer gluten detector. Food Chem 2019;275:446–56.
[37] Wieser H, Koehler P. Is the calculation of the gluten content by multiplying the prolamin content by a factor of 2 valid? Eur Food Res Technol 2009;229:9–13.
[38] Röckendorf N, Meckelein B, Scherf KA, Schalk K, Koehler P, Frey A. Identification of novel antibody-reactive detection sites for comprehensive gluten monitoring. PLoS One 2017;12:e0181566.
[39] Garrido-Maestu A, Azinheiro S, Fucinos P, Carvalho J, Prado M. Highly sensitive detection of gluten-containing cereals in food samples by real-time loop-mediated isothermal amplification (qLAMP) and real-time polymerase chain reaction (qPCR). Food Chem 2018;246:156–63.
[40] Fiedler KL, McGrath SC, Callahan JH, Ross MM. Characterization of grain-specific peptide markers for the detection of gluten by mass spectroscopy. J Agric Food Chem 2014;62:5835–44.

[41] Colgrave ML, Goswami H, Byrne K, Blundell M, Howitt CA, Tanner GJ. Proteomic profiling of 16 cereal grains and the application of targeted proteomics to detect wheat contamination. J Proteome Res 2015;14:2659–68.

[42] Schalk K, Koehler P, Scherf KA. Targeted liquid chromatography tandem mass spectrometry to quantitate wheat gluten using well-defined reference proteins. PLoS One 2018;13:e0192804.

[43] Schalk K, Koehler P, Scherf KA. Quantitation of specific barley, rye, and oat marker peptides by targeted liquid chromatography-mass spectrometry to determine gluten concentrations. J Agric Food Chem 2018;66:3581–92.

[44] Manfredi A, Mattarozzi M, Gianetto M, Careri M. Multiplex liquid chromatography-tandem mass spectrometry for the detection of wheat, oat, barley and rye prolamins towards the assessment of gluten-free product safety. Anal Chim Acta 2015;895:62–70.

[45] White LE, Bannerman E, Gillett PM. Coeliac disease and the gluten-free diet: a review of the burdens; factors associated with adherence and impact on health-related quality of life, with specific focus on adolescence. J Hum Nutr Diet 2016;29:593–606.

[46] Burger JP, de Brouwer B, IntHout J, Wahab PJ, Tummers M, Drenth JP. Systematic review with meta-analysis: dietary adherence influences normalization of health-related quality of life in coeliac disease. Clin Nutr 2017;36:399–406.

[47] Muhammad H, Reeves S, Jeanes YM. Identifying and improving adherence to the gluten-free diet in people with coeliac disease. Proc Nutr Soc 2019;11:1–8.

[48] Lee HJ, Anderson Z, Ryu D. Gluten contamination in foods labeled as "gluten free" in the United States. J Food Protect 2014;77:1830–3.

[49] Syage JA, Kelly CP, Dickason MA, Ramirez AC, Leon F, Dominguez R, Sealey-Voyksner JA. Determination of gluten consumption in celiac disease patients on a gluten-free diet. Am J Clin Nutr 2018;107:201–7.

[50] Moreno ML, Cebolla A, Munoz-Suano A, Carrillo-Carrion C, Comino I, Pizzarro A, et al. Detection of gluten immunogenic peptides in the urine of patients with coeliac disease reveals transgressions in the gluten-free diet and incomplete mucosal healing. Gut 2017;66:250–7.

[51] Husby S, Bai JC. Follow-up of celiac disease. Gastroenterol Clin North Am 2019;48:127–36.

[52] Kelly CP, Bai JC, Liu E, Leffler DA. Advances in diagnosis and management of celiac disease. Gastroenterology 2015;148:1175–86.

[53] Samasca G, Lerner A, Girbovan A, Sur G, Lupan I, Makovicky P, et al. Challenges in gluten-free diet in coeliac disease: Prague consensus. Eur J Clin Invest 2017;47:394–7.

[54] Valitutti F, Trovato CM, Montuori M, Cucchiara S. Pediatric celiac disease: follow-up in the spotlight. Adv Nutr 2017;8:356–61.

[55] Saturni L, Ferretti G, Bacchetti T. The gluten-free diet. Safety and nutritional quality. Nutrients 2010;2:16–34.

[56] Vici G, Belli L, Biondi M, Polzonetti V. Gluten free diet and nutrient deficiencies: a review. Clin Nutr 2016;35:1236–41.

[57] Naik RD, Seidner DL, Adams DW. Nutritional consideration in celiac disease and nonceliac gluten sensitivity. Gastroenterol Clin North Am 2018;47:139–54.

[58] Dennis M, Lee AR, McCarthy T. Nutritional considerations of the gluten-free diet. Gastroenterol Clin North Am 2019;48:53–72.

[59] Babio N, Alcazar M, Castillejo G, Recasens M, Martínez-Cerezo F, Gutiérrez-Pensado V, et al. Patients with celiac disease reported higher consumption of added sugar and total fat than healthy individuals. J Pediatr Gastroenterol Nutr 2017;64:63–9.

[60] Maladrino N, Capristo E, Farnetti S, Leggio L, Abenavoli L, Addolorato G, Gasbarrini G. Metabolic and nutritional features in adult celiac patients. Dig Dis 2008;26:128.

[61] Anania C, Pacifico L, Olivero F, Perla FM, Chiesa C. Cardiometabolic risk factors in children with celiac disease on a gluten-free diet. World J Clin Pediatr 2017;6:143–8.

[62] Kurada S, Yadav A, Leffler DA. Current and novel therapeutic strategies in celiac disease. Expert Rev Clin Pharmacol 2016;9:1211–23.

[63] Serena G, Kelly CP, Fasano A. Nondietary therapies for celiac disease. Gastroenterol Clin North Am 2019;48:145–63.

[64] Gass J, Bethune MT, Siegel M, Spencer A, Khosla C. Combination enzyme therapy for gastric digestion of dietary gluten in patients with celiac sprue. Gastroenterology 2007;133:472–80.

[65] Lähdeaho ML, Kaukinen K, Laurila K, Vuotikka P, Koivurova OP, Kärjä-Lahdensuu T, et al. Glutenase ALV003 attenuates gluten-induced mucosal injury in patients with celiac disease. Gastroenterology 2014;146:1649–58.

[66] Tack GJ, van de Water JM, Bruins MJ, Kooy-Winkelaar EM, van Bergen J, Bonnet P, et al. Consumption of gluten with gluten-degrading enzyme by celiac patients: a pilot-study. World J Gastroenterol 2013;19:5837–47.

[67] Montserrat V, Bruins MJ, Edens L, Koning F. Influence of dietary components on *Aspergillus niger* prolyl endoprotease mediated gluten degradation. Food Chem 2015;174:440–5.

[68] Ehren J, Moron B, Martin E, Bethune MT, Gray GM, Khosla C. A food-grade enzyme preparation with modest gluten detoxification properties. PLoS One 2009;4:e6313.

[69] Chander AM, Yadav H, Jain S, Bhadada SK, Dhawan DK. Cross-talk between gluten, intestinal microbiota and intestinal mucosa in celiac disease: recent advances and basis of autoimmunity. Front Microbiol 2018;9:2597.

[70] Valitutti F, Fasano A. Breaking down barriers: how understanding celiac disease pathogenesis informed the development of novel treatments. Dig Dis Sci 2019;64:1748–58.

[71] Kelly CP, Green PH, Murray JA, Dimarino A, Colatrella A, Leffler DA, et al. Larazotide acetate in patients with coeliac disease undergoing a gluten challenge: a randomised placebo-controlled study. Aliment Pharmacol Ther 2013;37:252–62.

[72] Leffler DA, Kelly CP, Green PHR, Fedorak RN, di Marino A, Perrow W, et al. Larazotide acetate for persistent symptoms of celiac disease despite a gluten-free diet: a randomized controlled trial. Gastroenterology 2015;148:1311–9.

[73] Brown GJ, Daveson J, Marjason JK, Ffrench RA, Smith D, Sullivan M, et al. A phase I study to determine safety, tolerability and bioactivity of Nexvax2 in HLA DQ2 + volunteers with celiac disease following a long-term, strict gluten-free diet. Gastroenterology 2011;140:S437–8.

[74] Di Sabatino A, Lenti MV, Corrazza GR, Gianfrani C. Vaccine immunotherapy for celiac disease. Front Med 2018;5:187.

[75] Goel G, King T, Daveson AJ, Adrews JM, Krishnarajah J, Krause R, et al. Epitope-specific immunotherapy targeting CD4-positive T cells in coeliac disease: two randomised, double-blind, placebo-controlled phase 1 studies. Lancet Gastroenterol Hepatol 2017;2:479–93.

[76] Croese J, Giacomin P, Navarro S, Clouston A, McCann L, Dougall A, et al. Experimental hookworm infection and gluten microchallenge promote tolerance in celiac disease. J Allergy Clin Immunol 2015;135:508–16.

[77] Hindryckx P, Levesque BG, Holvoet T, Durand S, Tang CM, Parker C, et al. Disease activity indices in coeliac disease: systematic review and recommendations for clinical trials. Gut 2018;67:61–9.

[78] Gottlieb K, Dawson J, Hussain F, Murray JA. Development of drugs for celiac disease: review of endpoints for Phase 2 and 3 trials. Gastroenterol Rep 2015;3:91–102.

[79] Ludvigsson JF, Ciacci C, Green PH, Kaukinen K, Korponay-Szabo IR, Kurppa K, et al. Outcome measures in coeliac disease trials: the Tampere recommendations. Gut 2018;67:1410–24.

Chapter 13

Future research on wheat hypersensitivities

The awareness of wheat-related disorders among the medical community and the general population has risen tremendously during the past decades and the popularity of a wheat (gluten)-free diet has remarkably increased in the Western world. According to the bibliographic database PubMed/Medline, the annual number of scientific medical papers on wheat hypersensitivities and gluten increased dramatically from year 1997 to 2017, regarding celiac disease (CD) from 353 to 901, wheat allergy (WA) from 0 to 32, non-celiac gluten sensitivity (NCGS) from 26 to 104 and gluten from 151 to 713. International multidisciplinary investigations have achieved more detailed insights into the basic principles of wheat hypersensitivities. However, many issues concerning the various features of wheat-related diseases are still unresolved and need to be studied by future research. Lack of knowledge is specifically observed in NCGS and irritable bowel syndrome (IBS). The aim of this chapter is to illustrate gaps in the background on epidemiology, genetics, triggering factors, environmental factors, pathogenic mechanisms, diagnosis, and treatment of wheat hypersensitivities.

13.1 Epidemiology

The acquisition of correct epidemiological data on wheat hypersensitivities is difficult, because representative and high numbers of affected patients and controls are necessary. Moreover, the diseases and their symptoms must be clearly defined and the diagnostic tools must be standardized. However, many published studies on the prevalence of wheat hypersensitivities do not fulfill these criteria. CD and WA frequencies have been evaluated in developed countries for quite a long time and average data are in a relatively small range ($\approx 1\%$ and 0.5%–2.0%, respectively). Frequencies in developing countries are uncertain and mostly underestimated due to the lack of awareness of the diseases and missing diagnostic facilities. In these countries, and especially in China and India with the highest population rates, epidemiological evaluation under standardized conditions should be urgently enforced. A number of studies revealed that cases of CD and allergies increased during the last decades. More research is needed to investigate the reasons for this increase to reduce the risk

Wheat—An Exceptional Crop. https://doi.org/10.1016/B978-0-12-821715-3.00013-7

of developing these disorders. In the cases of NCGS and IBS, wide ranges of frequencies (0.6%–6% and 5%–20%, respectively) have been reported. This lack of precise data is a global problem and the clinical pictures and the diagnostic procedures have to be defined, before prevalences can be determined more accurately.

13.2 Genetics

Allergies and CD are known to be closely related to genetics. Affected parents are more likely to have affected children and monozygotic twins have the same manifestations in the majority of cases. HLA-DQ2/8 genes have been proven to be strongly associated with CD and recent studies additionally identified a number of non-HLA-genes, but these findings explain only about 50% of the genetic susceptibility to CD. Thus identification of further non-HLA-genes related to CD is necessary. Many non-HLA-genes typical of CD are shared with autoimmune diseases. Therefore the genetic background should be investigated furthermore in order to get more insight into the associations between CD and autoimmune diseases. The genetic determinants of other wheat hypersensitivities have not been studied in detail and should be a task of future research. Moreover, epigenetics as well as gene-gene and gene-environment interactions need to be investigated.

13.3 Triggering factors

Only common wheat has been intensely examined for constituents triggering hypersensitivities. Testing the other wheat species (spelt, durum wheat, emmer and einkorn) has been minimal. In case of CD, gluten proteins have been ambiguously identified as precipitating factors. There is no doubt that all species are harmful to CD patients due to similar quantities and compositions of gluten proteins. Some older studies on "ancient" wheat species (emmer and einkorn) proclaimed reduced CD toxicity, but detailed clinical evaluations are still missing.

Numerous proteins from common wheat, including both gluten proteins and non-gluten proteins, have been identified as culprit agents for dietary WA, respiratory allergy, and contact allergy. Regarding gluten proteins, wheat species and varieties differ only in a small range and gluten-sensitive allergic people therefore have to avoid all sources of wheat. In contrast, wheat species and their varieties distinctly vary in the composition of non-gluten proteins. Allergologists and cereal chemists should cooperate in identifying varieties with low allergenic potential. In particular, the question whether "ancient" wheat species are less allergenic or even not allergenic has to be clarified. Moreover, the influence of food processing such as cooking, baking or fermentation on allergic responses should be investigated in more details. Genetic engineering could be a promising approach to modify the allergenic epitopes of proteins.

A gluten-free diet (GFD) has been considered a useful treatment of NCGS. Therefore gluten proteins were initially supposed to be the culprit agents. Later studies indicated that other wheat constituents apart from gluten proteins such as fermentable oligo-, di-, and monosaccharides and polyols (FODMAPs) and amylase/trypsin inhibitors (ATIs) might contribute to the development of NCGS. Regarding diarrhea-predominant IBS (D-IBS), wheat has been proposed as one of the most relevant triggers, but which components of wheat are involved remains unclear. To clarify the role of gluten proteins, ATIs, and FODMAPs in both NCGS and D-IBS, multicenter placebo-controlled oral challenge tests with commonly accepted reference preparations are urgently needed.

Knowledge on the relationship between the structure of triggering agents and immune responses in wheat hypersensitivities is incomplete. Whereas a number of toxic or allergenic epitopes (amino acid sequences) of proteins have been identified in CD and WA, such information is missing in NCGS and D-IBS. Only few attempts were made to quantify the toxic or allergic effects of harmful components and dose-response relationships are widely unknown. Corresponding standardized methods to determine the degree of toxicity/immunogenicity have not yet been developed.

13.4 Environmental factors

Wheat hypersensitivities can be initiated in predisposed subjects at any age. Therefore environmental factors have been proposed to be associated with the development of hypersensitivities in addition to triggering proteins and genetic factors. Allergies are more common in industrialized countries than in developing countries and more common in urban populations than in rural populations. Based on these observations, a high level of hygiene has been suggested to be responsible for the increased sensitivity toward allergens. The hygiene hypothesis is also under discussion in the case of CD, but much more investigation and statistical evaluation are necessary to clarify, whether cleaner hygienic conditions boost the development of hypersensitivities.

Increased permeability of the small intestine ("leaky gut") is a precondition for triggering agents to reach the intestinal lymphatic tissue and the blood. Intestinal infection and inflammation but also physical activity, intake of various medications, extensive alcohol consumption or psychological stress have been considered responsible for an increased permeability. However, statistical evidence of the association between permeability and hypersensitivities is not yet available. More studies are therefore needed to judge the influence of increased permeability on the activation of the disorders. In particular, it is important to develop methods for the measurement of intestinal permeability that are more suitable than the classical lactulose/mannitol (LAMA) test.

It was believed for many years that a moderate introduction of gluten to the diet of children between 4 and 7 months and breastfeeding at that time may prevent the development of CD. However, recent multicenter studies revealed contrasting results. It is still an open question whether the time of gluten introduction, breastfeeding, and repeated gastrointestinal infections in early life have any influence on CD development and this question should be urgently answered.

Imbalanced intestinal microbiota and bacterial overgrowth have been reported as triggering factors for intestinal complaints induced by wheat hypersensitivities. It has been hypothesized that "good" (commensal) bacteria are decreased and "bad" (pathogenic) bacteria are increased leading to more infections and altered intestinal permeability. The increased exposure to antibiotics may contribute to these unfavorable conditions, but this has not been proven until now. Children born by cesarean section were proposed to have an increased risk of later CD due to an altered microbiota compared to children born vaginally. Whether the type of birth has an influence on the development of other wheat hypersensitivities, is still an open question. In general, more studies are needed to judge the relationship between intestinal microbiota and wheat hypersensitivities. Moreover, basic knowledge of this relationship will be essential to modulate the microbiota as a therapeutic approach. For example, the effects of probiotics and prebiotics should be studied in more detail.

13.5 Pathogenic mechanisms

The pathomechanisms of wheat hypersensitivities are complex and in parts not fully understood. Gluten proteins are the precipitating factors of CD, while both gluten and non-gluten proteins trigger the development of other wheat-related diseases. However, the pivotal structures (amino acid sequences) of epitopes, present in precipitating proteins, and their binding to specific receptors are frequently not known and should be investigated in detail. Regarding dietary hypersensitivities, the pathways of harmful proteins and peptides are partially unclear and further studies should include the investigation of gastrointestinal digestion, intestinal passage, and subsequent cascades of biochemical reactions.

Adaptive and innate immune responses are central points of the CD pathogenesis. Within the past 20 years, remarkable advances were achieved regarding the immune responses typical of the adaptive immunity. Further progress in determining details of the adaptive immune reactions can be expected for the near future. The role of the innate immune response and the underlying mechanisms are largely unclear and extensive studies are necessary. Moreover, unanswered questions are how innate and adaptive immunities play together and whether other wheat proteins apart from gluten (e.g., ATIs) contribute to innate responses. IgA and IgG antibodies and autoantibodies in the blood and the intestinal tissue are important markers of CD. It remains to be elucidated whether these antibodies account for the progression of CD at intestinal and extraintestinal levels. In addition, the mechanistic background of neurological and psychological manifestations, including the role of the blood-brain barrier, should be in the focus of future research.

The principles of IgE-mediated wheat allergy are well defined. However, some details of the pathogenesis are still unclear. For example, knowledge on the binding sites of immunogenic proteins that are recognized by antibodies is incomplete The scientific background of harmful effects of cofactors in wheat-dependent exercise-induced anaphylaxis (WDEIA) such as physical exercise and acetylsalicylic acid have not yet been clarified. In contrast to IgE-mediated allergies, the pathogenic mechanism of non-IgE-mediated allergy is widely unknown and in need of elucidation.

The existence of NCGS as a unique wheat hypersensitivity is now widely accepted by the medical community. However, the scientific understanding is still restricted and has only few certainties and many unresolved issues. Accurate information on the pathogenic mechanisms and potential triggers are missing. In particular, it must be clarified whether gluten alone or other wheat constituents are triggering factors. Thus extensive future research is necessary to generate more knowledge regarding NCGS.

The pathomechanism of IBS is largely obscure. Factors that seem to be important such as bacterial overgrowth or intestinal permeability require much more research. In cases of D-IBS patients, who often are sensitive to gluten, serum antibodies need to be comprehensively investigated with regard to their specificities toward single gluten protein types. Considering the high prevalence of IBS, well-structured multicenter studies are urgently needed to get insight into the pathogenic mechanisms of this disease.

13.6 Diagnosis

The diagnostic scheme for CD consists of symptomatology, serology, small intestinal histology, HLA status, and response to a GFD. The most recent diagnosis guidelines for symptomatic children suggest that the intestinal biopsy can be omitted, if CD-specific antibody titers are above 10 times the upper limit of normal. Further studies should focus on determining the adult cases, who may be spared from duodenal biopsy.

Diagnostic methods to identify IgE-mediated WA include blood and oral challenge tests and are well established. WDEIA may be an exception: The linkage between gluten intake, allergic reactions, and triggering cofactors is difficult to identify. Establishing a well-defined standardized oral challenge test with gluten in combination with exercise for example may contribute to a more accurate diagnosis. The diagnosis of non-IgE-mediated WA is difficult and requires further development. Differentiation between NCGS and IBS is particularly challenging due to similar intestinal symptoms and missing laboratory indicators. Future research is therefore needed to find out specific metabolic biomarkers in blood, urine and/or feces that might help identifying hypersensitivities in diagnostically difficult cases.

Serum antibody tests are often on the first line of diagnosis. However, laboratory tests vary in their performance due to different qualities of antigens and cut-off values. Collaborative studies are therefore needed to compare and evaluate commercial diagnostic kits regarding specificity and sensitivity with the aim to reach common recommendations. Recently introduced simple and rapid analytical methods using immunosensors should be further developed and tested so that they may replace time- and labor-consuming enzyme-linked immunosorbent assays (ELISA) methods in the future.

The diagnosis of WA, NCGS, and IBS frequently requires oral provocation tests. Numerous different poorly characterized wheat or gluten preparations have been applied for challenge. However, they did not allow for comparison of results between different clinical and scientific studies. There is therefore an urgent need to develop a commonly accepted well-defined gluten preparation for oral challenge. Its overall composition should be known and it should be safe for patients, easily applicable, and not distinguishable from a placebo.

13.7 Therapies

Avoidance of wheat (gluten)-containing foods by a GFD is currently the only effective treatment of wheat hypersensitivity, a therapy without side effects and clinical complications. The question how much wheat/gluten can be tolerated by patients suffering from wheat sensitivities has been answered only for CD: More than 20 mg gluten per day are not considered to be safe. In the cases of WA, NCGS, and IBS, true safe exposure thresholds are not known and should be defined by future studies in order to provide patients with information how strict the diet must be. Numerous patients are still unsatisfied with dietetic gluten-free foods due to inferior quality compared to their gluten-containing counterparts. Improving the quality of dietetic food products will be an ongoing task for food technologists.

The conventional dietary treatment is a big challenge for patients due to inferior quality, restricted availability, and higher costs of dietetic products. Therefore there is an urgent need to develop safe and cost-effective alternatives. In the case of WA, the oral, sublingual, or epicutaneous immunotherapy is a promising approach to induce tolerance. However, more studies are needed to assess safety and efficacy of this approach. Gluten is the most important trigger of wheat hypersensitivities caused by food intake. Therefore a number of therapeutic strategies have been developed to eliminate the harmful effects of gluten proteins. Most of them consist of detoxification of gluten by enzymatic degradation or inhibition of specific immune reactions. Clinical evaluations of corresponding medications are already ongoing and results appear promising. However, studies with longer duration periods and greater significance will be needed to assess both the efficacy and long-term safety. An important point to address is the amount of gluten that can be detoxified in vivo. The patient should know whether the preparation allows to consume gluten *ad libitum*, to ingest sparing quantities only or to treat traces of gluten inadvertently present in gluten-free foods.

13.8 Gluten analysis

The accurate quantitation of gluten in foods is essential for patients with wheat-mediated disorders, the industry producing gluten-free foods, and the food control authorities. Numerous immunochemical and nonimmunochemical methods have been developed by different research groups and companies and a number of assays are now commercially available. Unfortunately, each test has its specific features such as different extraction procedures, reference materials, and principle of gluten detection. It is therefore, not surprising that the comparison of results from different test kits for the same sample shows systematic deviations. Future research should enable the development of an immunological method and a complementary nonimmunochemical reference method that achieve consistent and accurate results for the entirety of gluten proteins. Appropriate extraction procedures should remove interfering substances and solubilize all gluten proteins and peptides from the food matrix independent of food processing (heat-treated, partial hydrolyzed or fermented). The reference materials used should include total gluten proteins (both prolamins and glutelins) as well as gluten hydrolysates.

The principle of conventional ELISA has been widely applied for the quantitation of gluten without reaching a worldwide acceptance regarding single test kits. Studies of the last decade have demonstrated that immunosensors are reliable, user-friendly, and cost-effective alternatives with short operation times. However, collaborative studies should clarify, whether the developed sensors are capable of substituting conventional ELISAs. Nonimmunochemical methods such as polymerase chain reaction (PCR) and liquid chromatography/mass spectrometry (LC/MS) may be promising alternatives to existing immunochemical tests like ELISAs in the future. However, performance criteria for specificity, sensitivity, accuracy, handling, time, and cost of analysis should be further established for each system and compared with corresponding ELISA kits. Finally, international collaborative studies should evaluate method performance and clarify, which of the developed sensors is the best candidate for substituting ELISA.

Index

Note: Page numbers followed by *f* indicate figures and *t* indicate tables.

A

Abdominal distension, 183, 184*f*
Abdominal pain, 183
Acetylsalicylic acid (ASA), 178–179, 226
Adaptive immune response, 168, 199*f*, 200–202, 236–237
Addison's disease, 195*t*, 197
Aegilops, 2–3, 10
Aflatoxins, 51
Agglutinins, 36
Albumins, 23, 24*t*, 25, 26*t*, 115–116, 116*t*, 142–143, 153, 165*t*, 173–174
Aleurone layer, 6–8, 7*f*, 20, 25, 37, 38*t*, 39–40, 128, 135, 137, 143
Alkylresorcinols, 42–43, 140
Allergens, 151, 221, 225, 228, 228*t*
Allergic rhinitis, 228
Allergies, 221–222, 228–229, 270–271
α-amylase, 33, 129–130
α-amylase inhibitors, 225
α-amylase/trypsin inhibitors (ATIs), 35, 166, 225, 236, 237*f*, 271
α-gliadins, 10, 28*f*, 29, 115–116, 170*t*, 174, 176*t*, 200–201, 201*f*, 225, 228*t*
α-tocopherol, 39–40, 40*t*, 42, 43*f*, 76–77, 78*t*, 127, 128*t*, 144
Alternative therapies, 263–264
Amino acids, 21, 21*f*, 22*t*, 26–28, 26–27*t*, 28*f*, 30, 115–116, 118–119, 119*f*, 142–143, 142*t*, 154, 169–170, 176*t*
Amylase trypsin inhibitors (ATIs), 203, 236–238, 237*f*, 271
Amylopectin, 16–18, 17–18*f*, 33, 74–75, 107, 136–137
Amylose, 16–18, 17–18*f*, 107, 136–137
Anaphylaxis, 222–224
Anemia, 185
Angioedema, 227, 227*f*
Animal feed, 97–98
Antibodies, 151, 179–180, 185–187, 189, 201–202, 222, 226, 235, 238–239, 246, 257
Antideamidated gliadin peptide antibodies (DGPAs), 179–180, 182, 205–206, 206*t*
Antiendomysium antibodies (EMAs), 204
Antigen-presenting cells (APCs), 168, 171, 175, 198–199, 199*f*, 201*f*
Antigliadin antibodies (AGAs), 205–206, 206*t*, 235
Antitransglutaminase antibodies (TGAs), 179–180, 182, 185–187, 197, 204, 206*t*
Anxiety, 192, 234–235
Arabic flat bread, 79
Arabinogalactan peptides, 20
Arabinose, 16, 19–20
Arabinoxylans, 19–20, 19*f*, 19*t*, 128, 137, 139–140, 145
Arteriosclerosis, 138
Aspirin. *See* Acetylsalicylic acid
A-starch, 16, 104–106
Attention-deficit/hyperactivity disorders (ADHD), 193
Autism spectrum disorders (ASD), 192–193
Autoantibodies, 200–202, 272
Autoimmune diseases, 195–198, 195*t*, 235, 270
Autoimmune hepatitis, 187, 195*t*, 197
Autoimmune thyroid disease, 195*t*, 196–197
Autophagy, 146
Avenins, 31–32, 36, 169, 170*t*, 174, 176*t*

B

Baguette, 76, 77*f*, 78
Bake-off technology, 72–73
Baker's asthma, 228–229, 228*t*
Baking process, 62–63, 71–72, 75–77
Baking test, 68
Barley, 63*f*, 113–114
Basophil activation test, 222, 239
Basophils, 150–151, 151*f*, 222, 235
Batter process, 104–105
Beer, 96–97, 251, 254
β-amylases, 32–33, 35, 112
β-carotene, 41–42, 41*f*, 87

β-glucans, 19*t*, 20, 140
Betaine, 146
β-sitosterol, 22*t*, 36–37, 42, 43*f*, 127, 145
Beverages, 96–97, 112–113
Bilobol, 42–43, 43*f*
Bile acids, 139–140
Bioavailability, 143–147
Biopsy, 163, 180, 194, 204–205, 208, 273
Biotin (B_7), 40*t*, 144
Blood-brain barrier (BBB), 157, 191
Blood tests, 198, 222–223, 262
B-lymphocytes (cells), 150–151, 199*f*
Bone mineral density, 210
Bran, 7, 55–56, 58, 92, 128–129, 128*t*, 137, 140
Bread, 63–80, 63*f*
  bake-off technology, 72–73
  baking process, 71–72
  flavor, 73–74, 74*t*
  quality, 75–76
  staling, 74–75
  storage, 74–75
  types, 76–80
  volume, 63–66, 68–69
Bread making recipes, 64–70
Bread rolls, 77*f*, 78–79
Breakfast cereals, 61, 92, 95–96, 121
Breeding, 1, 8–11, 54, 164–166, 173
Brioche, 82–83
B-starch, 16, 105
Bulgur, 92
Butyric acid, 139–140
B-vitamins, 39–40, 39*f*, 58–59

C

Cakes, 62, 80–82, 82*f*, 106
Calcium, 40, 40*t*, 78*t*, 128*t*, 143, 185
Campesterol, 22*t*, 36–37, 42–44, 145
Campestanol, 42, 43*f*
Carbohydrates, 13–20, 15*f*, 78*t*, 137, 168, 238–239, 245
Cardiovascular diseases (CVD), 140, 188
Carotenoids, 34, 36, 41–42, 87
CD-toxic cereals, 169, 170*t*
CD-toxic proteins and peptides, 175, 176*t*
Celiac disease (CD)
  associated diseases, 195–198, 195*t*
  causative proteins, 168–177
  clinical features, 181–195
  diagnosis, 203–210, 204*f*
  environmental factors, 177–179
  epidemiology, 163–167, 164–165*t*
  genetics, 167–168
  history, 161–163
  pathology, 179–181
  pathomechanism, 198–203
  prevalence, 163–164, 164*t*
  serology, 179–181
Cellulose, 17*f*, 18–20, 19*t*, 128, 130, 137–138
Cereal(s), 1, 2*f*, 13, 47–48, 61, 63–64, 63*f*, 169–170, 172–173, 251
Cereal proteins, 168, 170*t*, 173–177
Cerebellar ataxia, 190, 235
Chapatti, 79
Chemokines, 150–151, 189–190, 199*f*, 201–202, 238
Cholesterol, 134, 140
Chorleywood bread process, 70
Ciabatta, 78
Claudins, 155–156, 235
Clustered Regularly Interspaced Short Palindromic Repeats (CRISPR), 10–11
Codex Alimentarius Commission, 137
Codex Alimentarius Standard for Gluten-Free Foods, 173–174
Colorectal cancer, 139
Common wheat, 1–5, 13, 14*t*, 24*t*, 25, 39–41, 48, 58*t*, 65, 86–87, 118*f*, 122, 164–166, 175–177, 237–238, 270
Constipation, 138–139, 245
Contact dermatitis, 229–230
Contact urticaria, 153, 229–230
Cookies, 84–85, 84*f*
Copper, 40, 40*t*, 143, 185
Corticosteroids, 195
Couscous, 91–92
Crackers, 85
Crepe, 83
Crisp bread, 79–80
Croissants, 82, 82*f*

D

Dark bread, 77
Deoxynivalenol (DON), 51
Depression, 192
Dermatitis herpetiformis (DH), 185–187, 186*f*
Dextrin, 69, 72, 129–130
Dextrose equivalent (DE), 110–112, 111*t*
Diabetes mellitus, 138, 195*t*, 196
Diarrhea, 161, 181–183, 234–235, 245
Diagnosis
  allergies, 166, 221–222, 226, 229
  celiac disease, 203–210, 204*f*
  irritable bowel syndrome, 245–247
  non-celiac gluten sensitivity, 238–241

Dietary fiber (DF), 14*t*, 58*t*, 78*t*, 100*t*, 128*t*, 135*t*, 137–141
cardiovascular diseases (CVD), 140
chemical composition, 14*t*, 58*t*, 78*t*, 116*t*, 128*t*
colonic functions, 138–139
definitions, 137
obesity, 141
post-prandial glycemic response, 141
Digestion, of proteins, 153–154
Disaccharides, 14–16, 32–33, 72, 110–111, 156
Disulfide bonds, 23, 29–31, 29*f*, 116–117, 120–121, 126, 169
Diverticular disease, 138–139
Donuts, 83, 106
Dough preparation, 66, 68–71, 79
Down's syndrome, 195*t*, 198
Dry baked products, 84–86, 84*f*
Durum wheat, 1–3, 5, 9, 21–22, 24*t*, 48, 55–56, 86–87, 92

E

Economics, 98–101
Einkorn, 2–3, 2*t*, 24*t*, 25, 47–48, 62, 237–238, 258, 270
Elasticity, 63–64, 113–114, 117–119, 251
ELISAs. *See* Enzyme-linked immunosorbent assays (ELISAs)
Emmer, 1–3, 2*t*, 24*t*, 47–48, 62, 173, 237–238, 270
Emulsifiers, 69–70, 81–82, 253
Endoscopy, 194, 207–210, 238–239
Endosperm, 3, 5–7, 7*f*, 18, 20, 26–28, 35, 37–39, 38*t*, 55–56, 87–88, 143, 237
Enterocytes, 154–156, 198–199, 202, 207–208, 208*f*
Enterodiol, 42, 43*f*
Environmental factors, 167, 177–179, 221, 271–272
Enzyme-linked immunosorbent assays (ELISAs), 205–206, 256–259, 257*f*
Enzymes, 8, 25, 32–34, 69, 73, 97, 112, 140, 144, 153–154, 175, 198, 200, 206–207, 225, 236–237, 240–241, 253, 257, 263
Eosinophils, 150, 152, 224–225, 235
Eosinophil esophagitis, 224–225
Eosinophilic gastritis, 224–225
Epidemiology, 163–167, 269–270
Epidermal transglutaminase (eTG), 185–187
Epilepsy, 190–191, 235
Epinephrine, 223, 226
Epithelial cells, 139–140, 154–155, 155*f*, 202
Epithelial passage, 154–155, 198–199
Essential amino acids, 25–28, 115–116, 142–143, 142*t*, 169–170
Ethanol production, 129–130
European Chemicals Agency (ECHA), 52
European Food Safety Authority (EFSA), 52, 135
European Society for Paediatric Gastroenterology Hepatology and Nutrition (ESPGHAN), 204
Exorphins, 157
Extensigram, 66–67, 67*f*, 118, 118*f*
Extraction rate, 57–59, 57–58*t*, 64–65, 79, 135
Extrusion, 84–87, 89, 96–98, 120

F

Farina, 55–56, 86–87, 89, 95, 122
Farinogram, 66, 67*f*
Farming, 47–54
Fatigue, 182, 185, 192
Fatty acids, 8, 34, 36–37, 37*f*, 127, 127–128*t*, 139–140, 143
Fermentable oligo-, di-, and monosaccharides and polyols (FODMAPs), 236, 271
Fertilization, 3, 21–22, 25, 50
Fertilizers, 4, 49–50, 53
Ferulic acid, 19–20, 19*f*, 42, 140, 145
Fettuccine, 90–91
Fine bakery goods, 80–84
Flakes, 96
Flat bread, 79–80
Flavones, 41–42
FODMAPs. *See* Fermentable oligo-, di-, and monosaccharides and polyols
Folic acid (B_9), 39–40, 39*f*, 40*t*, 78*t*, 128*t*, 144, 183–184
Food allergies, 221–223
Food and Agriculture Organization (FAO), 48, 63, 136
Food and Drug Administration (FDA), 135
Frekeeh, 92
Fructans, 18–20, 19*t*, 137
Fructo-oligosaccharides, 14–16, 137
Fructose, 14–16, 73, 110–112
Fungicides, 49–51

G

Galactose, 15*f*, 16, 36–37
γ-gliadins, 23, 25, 28–30, 28–29*f*, 169, 170*t*, 174, 176*t*, 225, 228*t*

Gelatinization, 17–18, 18*f*, 110, 136–137
Genetic mutation, 9–11
Genetic engineering, 8–11, 54, 174, 270
Genetics, 270
Genetic tests, 208–209
Genome-wide association studies (GWAS), 167–168
Germ, 5–8, 25, 32, 36–37, 38*t*, 39–40, 57–58, 126–127, 144
Germ oil, 126–127, 127*t*
Gibberellins, 52–53
Gingerbread, 68, 83–84
Gliadins, 23, 24*t*, 25–26, 27*t*, 28, 28*f*, 115–118, 121*f*, 152–153, 164–166, 170*t*, 174, 176*t*, 179–180, 201*f*, 225, 228*t*, 236
Gliadorphins, 157, 193
Globulins, 23, 115–116, 142–143, 153
Glucoamylase, 32–33, 112, 129–130
Glucose, 8, 16, 33, 36–37, 73, 108, 110–112, 111*t*, 136, 141
Glucose syrup, 112–113
Glutamine, 22*t*, 25–29, 117–119, 126, 175, 200, 202
Glutelins, 113–114, 169, 174, 258–260
Gluten, 31–32, 32*f*
  chemical composition, 115–117, 116*t*
  definition, 113–114
  film-forming, 113–114, 119, 126*f*
  food applications, 121–124
  hidden, 251, 252*t*
  modification, 119–121, 121*f*
  nonfood applications, 124–126, 126*f*
  physical properties, 117–119, 118–119*f*
  production, 114–115, 115*f*
Gluten analysis, 275
  extraction, 255
  immunochemical methods, 256–260
  nonimmunochemical methods, 260–261
  reference material, 256
Gluten ataxia (GA), 235
Gluten challenge, 182, 239–241
Gluten detoxification, 264
Gluten-free bread, 252–253
Gluten-free diet (GFD), 161, 209, 226, 235, 239–240, 249–251, 261–262, 271
Gluten-free products, 251–255, 252*t*
Glutenins, 23, 25–28, 27*t*, 28*f*, 30, 66, 115–118, 119*f*, 152–153, 169, 174, 225, 228*t*, 236
Gluten peptides, 175, 200
Gluten proteins, 161, 272
  chemistry, 169–170
  classification, 170*t*
  disulfide structures, 29, 29*f*
  molecular weights, 169–170
  physical and chemical treatments of, 174
  proportions, 170*t*
  repetitive sequences, 169–170, 171*t*
Glutomatic, 114
Glycolipids, 18, 35–37, 38*t*, 126–127, 143
Glyphosate, 52, 52*f*
Graham flour, 58
Grain, 54–55
  morphology, 19–20
  storage, 54–55
  yield, 8–10, 48–50, 54

H

Hammer mills, 56–58
Health claims, 135, 138, 140–141, 145–146
Herbicides, 49–52
Histamine, 203–204, 222, 227
HLA genes, 167–168
HLA-DQ2/8 heterodimers, 198–199
HMW glutenin subunits (HMW-GS), 23, 25, 28*f*, 30, 66, 118–119, 119*f*, 169, 170*t*, 176*t*
Hookworm therapy, 178, 264
Hordeins, 31–32, 169, 170*t*, 174, 176*t*, 258, 260–261
Human immunity, 149–152
Human leukocyte antigen (HLA), 167–168
Hybridization, 9
Hybrid wheat, 9–11
Hydrolyzed wheat protein (HWP), 225
Hygiene hypothesis, 166, 178, 271

I

ICC methods, 166
IgA antibodies, 150–151, 185–187, 201–202, 206, 206*t*, 235, 272
IgA deficiency, 180, 195*t*, 198, 206, 206*t*
IgE antibodies, 151, 222
IgE-mediated wheat allergy (IgE-WA), 223–224
IgG antibodies, 150–151, 180, 198, 201–202, 206, 206*t*, 272
Immunoglobulin (Ig), 150–151, 180, 199*f*
Immunosensors, 206–207, 259, 273–275
Infections, 188–189
Infertility, 188
Innate immune response, 202–203, 237–238, 272
Insecticides, 49–51
Interferon-γ (IFN-γ), 171, 199*f*, 201–202

Interleukins (ILs), 150–151, 171, 189, 202, 224–225, 238
International Association for Cereal Science and Technology (ICC), 65
International Wheat Genome Sequencing Consortium (IWGSC), 4
Intestinal
biopsy, 163, 170, 182, 204, 207–208, 273
histology, 207–208, 208*f*
inflammation, 189–190
microbiota, 177, 272
permeability, 154–156
tract, 154–155
Intraepithelial lymphocytes (IELs), 180–181, 202, 235
Iron, 10, 40, 40*t*, 78*t*, 79, 128*t*, 143, 172, 185, 262
Irritable bowel syndrome (IBS), 233, 239
diagnosis, 245–246
pathomechanism, 246
prevalence, 245
symptoms, 245
treatment, 247

J
Junctional adhesion molecule A, 155–156, 155*f*

K
Kidney diseases, 187–188
Khorasan, 3, 253

L
Lactulose/mannitol (LAMA) test, 156, 271
Larazotide acetate, 263–264
Lasagne, 90*f*, 91
Leavened bread, 76–79, 77*f*
Leavening agents, 68–69
Linoleic acid, 36, 127, 127–128*t*, 143
Linolenic acid, 36, 127, 127–128*t*, 143
Lignans, 42, 140
Lignin, 19*t*, 20, 138
Lipases, 34
Lipids, 8, 14*t*, 18, 35–39, 38*t*, 58, 58*t*, 78*t*, 104, 116*t*, 126–128, 128*t*, 133–134, 135*t*, 143–145, 240, 262
Liquid chromatography-mass spectrometry (LC-MS), 260–261
Liver disorders, 187
LMW glutenin subunits (LMW-GS), 23, 25, 30–31, 115–116, 169, 170*t*, 174, 176*t*, 201
Lutein, 41–42, 41*f*, 87
Lymphoma, 193–195

M
Macaroni, 86–88, 90*f*, 91
Magnesium, 40, 40*t*, 78*t*, 128*t*, 143–144, 172, 185
Maillard reaction, 72–73, 258–259
Malabsorption, 183–185
Malignancy, 193–195
Maltose, 14–16, 32–33, 68–69, 73, 110–112, 111*t*
Manganese, 40, 40*t*, 128*t*, 143, 187
Mannitol, 14–16, 113, 156, 182
Marsh-Oberhuber classification, 208
Martin process, 104–105
Mass screening, 209–210
Mast cells, 150–152, 151*f*, 227
33-Mer peptide, 175–177
Metabolic proteins, 32–34
Microbiota, 42, 137–138, 150, 156, 166, 177–178, 272
Milling process, 55–59, 146–147
Mineral deficiency, 34, 183–184
Minerals, 14*t*, 39–40, 40*t*, 57–59, 58*t*, 76–77, 78*t*, 96, 128, 128*t*, 135, 135*t*, 143–145, 185, 210
Monosaccharides, 13–16, 15*f*, 20, 110–111
Mortality, 193–195
Mycotoxins, 51, 54, 97

N
Naltrexone, 193
Neurological disorders, 189–191
Nicotinamide (B_3), 39–40, 40*t*, 76–77, 78*t*, 128, 128*t*, 144–145
Nitrogen (N), 50
Nonceliac gluten sensitivity (NCGS), 190
clinical features, 234–235
diagnosis, 238–241
pathogenesis, 235–238
prevalence, 233–234
symptoms, 234–235
triggering factors, 235–238
Non-HLA genes, 167–168, 197–198, 270
Non-IgE-mediated wheat allergy (Non-IgE-WA), 152, 224–225
Nonpolar lipids, 36–39, 38*t*, 126–127, 143
Nonstarch polysaccharides (NSP), 13–16, 18–20, 19*t*, 128, 137
Non-steroidal antiinflammatory drugs (NSAIDs), 178–179, 226
Noodles, 92–94

Nucellar epidermis, 5, 7, 7*f*
Nutrients, 135–136, 135*t*
Nutritional deficiencies, 133–134, 183–184, 191, 223, 250–251, 262
Nutritional value, 59, 77, 133–147, 254

O

Oats, 1–2, 63*f*, 113–114, 161, 169, 170*t*, 172–174, 250, 254
Obesity, 141
Occludin, 155–156
Oleic acid, 36, 127–128, 127–128*t*, 143
Oligosaccarides, 14–16, 32–33, 73, 110–111, 111*t*, 116–117
ω1, 2-gliadins, 15*f*, 24*t*, 115–116, 169, 170*t*, 176*t*, 225
ω5-gliadins, 28–29, 28*f*, 169, 170*t*, 225
Opioid peptides, 157
Oral provocation test, 222–223
Oral enzyme therapy, 263
Organ culture test, 171–172, 175, 202
Organic farming, 53
Osborne fractions, 23, 25, 26*t*
Osteoporosis, 161, 185

P

Palmitic acid, 127*t*
Pancake, 82*f*, 83
Pantothenic acid (B_5), 39*f*, 40*t*, 78*t*, 128*t*, 144
Paracellular pathway, 154–155
Pasta, 86–91
  preparation, 88–89
  quality, 90
  types, 90–91
Paste products, 86–94
Pathomechanisms
  allergy, 152
  celiac disease, 198–203
  irritable bowel syndrome, 246
  nonceliac gluten sensitivity, 246
Peptidases, 33
Pericarp, 6–7, 7*f*, 39–40
Peripheral neuropathy, 190
Pesticides, 51, 53
Phagocytes, 150
Phenolic acids, 145
Phosphate, 143
Phospholipids, 37
Phosphorus, 40*t*, 57*t*, 128*t*
Phytic acid, 143–144
Phytochemicals, 41–44, 145–146
Phytosterols, 145
Pin mills, 56–57
Pizza, 80
Plant growth regulators (PGRs), 52–53
Poaceae, 1–2, 2*f*
Polymerase chain reaction (PCR), 209, 260–261, 275
Polymerization, 30–31
Polypeptides, 154
Polysaccharides, 16
Potassium, 40*t*, 78*t*, 128*t*
Pretzels, 77*f*, 79
Prolamins, 169
Proline, 22*t*, 26–28, 28*f*
Prolyl endopeptidases (PEPs), 263
Protective proteins, 35
Proteins, 142–143, 152–153, 225
  amino acids, 21, 21*f*, 28*f*
  classification, 153
  digestion, 153–154
  distribution, 22–23
  disulfide bonds, 30–31
Proton pump inhibitors, 166–167
Pseudocereals, 251
Psoriasis, 187
Psychiatric disorders, 191–193
Puff pastry, 82
Puroindolines, 35–36
Pyridoxine (B_6), 39*f*, 40*t*, 78*t*, 128*t*, 144

Q

Quality of life (QOL), 261

R

Raffinose, 14–16
Ravioli, 94
Refractory CD (RCD), 194–195
Repetitive amino acid sequences, 119*f*, 171*t*
Reproductive disorders, 188–189
Resistant starch, 137
Respiratory allergy, 228–229
Retrogradation, 17, 18*f*
Riboflavin (B_2), 39*f*, 40*t*, 78*t*, 128*t*, 144
Ring dryer, 114–115, 115*f*
Roller mills, 55–56, 56*f*
Rye, 62, 63*f*, 169, 172–173, 258

S

Saccharomyces cerevisiae, 68–69
Saccharose, 74*t*
Salt, 64, 93
Salted pretzels, 86

Salt sticks, 84*f*, 86
Schizophrenia, 192
Secalins, 31–32, 169
Seitan, 123–124
Selenium, 143
Semolina, 87–88, 91–92
Serological tests, 206*t*
Serology, 179–181
Sitostanol, 42
Sjögren's syndrome, 195*t*, 196–197
Skin tests, 222
Small intestine, 153–154, 271
Sodium, 40*t*, 78*t*
Sorbitol, 113
Sourdough, 68
Spaghetti, 90, 90*f*
Spelt, 2*t*, 4, 24*t*
Spermidine, 43–44, 43*f*, 146
Spirits, 97
Stanols, 42
Starch, 14*t*, 16–18, 17*f*, 38*t*, 58*t*, 78*t*, 109*f*, 116*t*, 128*t*, 135*t*, 136–137, 252–253
  applications, 104
  damage, 56–57, 66
  granules, 103–104
  lipids, 18
  modified, 108–110
  native, 105–107
  production, 104–105
  properties, 103–104
Starch syrup, 111*t*
Starchy endosperm, 7*f*
Stearic acid, 37*f*, 127–128*t*
Steatorrhea, 183
Sterols, 145
Stigmasterol, 42
Storage proteins, 169
Straw, 129
Sugars, 14*t*, 111*t*, 135*t*
Sulfur (S), 50
Sweeteners, 110–113

T

Tagliatelle, 91
T-cell receptors (TCRs), 201*f*
T-cell test, 171
Terpenoids, 145
Testa, 7, 7*f*
Thiamin (B_1), 39*f*, 40*t*, 78*t*, 144
Therapy
  allergies, 152
  celiac disease, 209
  irritable bowel syndrome, 247
  nonceliac gluten sensitivity, 239–240
Thousand kernel weight (TKW), 5–6
Tight junctions, 155–156, 155*f*
T-lymphocytes (T cells), 150–151, 168, 171, 201*f*
Toast bread, 77
Tocopherols, 145
Tocotrienols, 145
Toll-like receptors (TLRs), 203, 235–236
Tortilla, 79
Toxicity, cereal proteins and peptides, 173–177
Transcellular pathway, 154–156
Transglutaminase 2 (TG2), 171, 179–180, 199*f*
Tricin, 41*f*, 42
Triglycerides, 37*f*, 127, 143
Triticale, 5
Triticum (T.) species, 2*t*
Tumor necrosis factor (TNF), 199*f*

V

Vaccination, 188
Video capsule endoscopy (VCE), 207–208
Villous atrophy, 194, 208*f*
Vital gluten, 114
Vitamin deficiency, 144
Vitamin E, 100*t*, 144, 183–184
Vitamins, 39–40, 143–145

W

Wafers, 85
Waffles, 84*f*, 85
Weight loss, 141
Wheat
  genome, 2*t*
  phylogeny, 2*f*
  production, 49*t*
  species, 2*t*
Wheat-dependent exercise-induced anaphylaxis (WDEIA), 226–227, 250–251, 273
  cofactors, 226–227
  diagnosis, 226
  prevalence, 226
  symptoms, 226
Wheat food allergy (dietary WA), 223–225
Wheat-related disorders, 149, 150*f*
White bread, 63*f*, 73–74, 76–77, 78*t*
White flour, 40*t*, 58, 64–65, 135*t*, 142*t*
Whole grain flour, 57–58, 137, 141–143, 142*t*
World Health Organization (WHO), 52

X

Xanthophylls, 41–42
Xylose, 170

Y

Yeast, 64, 68–69

Z

Zinc, 40*t*, 143
Zonulin, 155*f*, 263–264
Zwieback, 84*f*, 85–86

Kolibri 360 GmbH
Ferdinand-Jühlke-Straße 7
99095 Erfurt, Deutschland
produktsicherheit@kolibri360.de